EX LIBRIS

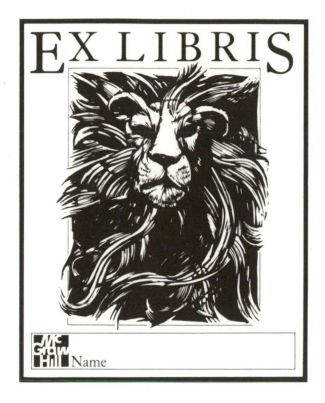

FOURIER SERIES
AND BOUNDARY
VALUE PROBLEMS

International Series in Pure and Applied Mathematics

Churchill/Brown Series

Joseph Fourier

FOURIER SERIES AND BOUNDARY VALUE PROBLEMS

Fifth Edition

James Ward Brown

Professor of Mathematics
The University of Michigan—Dearborn

Ruel V. Churchill

Late Professor of Mathematics
The University of Michigan

McGraw-Hill, Inc.

New York   St. Louis   San Francisco   Auckland   Bogotá
Caracas   Lisbon   London   Madrid   Mexico   Milan   Montreal
New Delhi   Paris   San Juan   Singapore   Sydney   Tokyo   Toronto

This book was set in Times Roman by Science Typographers, Inc.
The editors were Richard H. Wallis, Maggie Lanzillo, and James W. Bradley;
the production supervisor was Denise L. Puryear.
The cover was designed by John Hite.
R. R. Donnelley & Sons Company was printer and binder.

FOURIER SERIES AND BOUNDARY VALUE PROBLEMS

1 2 3 4 5 6 7 8 9 0 DOC 9 0 9 8 7 6 5 4 3 2

ISBN 0-07-008202-2

Library of Congress Cataloging-in-Publication Data

Brown, James Ward.
    Fourier series and boundary value problems / James Ward Brown,
  Ruel V. Churchill.—5th ed.
        p.   cm.
    Churchill's name appears first on the earlier editions.
    Includes bibliographical references and index.
    ISBN 0-07-008202-2
    1. Fourier series.   2. Functions, Orthogonal.   3. Boundary value
  problems.   I. Churchill, Ruel Vance, (date).   II. Title.
  QA404.B76   1993
  515′.2433—dc20   92-12194

ABOUT THE AUTHORS

JAMES WARD BROWN is Professor of Mathematics at The University of Michigan—Dearborn. He earned his A.B. in physics from Harvard University and his A.M. and Ph.D. in mathematics from The University of Michigan in Ann Arbor, where he was an Institute of Science and Technology Predoctoral Fellow. He was coauthor with Dr. Churchill of *Complex Variables and Applications*, now in its fifth edition. He has received a research grant from the National Science Foundation as well as a Distinguished Faculty Award from the Michigan Association of Governing Boards of Colleges and Universities. Dr. Brown is listed in *Who's Who in America*.

RUEL V. CHURCHILL was, at the time of his death in 1987, Professor Emeritus of Mathematics at The University of Michigan, where he began teaching in 1922. He received his B.S. in physics from the University of Chicago and his M.S. in physics and Ph.D. in mathematics from The University of Michigan. He was coauthor with Dr. Brown of *Complex Variables and Applications*, a classic text that he first wrote over 45 years ago. He was also the author of *Operational Mathematics*, now in its third edition. Dr. Churchill held various offices in the Mathematical Association of America and in other mathematical societies and councils.

To the memory of my father,
George H. Brown,

and of my long-time friend and coauthor,
Ruel V. Churchill.

These distinguished men of science for years influenced
the careers of many people, including myself.

J. W. B.

JOSEPH FOURIER

JEAN BAPTISTE JOSEPH FOURIER was born in Auxerre, about 100 miles south of Paris, on March 21, 1768. His fame is based on his mathematical theory of heat conduction, a theory involving expansions of arbitrary functions in certain types of trigonometric series. Although such expansions had been investigated earlier, they bear his name because of his major contributions. Fourier series are now fundamental tools in science, and this book is an introduction to their theory and applications.

Fourier's life was varied and difficult at times. Orphaned by the age of 9, he became interested in mathematics at a military school run by the Benedictines in Auxerre. He was an active supporter of the Revolution and narrowly escaped imprisonment and execution on more than one occasion. After the Revolution, Fourier accompanied Napoleon to Egypt in order to set up an educational institution in the newly conquered territory. Shortly after the French withdrew in 1801, Napoleon appointed Fourier prefect of a department in southern France with headquarters in Grenoble.

It was in Grenoble that Fourier did his most important scientific work. Since his professional life was almost equally divided between politics and science and since it was intimately geared to the Revolution and Napoleon, his advancement of the frontiers of mathematical science is quite remarkable.

The final years of Fourier's life were spent in Paris, where he was Secretary of the Académie des Sciences and succeeded Laplace as President of the Council of the Ecole Polytechnique. He died at the age of 62 on May 16, 1830.

CONTENTS

PREFACE

This is an introductory treatment of Fourier series and their applications to boundary value problems in partial differential equations of engineering and physics. It is designed for students who have completed a first course in ordinary differential equations and the equivalent of a term of advanced calculus. In order that the book be accessible to as great a variety of students as possible, there are footnotes referring to texts which give proofs of the more delicate results in advanced calculus that are occasionally needed. The physical applications, explained in some detail, are kept on a fairly elementary level.

The *first objective* of the book is to introduce the concept of orthogonal sets of functions and representations of arbitrary functions in series of functions from such sets. Representations of functions by Fourier series, involving sine and cosine functions, are given special attention. Fourier integral representations and expansions in series of Bessel functions and Legendre polynomials are also treated.

The *second objective* is a clear presentation of the classical method of separation of variables used in solving boundary value problems with the aid of those representations. Some attention is given to the verification of solutions and to uniqueness of solutions; for the method cannot be presented properly without such considerations. Other methods are treated in the authors' book *Complex Variables and Applications* and in Professor Churchill's book *Operational Mathematics*.

This book is a revision of the 1987 edition. The first two editions, published in 1941 and 1963, were written by Professor Churchill alone. While improvements appearing in earlier editions have been retained with this one, there are a number of major changes in this edition that should be mentioned.

The introduction of orthonormal sets of functions is now blended in with the treatment of Fourier series. Orthonormal sets are thus instilled earlier and are reinforced immediately with available examples. Also, much more attention is now paid to solving boundary value problems involving nonhomogeneous partial differential equations, as well as problems whose nonhomogeneous boundary conditions prevent direct application of the method of separation of variables. To be specific, considerable use is made, both in examples and in problem sets, of the method of variation of parameters, where the coefficients in certain eigenfunction expansions are found by solving ordinary differential equations.

Other improvements include a simpler derivation of the heat equation that does not involve vector calculus, a new section devoted exclusively to examples of eigenfunction expansions, and many more figures and problems to be worked out by the reader. There has been some rearrangement of the early material on separation of variables, and the exposition has been improved throughout.

The chapters on Bessel functions and Legendre polynomials, Chapters 7 and 8, are essentially independent of each other and can be taken up in either order. The last three sections of Chapter 2, on further properties of Fourier series, and Chapter 9, on uniqueness of solutions, can be omitted to shorten the course; this also applies to some sections of other chapters.

The preparation of this edition has benefited from the continued interest of various people, many of whom are colleagues and students. They include Jacqueline R. Brown, Michael A. Lachance, Ronald P. Morash, Joyce A. Moss, Frank J. Papp, Richard L. Patterson, Mark A. Pinsky, and Sandra M. Razook. Ralph P. Boas, Jr., and George H. Brown furnished some of the references that are cited in the footnotes; and the derivation of the laplacian in spherical coordinates that is given was suggested by a note of R. P. Agnew's in the *American Mathematical Monthly*, vol. 60 (1953). Finally, it should be emphasized that this edition could not have been possible without the enthusiastic editorial support of people at McGraw-Hill, most especially Richard H. Wallis and Maggie Lanzillo. They, in turn, obtained the following reviewers of both the last edition and the present one in manuscript form: Joseph M. Egar, Cleveland State University; K. Bruce Erickson, University of Washington; William W. Farr, Worcester Polytechnic Institute; Thomas L. Jackson, Old Dominion University; Charles R. MacCluer, Michigan State University; Robert Piziak, Baylor University; and Donald E. Ryan, Northwestern State University of Louisiana.

James Ward Brown

FOURIER SERIES
AND BOUNDARY
VALUE PROBLEMS

CHAPTER
1

PARTIAL
DIFFERENTIAL
EQUATIONS
OF PHYSICS

This book is concerned with two general topics:

(*a*) One is the representation of a given function by an infinite series involving a prescribed set of functions.

(*b*) The other is a method of solving boundary value problems in partial differential equations, with emphasis on equations that are prominent in physics and engineering.

Representations by series are encountered in solving such boundary value problems. The theories of those representations can be presented independently. They have such attractive features as the extension of concepts of geometry, vector analysis, and algebra into the field of mathematical analysis. Their mathematical precision is also pleasing. But they gain in unity and interest when presented in connection with boundary value problems.

The set of functions that make up the terms in the series representation is determined by the boundary value problem. Representations by Fourier series, which are certain types of series of sine and cosine functions, are associated with a large and important class of boundary value problems. We shall give special attention to the theory and application of Fourier series. But we shall also consider extensions and generalizations of such series, concentrating on Fourier integrals and series of Bessel functions and Legendre polynomials.

1

A boundary value problem is correctly set if it has one and only one solution within a given class of functions. Physical interpretations often suggest boundary conditions under which a problem may be correctly set. In fact, it is sometimes helpful to interpret a problem physically in order to judge whether the boundary conditions may be adequate. This is a prominent reason for associating such problems with their physical applications, aside from the opportunity to illustrate connections between mathematical analysis and the physical sciences.

The theory of partial differential equations gives results on the existence and uniqueness of solutions of boundary value problems. But such results are necessarily limited and complicated by the great variety of types of differential equations and domains on which they are defined, as well as types of boundary conditions. Instead of appealing to general theory in treating a specific problem, our approach will be to actually find a solution, which can often be verified and shown to be the only one possible.

1.   LINEAR BOUNDARY VALUE PROBLEMS

In the theory and application of ordinary or partial differential equations, the dependent variable, denoted here by u, is usually required to satisfy some conditions on the boundary of the domain on which the differential equation is defined. The equations that represent those boundary conditions may involve values of derivatives of u, as well as values of u itself, at points on the boundary. In addition, some conditions on the continuity of u and its derivatives within the domain and on the boundary may be required.

Such a set of requirements constitutes a *boundary value problem* in the function u. We use that terminology whenever the differential equation is accompanied by some boundary conditions, even though the conditions may not be adequate to ensure the existence of a unique solution of the problem.

EXAMPLE 1. The three equations

(1)
$$u''(x) - u(x) = -1 \qquad (0 < x < 1),$$
$$u'(0) = 0, \qquad u(1) = 0$$

make up a boundary value problem in ordinary differential equations. The differential equation is defined on the domain $0 < x < 1$, whose boundary points are $x = 0$ and $x = 1$. A solution of this problem which, together with its derivative, is continuous on the closed interval $0 \leqslant x \leqslant 1$ is

(2)
$$u(x) = 1 - \frac{\cosh x}{\cosh 1}.$$

Solution (2) is easily verified by direct substitution.

Frequently, it is convenient to indicate partial differentiation by writing independent variables as subscripts. If, for instance, u is a function of x and y, we may write

$$u_x \text{ or } u_x(x, y) \text{ for } \frac{\partial u}{\partial x}, \qquad u_{xx} \text{ for } \frac{\partial^2 u}{\partial x^2}, \qquad u_{xy} \text{ for } \frac{\partial^2 u}{\partial y\, \partial x},$$

etc. We shall always assume that the partial derivatives of u satisfy conditions allowing us to write $u_{yx} = u_{xy}$. Also, we shall be free to use the symbol $u_x(c, y)$, for example, to denote values of the function $\partial u / \partial x$ on the line $x = c$.

EXAMPLE 2. The problem consisting of the partial differential equation

$$(3) \qquad u_{xx}(x, y) + u_{yy}(x, y) = 0 \qquad (x > 0, y > 0)$$

and the two boundary conditions

$$(4) \qquad \begin{aligned} u(0, y) &= u_x(0, y) & (y > 0), \\ u(x, 0) &= \sin x + \cos x & (x \geqslant 0) \end{aligned}$$

is a boundary value problem in partial differential equations. The differential equation is defined in the first quadrant of the xy plane. As the reader can readily verify, the function

$$(5) \qquad u(x, y) = e^{-y}(\sin x + \cos x)$$

is a solution of this problem. The function (5) and its partial derivatives of the first and second order are continuous in the region $x \geqslant 0$, $y \geqslant 0$.

A differential equation in a function u, or a boundary condition on u, is *linear* if it is an equation of the first degree in u and derivatives of u. Thus the terms of the equation are either prescribed functions of the independent variables alone, including constants, or such functions multiplied by u or a derivative of u. Note that the general linear partial differential equation of the second order in $u = u(x, y)$ has the form

$$(6) \qquad Au_{xx} + Bu_{xy} + Cu_{yy} + Du_x + Eu_y + Fu = G,$$

where the letters A through G denote either constants or functions of the independent variables x and y only. The differential equations and boundary conditions in Examples 1 and 2 are all linear. The differential equation

$$(7) \qquad zu_{xx} + xy^2 u_{yy} - e^x u_z = f(y, z)$$

is linear in $u = u(x, y, z)$, but the equation $u_{xx} + uu_y = x$ is nonlinear in $u = u(x, y)$ because the term uu_y is not of the first degree as an algebraic expression in the two variables u and u_y [compare equation (6)].

A boundary value problem is *linear* if its differential equation and all its boundary conditions are linear. The boundary value problems in Examples 1 and 2 are, therefore, linear. The method of solution presented in this book does not apply to nonlinear problems.

A linear differential equation or boundary condition in u is *homogeneous* if each of its terms, other than zero itself, is of the first degree in the function u and its derivatives. Homogeneity will play a central role in our treatment of linear boundary value problems. Observe that equation (3) and the first of conditions (4) are homogeneous but that the second of those conditions is not. Equation (6) is homogeneous in a domain of the xy plane only when the function G is identically zero ($G \equiv 0$) throughout that domain; and equation (7) is nonhomogeneous unless $f(y, z) \equiv 0$ for all values of y and z being considered.

2. CONDUCTION OF HEAT

Thermal energy is transferred from warmer to cooler regions interior to a solid body by means of conduction. It is convenient to refer to that transfer as a *flow of heat*, as if heat were a fluid or gas that diffused through the body from regions of high concentration into regions of low concentration.

Let P_0 denote a point (x_0, y_0, z_0) interior to the body and S a plane or smooth curved surface through P_0. Also, let $\mathbf{n}$ be a unit vector that is normal to S at the point P_0 (Fig. 1). At time t, the *flux of heat* $\Phi(x_0, y_0, z_0, t)$ across S at P_0 in the direction of $\mathbf{n}$ is the quantity of heat per unit area per unit time that is being conducted across S at P_0 in that direction. Flux is, therefore, measured in such units as calories per square centimeter per second.

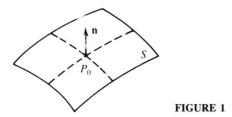

FIGURE 1

If $u(x, y, z, t)$ denotes temperatures at points of the body at time t and if n is a coordinate that represents distance in the direction of $\mathbf{n}$, the flux $\Phi(x_0, y_0, z_0, t)$ is positive when the directional derivative du/dn is negative at P_0 and negative when du/dn is positive there. A fundamental postulate, known as *Fourier's law*, in the mathematical theory of heat conduction states that the magnitude of the flux $\Phi(x_0, y_0, z_0, t)$ is proportional to the magnitude of the directional derivative du/dn at P_0 at time t. That is, there is a coefficient K, known as the *thermal conductivity* of the material, such that

$$(1) \qquad\qquad \Phi = -K \frac{du}{dn} \qquad\qquad (K > 0)$$

at P_0 and time t.

Another thermal coefficient of the material is its *specific heat* σ. This is the quantity of heat required to raise the temperature of a unit mass of the material one unit on the temperature scale. Unless otherwise stated, we shall always assume that the coefficients K and σ are constants and that the same is true of δ, the mass per unit volume of the material. With these assumptions, a second postulate in the mathematical theory is that conduction leads to a temperature function u which, together with its derivative u_t and those of the first and second order with respect to x, y, and z, is continuous throughout each domain interior to a solid body in which no heat is generated or lost.

Suppose now that heat flows only parallel to the x axis in the body, so that flux Φ and temperatures u depend on only x and t. Thus $\Phi = \Phi(x,t)$ and $u = u(x,t)$. We assume at present that heat is neither generated nor lost within the body and hence that heat enters or leaves only through the surface. We then construct a small rectangular parallelepiped, lying in the interior of the body, with one vertex at a point (x, y, z) and with faces parallel to the coordinate planes. The lengths of the edges are Δx, Δy, and Δz, as shown in Fig. 2. Observe that, since the parallelepiped is small, the continuous function u_t varies little in that region and has approximately the value $u_t(x,t)$ throughout it. This approximation improves, of course, as Δx tends to zero.

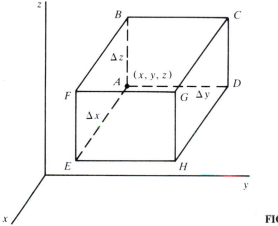

FIGURE 2

The mass of the element of material occupying the parallelepiped is $\delta \, \Delta x \, \Delta y \, \Delta z$. So, in view of the definition of specific heat σ stated above, we know that one measure of the quantity of heat entering that element per unit time at time t is approximately

(2) $$\sigma(\delta \, \Delta x \, \Delta y \, \Delta z) u_t(x,t).$$

Another way to measure that quantity is to observe that, since the flow of heat is parallel to the x axis, heat crosses only the surfaces $ABCD$ and $EFGH$ of the element, which are parallel to the yz plane. If the direction of the flux $\Phi(x,t)$

is in the *positive* direction of the x axis, it follows that the quantity of heat per unit time crossing the surface $ABCD$ into the element at time t is $\Phi(x, t)\,\Delta y\,\Delta z$. Because of the heat leaving the element through the face $EFGH$, the net quantity of heat entering the element per unit time is, then,

$$\Phi(x, t)\,\Delta y\,\Delta z - \Phi(x + \Delta x, t)\,\Delta y\,\Delta z.$$

In view of Fourier's law (1), this expression can be written

(3) $$K\left[u_x(x + \Delta x, t) - u_x(x, t)\right]\Delta y\,\Delta z.$$

Equating expressions (2) and (3) for the quantity of heat entering the element per unit time and then dividing by $\sigma\delta\,\Delta x\,\Delta y\,\Delta z$, we find that

$$u_t(x, t) = \frac{K}{\sigma\delta}\lim_{\Delta x \to 0}\frac{u_x(x + \Delta x, t) - u_x(x, t)}{\Delta x} = \frac{K}{\sigma\delta}u_{xx}(x, t).$$

The temperatures in the solid body, when heat flows only parallel to the x axis, thus satisfy the one-dimensional *heat equation*

(4) $$u_t(x, t) = ku_{xx}(x, t) \qquad \left(k = \frac{K}{\sigma\delta}\right).$$

The constant k here is called the *thermal diffusivity* of the material.

In the derivation of equation (4), we assumed that there is no source (or sink) of heat within the solid body, but only heat transfer by conduction. If there is a uniform source throughout the body that generates heat at a constant rate Q per unit volume, where Q denotes the quantity of heat generated per unit volume per unit time, it is easy to modify the derivation to obtain the nonhomogeneous heat equation

(5) $$u_t(x, t) = ku_{xx}(x, t) + q \qquad \left(q = \frac{Q}{\sigma\delta}\right).$$

This is accomplished by simply adding the term $Q\,\Delta x\,\Delta y\,\Delta z$ to expression (3) and proceeding in the same way as before. The rate Q per unit volume at which heat is generated may, in fact, be any continuous function of x and t, in which case the term q in equation (5) also has that property.

The heat equation describing flow in two and three dimensions is discussed in Sec. 3.

3. HIGHER DIMENSIONS AND BOUNDARY CONDITIONS

When the direction of heat flow in a solid body is not restricted to be simply parallel to the x axis, temperatures u in the body depend, in general, on all the space variables, as well as t. By considering the rate of heat passing through each of the six faces of the element in Fig. 2 (Sec. 2), one can derive (see

Problem 6, Sec. 4) the *three-dimensional* heat equation, satisfied by $u = u(x, y, z, t)$:

$$(1) \qquad u_t = k(u_{xx} + u_{yy} + u_{zz}).$$

The constant k is the thermal diffusivity of the material, appearing in equation (4), Sec. 2. When the *laplacian*

$$(2) \qquad \nabla^2 u = u_{xx} + u_{yy} + u_{zz}$$

is used, equation (1) takes the compact form

$$(3) \qquad u_t = k\nabla^2 u.$$

Note that when there is no flow of heat parallel to the z axis, so that $u_{zz} = 0$ and $u = u(x, y, t)$, equation (1) reduces to the heat equation for *two-dimensional* flow parallel to the xy plane:

$$(4) \qquad u_t = k(u_{xx} + u_{yy}).$$

The one-dimensional heat equation $u_t = ku_{xx}$ in Sec. 2 for temperatures $u = u(x, t)$ follows, of course, from this when there is, in addition, no flow parallel to the y axis. If temperatures are in a *steady state*, in which case u does not vary with time, equation (1) becomes *Laplace's equation*

$$(5) \qquad u_{xx} + u_{yy} + u_{zz} = 0.$$

Equation (5) is often written as $\nabla^2 u = 0$.

The derivation of equation (1) in Problem 6, Sec. 4, takes into account the possibility that heat may be generated in the solid body at a constant rate Q per unit volume, and the generalization

$$(6) \qquad u_t = k\nabla^2 u + q$$

of equation (5), Sec. 2, is obtained. If the rate Q is a continuous function of the space variables x, y, and z and temperatures are in a steady state, equation (6) becomes *Poisson's equation*

$$(7) \qquad \nabla^2 u = f(x, y, z),$$

where $f(x, y, z) = -q(x, y, z)/k$.

Equations that describe thermal conditions on the surfaces of the solid body and initial temperatures throughout the body must accompany the heat equation if we are to determine the temperature function u. The conditions on the surfaces may be other than just prescribed temperatures. Suppose, for example, that the flux Φ into the solid at points on a surface S is some constant Φ_0. That is, at each point P on S, Φ_0 units of heat per unit area per unit time flow across S in the *opposite direction* of an outward unit normal vector $\mathbf{n}$ at P. From Fourier's law (1) in Sec. 2, we know that if du/dn is the directional derivative of u at P in the direction of $\mathbf{n}$, the flux into the solid across S at P is the value of $K\,du/dn$ there. Hence

$$(8) \qquad\qquad K\frac{du}{dn} = \Phi_0$$

on the surface S. Observe that if S is perfectly insulated, $\Phi_0 = 0$ at points on S; and condition (8) becomes

$$(9) \qquad\qquad \frac{du}{dn} = 0.$$

On the other hand, there may be surface heat transfer between a boundary surface and a medium whose temperature is a constant T. The inward flux Φ, which can be negative, may then vary from point to point on S; and we assume that, at each point P, the flux is proportional to the difference between the temperature of the medium and the temperature at P. Under this assumption, which is sometimes called *Newton's law of cooling*, there is a positive constant H, known as the *surface conductance* of the material, such that $\Phi = H(T - u)$ at points on S. Condition (8) is then replaced by the condition

$$(10) \qquad\qquad K\frac{du}{dn} = H(T - u),$$

or

$$(11) \qquad\qquad \frac{du}{dn} = h(T - u) \qquad\qquad \left(h = \frac{H}{K}\right).$$

EXAMPLE. Consider a semi-infinite slab occupying the region $0 \leqslant x \leqslant c$, $y \geqslant 0$ of three-dimensional space. Figure 3 shows the cross section of the slab in the xy plane. Suppose that there is a constant flux Φ_0 into the slab at points on the face in the plane $x = 0$ and that there is surface heat transfer (possibly inward) between the face in the plane $x = c$ and a medium at temperature zero. Also, the surface in the plane $y = 0$ is insulated. Since $du/dn = -\partial u/\partial x$ and $du/dn = \partial u/\partial x$ on the faces in the planes $x = 0$ and $x = c$, respectively, a

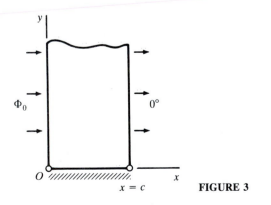

$x = c$ **FIGURE 3**

temperature function $u(x, y, z, t)$ evidently satisfies the boundary conditions

$$-Ku_x(0, y, z, t) = \Phi_0, \qquad u_x(c, y, z, t) = -hu(c, y, z, t).$$

The insulated surface gives rise to the boundary condition $u_y(x, 0, z, t) = 0$.

It should be emphasized that the various partial differential equations in this section are important in other areas of applied mathematics. In simple diffusion problems, for example, Fourier's law $\Phi = -K\,du/dn$ applies to the flux Φ of a substance that is diffusing within a porous solid. In that case, Φ represents the mass of the substance that is diffused per unit area per unit time through a surface, u denotes *concentration* (the mass of the diffusing substance per unit volume of the solid), and K is the *coefficient of diffusion*. Since the mass of the substance entering the element of volume in Fig. 2 (Sec. 2) per unit time is $\Delta x\,\Delta y\,\Delta z\,u_t$, one can replace the product $\sigma\delta$ in the derivation of the heat equation by unity to see that the concentration satisfies the *diffusion equation*

$$(12) \qquad\qquad u_t = K\nabla^2 u.$$

A function $u = u(x, y, z)$ that is continuous, together with its partial derivatives of the first and second order, and satisfies Laplace's equation (5) is called a *harmonic function*. We have seen in this section that the steady-state temperatures at points interior to a solid body in which no heat is generated are represented by a harmonic function. The steady-state concentration of a diffusing substance is also represented by such a function.

Among the many physical examples of harmonic functions, the velocity potential for the steady-state irrotational motion of an incompressible fluid is prominent in hydrodynamics and aerodynamics. An important harmonic function in electrical field theory is the electrostatic potential $V(x, y, z)$ in a region of space that is free of electric charges. The potential may be caused by a static distribution of electric charges outside that region. The fact that V is harmonic is a consequence of the inverse-square law of attraction or repulsion between charges. Likewise, gravitational potential is a harmonic function in regions of space not occupied by matter.

In this book, the physical problems involving the laplacian, and Laplace's equation in particular, are limited mostly to those for which the differential equations are derived in this chapter. Derivations of such differential equations in other areas of applied mathematics can be found in books on hydrodynamics, elasticity, vibrations and sound, electrical field theory, potential theory, and other branches of continuum mechanics. A number of such books are listed in the Bibliography at the back of this book.

4.   THE LAPLACIAN IN CYLINDRICAL AND SPHERICAL COORDINATES

We recall that the heat equation, derived in Sec. 2, and its modifications (Sec. 3), including Laplace's equation, can be written in terms of the laplacian

$$(1) \qquad\qquad \nabla^2 u = u_{xx} + u_{yy} + u_{zz}.$$

Often, because of the geometric configuration of the physical problem, it is more convenient to use the laplacian in other than rectangular coordinates. In this section, we show how the laplacian can be expressed in terms of the variables of two important coordinate systems already encountered in calculus.

The *cylindrical coordinates* ρ, ϕ, and z determine a point $P(\rho, \phi, z)$ whose rectangular coordinates are (Fig. 4)[†]

$$(2) \qquad x = \rho \cos \phi, \qquad y = \rho \sin \phi, \qquad z = z.$$

Thus ρ and ϕ are the *polar coordinates* in the xy plane of the point Q, where Q is the projection of P onto that plane. Relations (2) can be written

$$(3) \qquad \rho = \sqrt{x^2 + y^2}, \qquad \phi = \tan^{-1} \frac{y}{x}, \qquad z = z,$$

where the quadrant to which the angle ϕ belongs is determined by the signs of x and y, not by the ratio y/x alone.

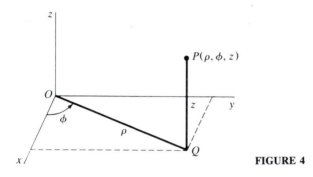

FIGURE 4

Let u denote a function of x, y, and z. Then, in view of relations (2), it is also a function of the three independent variables ρ, ϕ, and z. If u is continuous and possesses continuous partial derivatives of the first and second orders, the following method, based on the chain rule for differentiating composite functions, can be used to express the laplacian (1) in terms of ρ, ϕ, and z.

Relations (3) enable us to write

$$\frac{\partial u}{\partial x} = \frac{\partial u}{\partial \rho} \frac{\partial \rho}{\partial x} + \frac{\partial u}{\partial \phi} \frac{\partial \phi}{\partial x} = \frac{\partial u}{\partial \rho} \frac{x}{\sqrt{x^2 + y^2}} - \frac{\partial u}{\partial \phi} \frac{y}{x^2 + y^2} = \frac{x}{\rho} \frac{\partial u}{\partial \rho} - \frac{y}{\rho^2} \frac{\partial u}{\partial \phi}.$$

Hence, by relations (2),

$$(4) \qquad \frac{\partial u}{\partial x} = \cos \phi \frac{\partial u}{\partial \rho} - \frac{\sin \phi}{\rho} \frac{\partial u}{\partial \phi}.$$

[†] In calculus, the symbols r and θ are often used instead of ρ and ϕ, but the notation used here is common in physics and engineering. The notation for spherical coordinates, treated later in this section, may also differ somewhat from that learned in calculus.

Replacing the function u in equation (4) by $\partial u/\partial x$, we see that

$$(5) \qquad \frac{\partial^2 u}{\partial x^2} = \cos\phi \frac{\partial}{\partial\rho}\left(\frac{\partial u}{\partial x}\right) - \frac{\sin\phi}{\rho}\frac{\partial}{\partial\phi}\left(\frac{\partial u}{\partial x}\right).$$

We may now use expression (4) to substitute for the derivative $\partial u/\partial x$ appearing on the right-hand side of equation (5):

$$\frac{\partial^2 u}{\partial x^2} = \cos\phi \frac{\partial}{\partial\rho}\left(\cos\phi \frac{\partial u}{\partial\rho} - \frac{\sin\phi}{\rho}\frac{\partial u}{\partial\phi}\right) - \frac{\sin\phi}{\rho}\frac{\partial}{\partial\phi}\left(\cos\phi \frac{\partial u}{\partial\rho} - \frac{\sin\phi}{\rho}\frac{\partial u}{\partial\phi}\right).$$

By applying rules for differentiating differences and products of functions and using the relation

$$\frac{\partial^2 u}{\partial\rho\,\partial\phi} = \frac{\partial^2 u}{\partial\phi\,\partial\rho},$$

which is ensured by the continuity of the partial derivatives, we find that

$$(6) \qquad \frac{\partial^2 u}{\partial x^2} = \cos^2\phi \frac{\partial^2 u}{\partial\rho^2} - \frac{2\sin\phi\cos\phi}{\rho}\frac{\partial^2 u}{\partial\phi\,\partial\rho} + \frac{\sin^2\phi}{\rho^2}\frac{\partial^2 u}{\partial\phi^2}$$
$$+ \frac{\sin^2\phi}{\rho}\frac{\partial u}{\partial\rho} + \frac{2\sin\phi\cos\phi}{\rho^2}\frac{\partial u}{\partial\phi}.$$

In the same way, one can show that

$$(7) \qquad \frac{\partial u}{\partial y} = \frac{y}{\rho}\frac{\partial u}{\partial\rho} + \frac{x}{\rho^2}\frac{\partial u}{\partial\phi},$$

or

$$(8) \qquad \frac{\partial u}{\partial y} = \sin\phi \frac{\partial u}{\partial\rho} + \frac{\cos\phi}{\rho}\frac{\partial u}{\partial\phi},$$

and also that

$$(9) \qquad \frac{\partial^2 u}{\partial y^2} = \sin^2\phi \frac{\partial^2 u}{\partial\rho^2} + \frac{2\sin\phi\cos\phi}{\rho}\frac{\partial^2 u}{\partial\phi\,\partial\rho} + \frac{\cos^2\phi}{\rho^2}\frac{\partial^2 u}{\partial\phi^2}$$
$$+ \frac{\cos^2\phi}{\rho}\frac{\partial u}{\partial\rho} - \frac{2\sin\phi\cos\phi}{\rho^2}\frac{\partial u}{\partial\phi}.$$

By adding corresponding sides of equations (6) and (9), we arrive at the identity

$$(10) \qquad \frac{\partial^2 u}{\partial x^2} + \frac{\partial^2 u}{\partial y^2} = \frac{\partial^2 u}{\partial\rho^2} + \frac{1}{\rho}\frac{\partial u}{\partial\rho} + \frac{1}{\rho^2}\frac{\partial^2 u}{\partial\phi^2}.$$

Since rectangular and cylindrical coordinates share the coordinate z, it follows that *the laplacian of u in cylindrical coordinates is*

(11)
$$\nabla^2 u = \frac{\partial^2 u}{\partial \rho^2} + \frac{1}{\rho}\frac{\partial u}{\partial \rho} + \frac{1}{\rho^2}\frac{\partial^2 u}{\partial \phi^2} + \frac{\partial^2 u}{\partial z^2}.$$

We can group the first two terms and use the subscript notation for partial derivatives to write this in the form

(12)
$$\nabla^2 u = \frac{1}{\rho}(\rho u_\rho)_\rho + \frac{1}{\rho^2}u_{\phi\phi} + u_{zz}.$$

Expression (10) gives us the two-dimensional laplacian in polar coordinates. Note that Laplace's equation $\nabla^2 u = 0$ in that coordinate system in the xy plane can be written

(13)
$$\rho^2 u_{\rho\rho} + \rho u_\rho + u_{\phi\phi} = 0.$$

Note, too, how it follows from expression (11) that when temperatures u in a solid body vary only with ρ, and not with ϕ and z, the heat equation $u_t = k\nabla^2 u$ becomes

(14)
$$u_t = k\left(u_{\rho\rho} + \frac{1}{\rho}u_\rho\right).$$

Equations (13) and (14) will be of particular interest in the applications.

The *spherical coordinates r*, ϕ, and θ of a point $P(r,\phi,\theta)$ (Fig. 5) are related to x, y, and z as follows:

(15) $x = r \sin\theta \cos\phi, \qquad y = r \sin\theta \sin\phi, \qquad z = r \cos\theta.$

The coordinate ϕ is common to cylindrical and spherical coordinates, and the coordinates in those two systems are related by the equations

(16) $z = r \cos\theta, \qquad \rho = r \sin\theta, \qquad \phi = \phi.$

Expression (11) for the laplacian can be transformed into spherical coordinates quite readily by means of the proper interchange of letters, *without any*

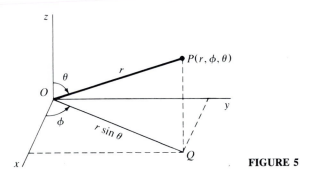

FIGURE 5

further application of the chain rule. This is accomplished in three steps, described below.

First, we observe that, except for the names of the variables involved, transformation (16) is the same as transformation (2). Since transformation (2) gave us equation (10), we know that

(17)
$$\frac{\partial^2 u}{\partial z^2} + \frac{\partial^2 u}{\partial \rho^2} = \frac{\partial^2 u}{\partial r^2} + \frac{1}{r}\frac{\partial u}{\partial r} + \frac{1}{r^2}\frac{\partial^2 u}{\partial \theta^2}.$$

Second, we note that when transformation (16) is used, the counterpart of equation (7) is

$$\frac{\partial u}{\partial \rho} = \frac{\rho}{r}\frac{\partial u}{\partial r} + \frac{z}{r^2}\frac{\partial u}{\partial \theta}.$$

With this and relations (16), we are able to write

(18)
$$\frac{1}{\rho}\frac{\partial u}{\partial \rho} + \frac{1}{\rho^2}\frac{\partial^2 u}{\partial \phi^2} = \frac{1}{r}\frac{\partial u}{\partial r} + \frac{\cot \theta}{r^2}\frac{\partial u}{\partial \theta} + \frac{1}{r^2 \sin^2 \theta}\frac{\partial^2 u}{\partial \phi^2}.$$

Third, by grouping the first and last terms in expression (11), and also the second and third terms there, we see that, according to equations (17) and (18), *the laplacian of u in spherical coordinates is*

(19)
$$\nabla^2 u = \frac{\partial^2 u}{\partial r^2} + \frac{2}{r}\frac{\partial u}{\partial r} + \frac{1}{r^2 \sin^2 \theta}\frac{\partial^2 u}{\partial \phi^2} + \frac{1}{r^2}\frac{\partial^2 u}{\partial \theta^2} + \frac{\cot \theta}{r^2}\frac{\partial u}{\partial \theta}.$$

Other forms of this expression are

(20)
$$\nabla^2 u = \frac{1}{r}(ru)_{rr} + \frac{1}{r^2 \sin^2 \theta} u_{\phi\phi} + \frac{1}{r^2 \sin \theta}(\sin \theta\, u_\theta)_\theta,$$

(21)
$$\nabla^2 u = \frac{1}{r^2}(r^2 u_r)_r + \frac{1}{r^2 \sin^2 \theta} u_{\phi\phi} + \frac{1}{r^2 \sin \theta}(\sin \theta\, u_\theta)_\theta.$$

Many of our applications later on will involve Laplace's equation $\nabla^2 u = 0$ in spherical coordinates when u is independent of ϕ. According to expression (20), that equation can then be written

(22)
$$r\frac{\partial^2}{\partial r^2}(ru) + \frac{1}{\sin \theta}\frac{\partial}{\partial \theta}\left(\sin \theta\,\frac{\partial u}{\partial \theta}\right) = 0.$$

PROBLEMS

1. Let $u(x)$ denote the steady-state temperatures in a slab bounded by the planes $x = 0$ and $x = c$ when those faces are kept at fixed temperatures $u = 0$ and $u = u_0$, respectively. Set up the boundary value problem for $u(x)$ and solve it to show that

$$u(x) = \frac{u_0}{c}x \quad \text{and} \quad \Phi_0 = K\frac{u_0}{c},$$

where Φ_0 is the flux of heat to the left across each plane $x = x_0$ $(0 \leqslant x_0 \leqslant c)$.

2. A slab occupies the region $0 \leqslant x \leqslant c$. There is a constant flux of heat Φ_0 into the slab through the face $x = 0$. The face $x = c$ is kept at temperature $u = 0$. Set up and solve the boundary value problem for the steady-state temperatures $u(x)$ in the slab.

$$\text{Answer: } u(x) = \frac{\Phi_0}{K}(c - x).$$

3. Let a slab $0 \leqslant x \leqslant c$ be subjected to surface heat transfer, according to Newton's law of cooling, at its faces $x = 0$ and $x = c$, the surface conductance H being the same on each face. Show that if the medium $x < 0$ has temperature zero and the medium $x > c$ has the constant temperature T, then the boundary value problem for steady-state temperatures $u(x)$ in the slab is

$$u''(x) = 0 \qquad\qquad (0 < x < c),$$

$$Ku'(0) = Hu(0), \qquad Ku'(c) = H[T - u(c)],$$

where K is the thermal conductivity of the material in the slab. Write $h = H/K$ and derive the expression

$$u(x) = \frac{T}{ch + 2}(hx + 1)$$

for those temperatures.

4. Let $u(r)$ denote the steady-state temperatures in a solid bounded by two concentric spheres $r = a$ and $r = b$ $(a < b)$ when the inner surface $r = a$ is kept at temperature zero and the outer surface $r = b$ is maintained at a constant temperature u_0. Show why Laplace's equation for $u = u(r)$ reduces to

$$\frac{d^2}{dr^2}(ru) = 0,$$

and then derive the expression

$$u(r) = \frac{bu_0}{b - a}\left(1 - \frac{a}{r}\right) \qquad\qquad (a \leqslant r \leqslant b).$$

Sketch the graph of $u(r)$ versus r.

5. In Problem 4, replace the condition on the outer surface $r = b$ with the condition that there is surface heat transfer into a medium at constant temperature T according to Newton's law of cooling. Then obtain the expression

$$u(r) = \frac{hb^2 T}{a + hb(b - a)}\left(1 - \frac{a}{r}\right) \qquad\qquad (a \leqslant r \leqslant b)$$

for the steady-state temperatures, where h is the ratio of the surface conductance H to the thermal conductivity K of the material.

6. Let $u = u(x, y, z, t)$ denote temperatures in a solid body throughout which there is a uniform heat source. Derive the heat equation

$$u_t = k\nabla^2 u + q$$

for those temperatures, where the constants k and q are the same ones as in equation (5), Sec. 2.

Suggestion: Modify the derivation of equation (5), Sec. 2, by also considering the net rate of heat entering the element in Fig. 2 (Sec. 2) through the faces parallel to the xz and xy planes. Since the faces are small, one may consider the needed flux at points on a given face to be constant over that face. Thus, for instance, the net rate of heat entering the element through the faces parallel to the xy plane is to be taken as

$$K\left[u_z(x, y, z + \Delta z, t) - u_z(x, y, z, t)\right]\Delta x\,\Delta y.$$

7. A slender wire lies along the x axis, and surface heat transfer takes place along the wire into the surrounding medium at a fixed temperature T. Modify the procedure in Sec. 2 to show that if $u = u(x, t)$ denotes temperatures in the wire, then

$$u_t = ku_{xx} + b(T - u),$$

where b is a positive constant.

Suggestion: Let r denote the radius of the wire, and apply Newton's law of cooling to see that the quantity of heat entering the element in Fig. 6 through its cylindrical surface per unit time is approximately $H[T - u(x, t)]2\pi r\,\Delta x$.

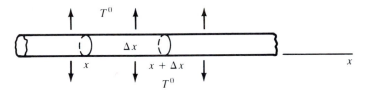

FIGURE 6

8. Suppose that the thermal coefficients K and σ are functions of x, y, and z. Modify the derivation in Problem 7 to show that the heat equation takes the form

$$\sigma\delta u_t = (Ku_x)_x + (Ku_y)_y + (Ku_z)_z$$

in a domain where all functions and derivatives involved are continuous.

9. Show that the substitution $\tau = kt$ can be used to write the two-dimensional heat equation $u_t = k(u_{xx} + u_{yy})$ in the form $u_\tau = u_{xx} + u_{yy}$, where $k = 1$.

10. Derive expressions (8) and (9) in Sec. 4 for $\partial u/\partial y$ and $\partial^2 u/\partial y^2$ in cylindrical coordinates.

11. In Sec. 4, show how expressions (20) and (21) for $\nabla^2 u$ in spherical coordinates follow from expression (19).

12. (*a*) Show that if u is a function of the polar coordinates ρ and ϕ, where $x = \rho\cos\phi$ and $y = \rho\sin\phi$, then

$$\frac{\partial u}{\partial \phi} = -y\frac{\partial u}{\partial x} + x\frac{\partial u}{\partial y}.$$

(*b*) Let $u(\rho, \phi)$ denote temperatures, independent of the cylindrical coordinate z, in a long rod, parallel to the z axis, whose cross section in the xy plane is the

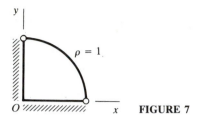

FIGURE 7

sector $0 \leqslant \rho \leqslant 1, 0 \leqslant \phi \leqslant \pi/2$ of a disk (Fig. 7). Use the result in part (a) to show that if the rod is insulated on its planar surfaces, where $\phi = 0$ and $\phi = \pi/2$, then u must satisfy the boundary conditions

$$u_\phi(\rho, 0) = 0, \qquad u_\phi\!\left(\rho, \frac{\pi}{2}\right) = 0 \qquad\qquad (0 < \rho < 1).$$

13. Suppose that temperatures u in a solid hemisphere $r \leqslant 1, 0 \leqslant \theta \leqslant \pi/2$ are independent of the spherical coordinate ϕ, so that $u = u(r, \theta)$, and that the base of the hemisphere is insulated (Fig. 8). Use transformation (16), Sec. 4, relating cylindrical to spherical coordinates, to show that

$$\frac{\partial u}{\partial \theta} = -\rho \frac{\partial u}{\partial z} + z \frac{\partial u}{\partial \rho}.$$

Thus show that u must satisfy the boundary condition $u_\theta(r, \pi/2) = 0$.

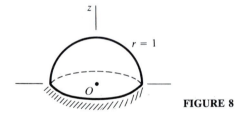

FIGURE 8

14. Show that the physical dimensions of thermal conductivity k (Sec. 2) are $L^2 T^{-1}$, where L denotes length and T time.

    Suggestion: Observe first that the dimensions of thermal conductivity K and specific heat σ are $AL^{-1}T^{-1}B^{-1}$ and $AM^{-1}B^{-1}$, respectively, where M denotes mass, A quantity of heat, and B temperature. Then recall that $k = K/(\sigma\delta)$, where δ is density (ML^{-3}).

5. A VIBRATING STRING

A tightly stretched string, whose position of equilibrium is some interval on the x axis, is vibrating in the xy plane. Each point of the string, with coordinates $(x, 0)$ in the equilibrium position, has a transverse displacement $y = y(x, t)$ at time t. We assume that the displacements y are small relative to the length of the string, that slopes are small, and that other conditions are such that the movement of each point is parallel to the y axis. Then, at time t, a point on the string has coordinates (x, y), where $y = y(x, t)$.

Let the tension of the string be great enough that the string behaves as if it were perfectly flexible. That is, at a point (x, y) on the string, the part of the string to the left of that point exerts a force **T**, in the tangential direction, on the part to the right; and any resistance to bending at the point is to be neglected. The magnitude of the x component of the tensile force **T** is denoted by H. See Fig. 9, where that x component has the same positive sense as the x axis. Our final assumption here is that H is constant. That is, the variation of H with respect to x and t can be neglected.

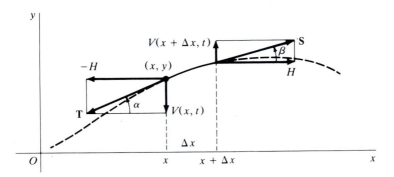

FIGURE 9

These idealizing assumptions are severe, but they are justified in many applications. They are adequately satisfied, for instance, by strings of musical instruments under ordinary conditions of operation. Mathematically, the assumptions will lead us to a partial differential equation in $y(x, t)$ that is linear.

Now let $V(x, t)$ denote the y component of the tensile force **T** exerted by the left-hand portion of the string on the right-hand portion at the point (x, y). We take the positive sense of V to be that of the y axis. If α is the angle of inclination of the string at the point (x, y) at time t, then

(1)
$$\frac{-V(x, t)}{H} = \tan \alpha = y_x(x, t).$$

This is indicated in Fig. 9, where $V(x, t) < 0$ and $y_x(x, t) > 0$. If $V(x, t) > 0$, then $\pi/2 < \alpha < \pi$ and $y_x(x, t) < 0$; and a similar sketch shows that

$$\frac{V(x, t)}{H} = \tan(\pi - \alpha) = -\tan \alpha = -y_x(x, t).$$

Hence relations (1) still hold. Note, too, that $y_x(x, t) = 0$ when $V(x, t) = 0$, since $\alpha = 0$ then. It follows from relations (1) that *the y component $V(x, t)$ of the force exerted at time t by the part of the string to the left of a point (x, y) on* *the part to the right is given by the equation*

(2) $$V(x,t) = -Hy_x(x,t) \qquad (H > 0),$$

which is basic for deriving the equation of motion of the string. Equation (2) is also used in setting up certain types of boundary conditions.

Suppose that all external forces such as the weight of the string and resistance forces, other than forces at the end points, can be neglected. Consider a segment of the string not containing an end point and whose projection onto the x axis has length Δx. Since x components of displacements are negligible, the mass of the segment is $\delta \Delta x$, where the constant δ is the mass per unit length of the string. At time t, the y component of the force exerted by the string on the segment at the left-hand end (x, y) is $V(x, t)$, given by equation (2). The tangential force **S** exerted on the other end of the segment by the part of the string to the right is also indicated in Fig. 9. Its y component $V(x + \Delta x, t)$ evidently satisfies the relation

$$\frac{V(x + \Delta x, t)}{H} = \tan \beta,$$

where β is the angle of inclination of the string at that other end of the segment. That is,

(3) $$V(x + \Delta x, t) = Hy_x(x + \Delta x, t) \qquad (H > 0).$$

Note that, except for a minus sign, this is equation (2) when the argument x there is replaced by $x + \Delta x$.

Now the acceleration of the end (x, y) in the y direction is $y_{tt}(x, t)$. Consequently, by Newton's second law of motion (mass times acceleration equals force), it follows from equations (2) and (3) that

(4) $$\delta \Delta x\, y_{tt}(x, t) = -Hy_x(x, t) + Hy_x(x + \Delta x, t),$$

approximately, when Δx is small. Hence

$$y_{tt}(x, t) = \frac{H}{\delta} \lim_{\Delta x \to 0} \frac{y_x(x + \Delta x, t) - y_x(x, t)}{\Delta x} = \frac{H}{\delta} y_{xx}(x, t)$$

whenever these partial derivatives exist. Thus the function $y(x, t)$, which represents the transverse displacements in a stretched string under the conditions stated above, satisfies the one-dimensional *wave equation*

(5) $$y_{tt}(x, t) = a^2 y_{xx}(x, t) \qquad \left(a^2 = \frac{H}{\delta}\right).$$

The constant a has the physical dimensions of velocity.

One can choose units for the time variable so that $a = 1$ in the wave equation. More precisely, if we make the substitution $\tau = at$, the chain rule shows that

$$\frac{\partial y}{\partial t} = a\frac{\partial y}{\partial \tau} \quad \text{and} \quad \frac{\partial^2 y}{\partial t^2} = a\frac{\partial}{\partial \tau}\left(a\frac{\partial y}{\partial \tau}\right) = a^2\frac{\partial^2 y}{\partial \tau^2}.$$

Equation (5) then becomes $y_{\tau\tau} = y_{xx}$. (A similar observation was made in Problem 9, Sec. 4, with regard to the heat equation.)

When external forces parallel to the y axis act along the string, we let F denote the force per unit length of string, the positive sense of F being that of the y axis. Then a term $F\Delta x$ must be added on the right-hand side of equation (4), and the equation of motion is

$$(6) \qquad\qquad y_{tt}(x,t) = a^2 y_{xx}(x,t) + \frac{F}{\delta}.$$

In particular, with the y axis vertical and its positive sense upward, suppose that the external force consists of the weight of the string. Then $F\Delta x = -\delta\Delta xg$, where the positive constant g is the acceleration due to gravity; and equation (6) becomes the linear nonhomogeneous equation

$$(7) \qquad\qquad y_{tt}(x,t) = a^2 y_{xx}(x,t) - g.$$

In equation (6), F may be a function of x, t, y, or derivatives of y. If the external force per unit length is a damping force proportional to the velocity in the y direction, for example, F is replaced by $-By_t$, where the positive constant B is a damping coefficient. Then the equation of motion is linear and homogeneous:

$$(8) \qquad\qquad y_{tt}(x,t) = a^2 y_{xx}(x,t) - by_t(x,t) \qquad\qquad \left(b = \frac{B}{\delta}\right).$$

If an end $x = 0$ of the string is kept fixed at the origin at all times $t \geqslant 0$, the boundary condition there is clearly

$$(9) \qquad\qquad y(0,t) = 0 \qquad\qquad\qquad (t \geqslant 0).$$

But if the end is permitted to slide along the y axis and is moved along that axis with a displacement $f(t)$, the boundary condition is the linear nonhomogeneous one

$$(10) \qquad\qquad y(0,t) = f(t) \qquad\qquad\qquad (t \geqslant 0).$$

Suppose that the left-hand end is attached to a ring which can slide along the y axis. When a force $F(t)$ ($t > 0$) in the y direction is applied to that end, $F(t)$ is the limit, as x tends to zero through positive values, of the force $V(x,t)$ described earlier in this section. According to equation (2), the boundary condition at $x = 0$ is then

$$-Hy_x(0,t) = F(t) \qquad\qquad\qquad (t > 0).$$

The minus sign disappears, however, if $x = 0$ is the *right-hand* end, in view of equation (3).

6.   VIBRATIONS OF BARS AND MEMBRANES

We describe here two other types of vibrations for which the displacements satisfy wave equations. We continue to limit our attention to fairly simple phenomena.

(a)   Longitudinal Vibrations of Bars

Let the coordinate x denote distances from one end of an elastic bar, in the shape of a cylinder or prism, to other cross sections when the bar is unstrained. Displacements of the ends or initial displacements or velocities in the bar, all directed lengthwise along it and uniform over each cross section involved, cause the sections to move parallel to the x axis. At time t, the longitudinal displacement of the section at a point x is denoted by $y(x, t)$. Thus the origin of the displacement y of the section at x is in a fixed coordinate system outside the bar, in the plane of the unstrained position of that section (Fig. 10).

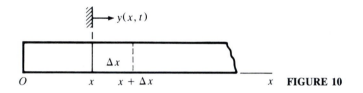

O x $x + \Delta x$ x **FIGURE 10**

At the same time, a neighboring section, labeled $x + \Delta x$ in Fig. 10 and to the right of the section at x, has a displacement $y(x + \Delta x, t)$. The element of the bar with natural length Δx is, then, stretched or compressed by the amount $y(x + \Delta x, t) - y(x, t)$. We assume that such an extension or compression of the element satisfies Hooke's law and that the effect of the inertia of the moving element is negligible. Hence the force exerted on the section at x by the part of the bar to the left of that section is

$$-AE\frac{y(x + \Delta x, t) - y(x, t)}{\Delta x},$$

where A is the area of each cross section, the positive constant E is *Young's modulus of elasticity*, and the ratio shown represents the relative change in length of the element. When Δx tends to zero, it follows that the longitudinal force $F(x, t)$ exerted on the element at its left-hand end is given by the basic equation

(1) $$F(x, t) = -AEy_x(x, t).$$

Similarly, the force on the right-hand end is

(2) $$F(x + \Delta x, t) = AEy_x(x + \Delta x, t).$$

Let the constant δ denote the mass per unit volume of the material. Then, applying Newton's second law to the motion of an element of the bar of length

Δx, we may write

(3) $\delta A\,\Delta x\, y_{tt}(x, t) = -AEy_x(x, t) + AEy_x(x + \Delta x, t).$

We find, after dividing by $\delta A\,\Delta x$ and letting Δx tend to zero, that

(4) $$y_{tt}(x, t) = a^2 y_{xx}(x, t) \qquad \left(a^2 = \frac{E}{\delta}\right).$$

Thus the longitudinal displacements $y(x, t)$ in an elastic bar satisfy the wave equation (4) when no external longitudinal forces act on the bar, other than at the ends. We have assumed that displacements are small enough that Hooke's law applies and that sections remain planar after being displaced. The elastic bar here may be replaced by a column of air, in which case equation (4) has applications in the theory of sound.

The boundary condition $y(0, t) = 0$ signifies that the end $x = 0$ of the bar is held fixed. If, instead, the end $x = 0$ is free when $t > 0$, then no force acts at that end; that is, $F(0, t) = 0$ and, in view of equation (1),

(5) $y_x(0, t) = 0$ $(t > 0).$

(b)   Transverse Vibrations of Membranes

Let $z(x, y, t)$ denote small displacements in the z direction, at time t, of points on a flexible membrane that is tightly stretched over a horizontal frame. In the equilibrium position, a point on the membrane has coordinates (x, y) in the xy plane. The plane through that point and parallel to the xz plane intersects the displaced membrane in a curve containing the points labeled A and B in Fig. 11. By making similar constructions, we can form the element $ABCD$ of the membrane that is also shown in Fig. 11. The projection of the element onto the xy plane is a small rectangle with edges of length Δx and Δy.

We now examine the internal tensile forces that are exerted on the element at points of the curve AB, those forces being tangent to the element and normal to AB. In analyzing such a force, we let H denote the magnitude *per unit length* along AB of the component parallel to the xy plane. We assume

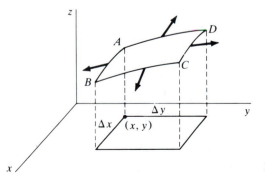

FIGURE 11

that H is constant, regardless of what point or curve on the membrane is being discussed. In view of expressions (2) and (3) in Sec. 5 for the forces V on the ends of a segment of a vibrating string, we know that the force in the z direction exerted over the curve AB is approximately $-Hz_y(x, y, t)\,\Delta x$ and that the corresponding force over the curve DC is approximately $Hz_y(x, y + \Delta y, t)\,\Delta x$. Similar expressions are found for the vertical forces exerted over AD and BC when the tensile forces on those curves are considered. It then follows that the sum of the vertical forces exerted over the entire boundary of the element is approximately

$$(6) \qquad \begin{aligned} &-Hz_y(x, y, t)\,\Delta x + Hz_y(x, y + \Delta y, t)\,\Delta x \\ &\quad -Hz_x(x, y, t)\,\Delta y + Hz_x(x + \Delta x, y, t)\,\Delta y. \end{aligned}$$

If Newton's second law is applied to the motion of the element in the z direction and if δ denotes the mass per unit area of the membrane, it follows from expression (6) for the total force on the element that $z(x, y, t)$ satisfies the *two-dimensional* wave equation

$$(7) \qquad\qquad z_{tt} = a^2(z_{xx} + z_{yy}) \qquad\qquad \left(a^2 = \frac{H}{\delta}\right).$$

Details of these final steps are left to the problems, where it is also shown that if an external transverse force $F(x, y, t)$ per unit area acts over the membrane, the equation of motion takes the form

$$(8) \qquad\qquad z_{tt} = a^2(z_{xx} + z_{yy}) + \frac{F}{\delta}.$$

Equation (8) arises, for example, when the z axis is directed vertically upward and the weight of the membrane is taken into account in the derivation of equation (7). Then $F/\delta = -g$, where g is the acceleration due to gravity.

From equation (7), one can see that the *static* transverse displacements $z(x, y)$ of a stretched membrane satisfy Laplace's equation (Sec. 3) in two dimensions. Here the displacements are the result of displacements, perpendicular to the xy plane, of parts of the frame that support the membrane when no external forces are exerted except at the boundary.

PROBLEMS

1. A stretched string, with its ends fixed at the points 0 and $2c$ on the x axis, hangs at rest under its own weight. The y axis is directed vertically upward. Point out how it follows from the nonhomogeneous wave equation (7), Sec. 5, that the static displacements $y(x)$ of points on the string must satisfy the differential equation

$$a^2 y''(x) = g \qquad\qquad \left(a^2 = \frac{H}{\delta}\right)$$

on the interval $0 < x < 2c$, in addition to the boundary conditions

$$y(0) = 0, \qquad y(2c) = 0.$$

By solving this boundary value problem, show that the string hangs in the parabolic arc

$$(x - c)^2 = \frac{2a^2}{g}\left(y + \frac{gc^2}{2a^2}\right) \qquad (0 \leqslant x \leqslant 2c)$$

and that the depth of the vertex of the arc varies directly with c^2 and δ and inversely with H.

2. Use expression (2), Sec. 5, for the vertical force V and the equation of the arc in which the string in Problem 1 lies to show that the vertical force exerted on that string by each support is δcg, half the weight of the string.

3. Let $z(\rho)$ represent static transverse displacements in a membrane, stretched between the two circles $\rho = 1$ and $\rho = \rho_0$ ($\rho_0 > 1$) in the plane $z = 0$, after the outer support $\rho = \rho_0$ is displaced by a distance $z = z_0$. State why the boundary value problem in $z(\rho)$ can be written

$$\frac{d}{d\rho}\left(\rho\frac{dz}{d\rho}\right) = 0 \qquad (1 < \rho < \rho_0),$$

$$z(1) = 0, \qquad z(\rho_0) = z_0,$$

and obtain the solution

$$z(\rho) = z_0\frac{\ln \rho}{\ln \rho_0} \qquad (1 \leqslant \rho \leqslant \rho_0).$$

4. Show that the steady-state temperatures $u(\rho)$ in a hollow cylinder $1 \leqslant \rho \leqslant \rho_0$, $-\infty < z < \infty$ also satisfy the boundary value problem written in Problem 3 if $u = 0$ on the inner cylindrical surface and $u = z_0$ on the outer one. Thus show that Problem 3 is a *membrane analogy* for this temperature problem. Soap films have been used to display such analogies.

5. Give needed details in the derivation of equation (6), Sec. 5, for the forced vibrations of a stretched string.

6. The physical dimensions of H, the magnitude of the x component of the tensile force in a string, are those of mass times acceleration: MLT^{-2}, where M denotes mass, L length, and T time. Show that, since $a^2 = H/\delta$, the constant a has the dimensions of velocity: LT^{-1}.

7. A strand of wire 1 ft long, stretched between the origin and the point 1 on the x axis, weighs 0.032 lb ($\delta g = 0.032$, $g = 32$ ft/s^2) and $H = 10$ lb. At the instant $t = 0$, the strand lies along the x axis but has a velocity of 1 ft/s in the direction of the y axis, perhaps because the supports were in motion and were brought to rest at that instant. Assuming that no external forces act along the wire, state why the displacements $y(x, t)$ should satisfy this boundary value problem:

$$y_{tt}(x, t) = 10^4 y_{xx}(x, t) \qquad (0 < x < 1, t > 0),$$

$$y(0, t) = y(1, t) = 0, \qquad y(x, 0) = 0, \qquad y_t(x, 0) = 1.$$

8. The end $x = 0$ of a cylindrical elastic bar is kept fixed, and a constant compressive force of magnitude F_0 units per unit area is exerted at all times $t > 0$ over the end $x = c$. The bar is initially unstrained and at rest, with no external forces acting along it. State why the function $y(x, t)$ representing the longitudinal displacements of cross sections should satisfy this boundary value problem, where $a^2 = E/\delta$:

$$y_{tt}(x, t) = a^2 y_{xx}(x, t) \qquad\qquad (0 < x < c, t > 0),$$

$$y(0, t) = 0, \qquad E y_x(c, t) = -F_0, \qquad y(x, 0) = y_t(x, 0) = 0.$$

9. The left-hand end $x = 0$ of a horizontal elastic bar is elastically supported in such a way that the longitudinal force per unit area exerted on the bar at that end is proportional to the displacement of the end, but opposite in sign. State why the end condition there has the form

$$y_x(0, t) = by(0, t) \qquad\qquad\qquad (b > 0).$$

10. Use expression (6), Sec. 6, to derive the nonhomogeneous wave equation (8), Sec. 6, for a membrane when there is an external transverse force $F(x, y, t)$ per unit area acting on it. [Note that if this force is zero $(F \equiv 0)$, the equation reduces to equation (7), Sec. 6.]

11. Let $z(x, y)$ denote the static transverse displacements in a membrane over which an external transverse force $F(x, y)$ per unit area acts. Show how it follows from the nonhomogeneous wave equation (8), Sec. 6, that $z(x, y)$ satisfies *Poisson's equation*:

$$z_{xx} + z_{yy} + f = 0 \qquad\qquad \left(f = \frac{F}{H}\right).$$

[Compare equation (7), Sec. 3.]

12. A uniform transverse force of F_0 units per unit area acts over a membrane, stretched between the two circles $\rho = 1$ and $\rho = \rho_0$ $(\rho_0 > 1)$ in the plane $z = 0$. From Problem 11, show that the static transverse displacements $z(\rho)$ satisfy the equation

$$(\rho z')' + f_0 \rho = 0 \qquad\qquad \left(f_0 = \frac{F_0}{H}\right),$$

and derive the expression

$$z(\rho) = \frac{f_0}{4}(\rho_0^2 - 1)\left(\frac{\ln \rho}{\ln \rho_0} - \frac{\rho^2 - 1}{\rho_0^2 - 1}\right) \qquad (1 \leqslant \rho \leqslant \rho_0).$$

7. TYPES OF EQUATIONS AND BOUNDARY CONDITIONS

The second-order linear partial differential equation (Sec. 1)

$$(1) \qquad\qquad A u_{xx} + B u_{xy} + C u_{yy} + D u_x + E u_y + F u = G$$

in $u = u(x, y)$, where $A, B, \ldots, G$ are constants or functions of x and y, is classified in any given region of the xy plane according to whether $B^2 - 4AC$ is positive, negative, or zero throughout that region. Specifically, equation (1) is

(*a*) *Hyperbolic* if $B^2 - 4AC > 0$;
(*b*) *Elliptic* if $B^2 - 4AC < 0$;
(*c*) *Parabolic* if $B^2 - 4AC = 0$.

For each of these categories, equation (1) and its solutions have distinct features. Some indication of this is given in Problems 15 and 16, Sec. 9. The terminology used here is suggested by the fact (Problem 6, Sec. 9) that when $A, B, \dots, F$ are constants and $G \equiv 0$, equation (1) always has solutions of the form $u = \exp(\lambda x + \mu y)$, where the constants λ and μ satisfy the algebraic equation

$$(2) \qquad\qquad A\lambda^2 + B\lambda\mu + C\mu^2 + D\lambda + E\mu + F = 0.$$

From analytic geometry, we know that such an equation represents a conic section in the $\lambda\mu$ plane and that the different types of conic sections arising are similarly determined by $B^2 - 4AC$.

EXAMPLES. Laplace's equation

$$u_{xx} + u_{yy} = 0$$

is a special case of equation (1) in which $A = C = 1$ and $B = 0$. Hence it is elliptic throughout the xy plane. Poisson's equation (Sec. 3)

$$u_{xx} + u_{yy} = f(x, y)$$

in two dimensions is elliptic in any region of the xy plane where $f(x, y)$ is defined. The one-dimensional heat equation

$$-ku_{xx} + u_t = 0$$

in $u = u(x, t)$ is parabolic in the xt plane, and the one-dimensional wave equation

$$-a^2 y_{xx} + y_{tt} = 0$$

in $y = y(x, t)$ is hyperbolic there.
Another special case of equation (1) is the *telegraph equation*[†]

$$v_{xx} = KLv_{tt} + (KR + LS)v_t + RSv.$$

Here $v(x, t)$ represents either the electrostatic potential or current at time t at a point x units from one end of a transmission line or cable that has

---

[†] A derivation of this equation is outlined in the book by Churchill (1972, pp. 272–273) that is listed in the Bibliography.

electrostatic capacity K, self-inductance L, resistance R, and leakage conductance S, all per unit length. The equation is hyperbolic if $KL > 0$. It is parabolic if either K or L is zero.

As indicated below, the three types of second-order linear equations just described require, in general, different types of boundary conditions in order to determine a solution.

Let u denote the dependent variable in a boundary value problem. A condition that prescribes the values of u itself along a portion of the boundary is known as a *Dirichlet condition*. The problem of determining a harmonic function on a domain such that the function assumes prescribed values over the entire boundary of that domain is called a *Dirichlet problem*. In that case, the values of the function can be interpreted as steady-state temperatures. Such a physical interpretation leads us to expect that a Dirichlet problem may have a unique solution if the functions considered satisfy certain requirements as to their regularity.

A *Neumann condition* prescribes the values of normal derivatives du/dn on a part of the boundary. Another type of boundary condition is a *Robin condition*. It prescribes values of $hu + du/dn$ at boundary points, where h is either a constant or a function of the independent variables.

If a partial differential equation in y is of the second order with respect to one of the independent variables t and if the values of both y and y_t are prescribed when $t = 0$, the boundary condition is one of *Cauchy type* with respect to t. In the case of the wave equation $y_{tt} = a^2 y_{xx}$, such a condition corresponds physically to that of prescribing the initial values of the transverse displacements y and velocities y_t in a stretched string. Initial values for both y and y_t appear to be needed if the displacements $y(x, t)$ are to be determined.

When the equation is Laplace's equation $u_{xx} + u_{yy} = 0$ or the heat equation $ku_{xx} = u_t$, however, conditions of Cauchy type on u with respect to x cannot be imposed without severe restrictions. This is suggested by interpreting u physically as a temperature function. When the temperatures u in a slab $0 \leqslant x \leqslant c$ are prescribed on the face $x = 0$, for example, the flux Ku_x to the left through that face is ordinarily determined by the values of u there and by other conditions in the problem. Conversely, if the flux Ku_x is prescribed at $x = 0$, the temperatures there are affected.

8. METHODS OF SOLUTION

Some boundary value problems in partial differential equations can be solved by a method corresponding to the one usually used to solve such problems in ordinary differential equations, namely the method of first finding the general solution of the differential equation.

EXAMPLE 1. Let us solve the boundary value problem

(1) $$u_{xx}(x, y) = 0, \qquad u(0, y) = y^2, \qquad u(1, y) = 1$$

on the domain $0 < x < 1$, $-\infty < y < \infty$.

Successive integrations of the equation $u_{xx} = 0$ with respect to x, with y kept fixed, lead to the equations $u_x = \phi(y)$ and

(2) $$u = x\phi(y) + \psi(y),$$

where ϕ and ψ are arbitrary functions of y. The boundary conditions in problem (1) require that

$$\psi(y) = y^2, \qquad \phi(y) + \psi(y) = 1.$$

Thus $\phi(y) = 1 - y^2$, and the solution of the problem is

(3) $$u(x, y) = x(1 - y^2) + y^2.$$

EXAMPLE 2. We next solve the wave equation

(4) $$y_{tt}(x, t) = a^2 y_{xx}(x, t) \qquad (-\infty < x < \infty, t > 0),$$

subject to the boundary conditions

(5) $$y(x, 0) = f(x), \qquad y_t(x, 0) = 0 \qquad (-\infty < x < \infty),$$

in terms of the constant a and the function f.

The differential equation (4) can be simplified as follows by introducing the new independent variables

(6) $$u = x + at, \qquad v = x - at.$$

According to the chain rule for differentiating composite functions,

$$\frac{\partial y}{\partial t} = \frac{\partial y}{\partial u} \frac{\partial u}{\partial t} + \frac{\partial y}{\partial v} \frac{\partial v}{\partial t}.$$

That is,

(7) $$\frac{\partial y}{\partial t} = a \frac{\partial y}{\partial u} - a \frac{\partial y}{\partial v}.$$

Replacing the function y by $\partial y / \partial t$ in equation (7) yields the expression

$$\frac{\partial^2 y}{\partial t^2} = a \frac{\partial}{\partial u} \left(\frac{\partial y}{\partial t} \right) - a \frac{\partial}{\partial v} \left(\frac{\partial y}{\partial t} \right);$$

and using equation (7) again, this time to substitute for $\partial y / \partial t$ on the right here, we see that

$$\frac{\partial^2 y}{\partial t^2} = a\frac{\partial}{\partial u}\left(a\frac{\partial y}{\partial u} - a\frac{\partial y}{\partial v}\right) - a\frac{\partial}{\partial v}\left(a\frac{\partial y}{\partial u} - a\frac{\partial y}{\partial v}\right),$$

or

(8)
$$\frac{\partial^2 y}{\partial t^2} = a^2\left(\frac{\partial^2 y}{\partial u^2} - 2\frac{\partial^2 y}{\partial v\,\partial u} + \frac{\partial^2 y}{\partial v^2}\right).$$

We have, of course, assumed that

$$\frac{\partial^2 y}{\partial u\,\partial v} = \frac{\partial^2 y}{\partial v\,\partial u}.$$

In like manner, one can show that

(9)
$$\frac{\partial^2 y}{\partial x^2} = \frac{\partial^2 y}{\partial u^2} + 2\frac{\partial^2 y}{\partial v\,\partial u} + \frac{\partial^2 y}{\partial v^2}.$$

In view of expressions (8) and (9), then, equation (4) becomes

(10)
$$y_{uv} = 0$$

with the change of variables (6).

Equation (10) can be solved by successive integrations to give $y_u = \phi'(u)$ and

$$y = \phi(u) + \psi(v),$$

where the arbitrary functions ϕ and ψ are twice-differentiable. The *general solution* of the wave equation (4) is, therefore,

(11)
$$y = \phi(x + at) + \psi(x - at).$$

In this example, the boundary conditions are simple enough that we can actually determine the functions ϕ and ψ. Observe that the function (11) satisfies conditions (5) when

$$\phi(x) + \psi(x) = f(x) \qquad \text{and} \qquad a\phi'(x) - a\psi'(x) = 0.$$

Thus $\phi(x) - \psi(x) = c$, where c is a constant; and it follows that

$$2\phi(x) = f(x) + c \qquad \text{and} \qquad 2\psi(x) = f(x) - c.$$

Consequently,

(12)
$$y(x,t) = \frac{1}{2}[f(x + at) + f(x - at)].$$

The solution (12) of the boundary value problem consisting of equations (4) and (5) is known as *d'Alembert's solution*. It is easily verified under the assumption that $f'(x)$ and $f''(x)$ exist for all x.

The method for solving boundary value problems illustrated in the two examples here has severe limitations. The general solutions (2) and (11),

involving arbitrary functions, were obtained by successive integrations, a procedure that applies to relatively few types of partial differential equations. But,
even in the exceptional cases in which such general solutions can be found, the
determination of the arbitrary functions directly from the boundary conditions is
often difficult.

Among a variety of other methods, the one to be developed in this book
will be suggested by the example in Sec. 9. That method, which is sometimes
called the *Fourier method*, is a classical and powerful one. Before turning to it,
however, we mention some other important ones. Methods based on Laplace,
Fourier, and other integral transforms, all included in the subject of operational
mathematics, are especially effective.[†] The classical method of conformal map-
ping in the theory of functions of a complex variable applies to a prominent
class of problems involving Laplace's equation in two dimensions.[‡] There are
still other ways of reducing or solving such problems, including applications of
so-called Green's functions and numerical, or computational, methods.

Even when a problem yields to more than one method, however, different
methods sometimes produce different forms of the solution; and each form may
have its own desirable features. On the other hand, some problems require
successive applications of two or more methods. Others, including some fairly
simple ones, have defied all known exact methods. The development of new
methods is an activity in present-day mathematical research.

9. ON THE SUPERPOSITION OF SEPARATED SOLUTIONS

The purpose of this section is to motivate the two main topics of the book,
indicated at the beginning of the chapter. Namely, in seeking a solution of the
boundary value problem in the example below, we shall find it necessary to
expand an arbitrary function in a series of trigonometric functions (Chap. 2), as
well as to formalize the Fourier method for solving boundary value problems in
partial differential equations (Chap. 3).

EXAMPLE. The Dirichlet problem (see Sec. 7)

$$(1) \qquad\qquad u_{xx}(x, y) + u_{yy}(x, y) = 0 \qquad\qquad (0 < x < 1, y > 0),$$

$$(2) \qquad\qquad u(0, y) = 0, \qquad u(1, y) = 0 \qquad\qquad (y > 0),$$

$$(3) \qquad\qquad u(x, 0) = f(x) \qquad\qquad (0 < x < 1)$$

is satisfied by steady-state temperatures $u(x, y)$, subject to the indicated bound-
ary conditions, in a semi-infinite slab occupying the region $0 \leqslant x \leqslant 1$, $y \geqslant 0$ of

---

[†] See the book by Churchill (1972), listed in the Bibliography.

[‡] See the authors' book (1990), also listed in the Bibliography.

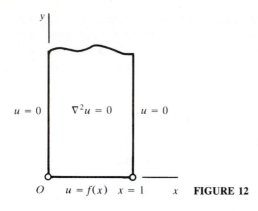

$O \quad u = f(x) \quad x = 1 \qquad x$ **FIGURE 12**

three-dimensional space (Fig. 12). We shall not be concerned here with precise conditions on the function f. We assume only that f is bounded and observe that it is then physically reasonable to seek solutions of equation (1) that tend to zero as y tends to infinity.

Since equation (1) has constant coefficients and is of the type treated below in Problem 6, we know from that problem that the function

$$(4) \qquad\qquad u(x, y) = e^{\lambda x + \mu y} = e^{\lambda x} e^{\mu y}$$

is a solution, where λ and μ are any constants (real or complex) related by the equation

$$(5) \qquad\qquad \lambda^2 + \mu^2 = 0.$$

Such a solution is *separated* in the sense that it is the product of two functions one of which depends only on x and the other of which depends only on y.[†] In anticipation of real-valued solutions of equation (1) that tend to zero as y tends to infinity, we stipulate that μ must be a negative real number and write $\mu = -\nu$, where $\nu > 0$. Then $\lambda = \pm \nu i$, where $i = \sqrt{-1}$, according to relation (5). So, in view of *Euler's formula*[‡]

$$e^{i\theta} = \cos \theta + i \sin \theta,$$

we have these two families of solutions:

$$U_1(x, y) = e^{-\nu y} e^{i\nu x} = e^{-\nu y}(\cos \nu x + i \sin \nu x),$$

$$U_2(x, y) = e^{-\nu y} e^{-i\nu x} = e^{-\nu y}(\cos \nu x - i \sin \nu x).$$

[†] This terminology is borrowed from the book by Pinsky (1991) that is listed in the Bibliography. While the method used here is well known, that book has an especially good variety of problems to which it is applied.

[‡] Complete justification of basic facts from complex analysis that are used in this section can be found in the authors' book (1990), listed in the Bibliography.

   Now the partial differential equation (1) is linear and homogeneous; and, as is the case with linear homogeneous *ordinary* differential equations, any linear combination of solutions of equation (1) is also a solution (see Problem 7). Accepting the fact that such a superposition principle remains valid when complex-valued functions and complex constants are involved, we arrive at the *real-valued* solutions

$$U_3(x, y) = \frac{U_1(x, y) + U_2(x, y)}{2} = e^{-\nu y} \cos \nu x,$$

$$U_4(x, y) = \frac{U_1(x, y) - U_2(x, y)}{2i} = e^{-\nu y} \sin \nu x$$

of equation (1), where ν has any positive value. It is, of course, a simple matter to verify directly that $U_3(x, y)$ and $U_4(x, y)$ actually satisfy equation (1).

   Turning now to the boundary conditions (2), we see that the solutions $U_3(x, y)$ cannot satisfy the first of those conditions since $\cos 0 = 1$. But the solutions $U_4(x, y)$ satisfy both conditions, provided that $\sin \nu = 0$. Since the only (positive) values of ν having that property are $\nu = n\pi$ ($n = 1, 2, \ldots$), it follows that the functions

(6) $u_n(x, y) = e^{-n\pi y} \sin n\pi x$ $(n = 1, 2, \ldots)$

all satisfy conditions (1) and (2). Referring once again to Problem 7, we see that any linear combination

(7) $u(x, y) = \sum_{n=1}^{N} b_n e^{-n\pi y} \sin n\pi x$

of the first N of the functions (6) satisfies equation (1). It is, moreover, obvious that this sum also satisfies conditions (2).

   As for condition (3), the constants b_n must be determined so that

$$f(x) = \sum_{n=1}^{N} b_n \sin n\pi x \qquad (0 < x < 1).$$

If the function $f(x)$ is itself a linear combination of the sine functions

$$\sin \pi x, \quad \sin 2\pi x, \quad \ldots, \quad \sin N\pi x,$$

the needed values of $b_1, b_2, \ldots, b_N$ are evident. If, for instance,

(8) $f(x) = 2 \sin \pi x + \sin 3\pi x,$

we may take $N = 3$ and write $b_1 = 2$, $b_2 = 0$, $b_3 = 1$. The function

$$u(x, y) = 2e^{-\pi y} \sin \pi x + e^{-3\pi y} \sin 3\pi x$$

thus satisfies all the conditions (1)–(3) when $f(x)$ is the particular function (8).

But suppose that $f(x)$ is an arbitrary function. The following generalization of the above method is suggested. We might replace the sum (7) by a generalized linear combination, or infinite series,

$$(9) \qquad u(x, y) = \sum_{n=1}^{\infty} b_n e^{-n\pi y} \sin n\pi x.$$

We must, of course, assume that this series converges and that the superposition principle verified in Problem 7 can be extended so as to apply. Then condition (3) requires us to find values of the constants b_n such that

$$(10) \qquad f(x) = \sum_{n=1}^{\infty} b_n \sin n\pi x \qquad (0 < x < 1).$$

The problem of finding such coefficients, as well as ones in series involving cosines, is the problem, noted just prior to this example, that is to be treated in Chap. 2. Separated solutions and a superposition principle, extended to infinite series, are the foundation of the Fourier method for solving boundary value problems that is described in Chap. 3.

PROBLEMS

1. Solve each of these boundary value problems by first finding the general solution of the partial differential equation involved.
   (a) $u_{xx}(x, y) = 6xy$ $(0 < x < 1, \ -\infty < y < \infty)$,
   $u(0, y) = y, \ u_x(1, y) = 0$;
   (b) $u_{xy}(x, y) = 2x$ $(x > 0, \ y > 0)$,
   $u(0, y) = 0, \ u(x, 0) = x^2$.
   Answers: (a) $u(x, y) = (x^3 - 3x + 1)y$; (b) $u(x, y) = x^2(1 + y)$.

2. Whether a second-order linear partial differential equation in $u = u(x, y)$ is hyperbolic, elliptic, or parabolic (Sec. 7) can vary from region to region in the xy plane when at least one of the coefficients is a nonconstant function of x and y. Classify each of the following differential equations in various regions, and sketch those regions.
   (a) $yu_{xx} + u_{yy} = 0$; (b) $u_{xx} + 2x^2 u_{xy} + yu_{yy} = 0$;
   (c) $xu_{xx} + yu_{yy} - 3u_y = 2$; (d) $u_{xx} - 2xu_{xy} + (1 - y^2)u_{yy} = 0$.
   Answers: (a) Parabolic on the x axis, elliptic above it, and hyperbolic below it;
   (b) parabolic on the curve $y = x^4$, elliptic above it, and hyperbolic below it;
   (d) parabolic on the circle $x^2 + y^2 = 1$, elliptic inside it, and hyperbolic outside it.

3. In Example 2, Sec. 8, d'Alembert's solution

$$y(x, t) = \frac{1}{2}[f(x + at) + f(x - at)]$$

represents transverse displacements in a stretched string of infinite length, initially released at rest from a position $y = f(x)$ $(-\infty < x < \infty)$. Use that solution to show how the instantaneous position of the string at time t can be displayed graphically by adding ordinates of two curves, one obtained by translating the curve $y = \frac{1}{2}f(x)$ to the right through the distance at, the other by translating it to the left through the same distance. As t varies, the curve $y = \frac{1}{2}f(x)$ moves in each direction as a wave with velocity a. Sketch some instantaneous positions when $f(x)$ is zero except on a small interval about the origin.

4. Use the general solution (11) in Example 2, Sec. 8, to solve the boundary value problem

$$y_{tt}(x,t) = a^2 y_{xx}(x,t) \qquad (-\infty < x < \infty, t > 0),$$

$$y(x,0) = 0, \qquad y_t(x,0) = g(x) \qquad (-\infty < x < \infty).$$

Suggestion: Note that one can write

$$\int g(x)\, dx = \int_0^x g(s)\, ds + C.$$

Answer: $y(x,t) = \dfrac{1}{2a} \displaystyle\int_{x-at}^{x+at} g(s)\, ds.$

5. Let $Y(x,t)$ denote d'Alembert's solution (12) in Example 2, Sec. 8, of the boundary value problem solved there, and let $Z(x,t)$ denote the solution found in Problem 4 for a related boundary value problem. Verify directly that the sum

$$y(x,t) = Y(x,t) + Z(x,t)$$

is a solution of the boundary value problem

$$y_{tt}(x,t) = a^2 y_{xx}(x,t) \qquad (-\infty < x < \infty, t > 0),$$

$$y(x,0) = f(x), \qquad y_t(x,0) = g(x) \qquad (-\infty < x < \infty).$$

Thus show that

$$y(x,t) = \frac{1}{2}[f(x+at) + f(x-at)] + \frac{1}{2a}\int_{x-at}^{x+at} g(s)\, ds$$

is a solution of the problem here. Interpret the problem physically (see Problem 3).

6. Let the coefficients $A, B, \ldots, F$ in the linear homogeneous partial differential equation

$$Au_{xx} + Bu_{xy} + Cu_{yy} + Du_x + Eu_y + Fu = 0$$

be constants, rather than more general functions of x and y. By substituting the exponential function $u = \exp(\lambda x + \mu y)$, where λ and μ are constants, into that differential equation, show that it is always a solution when λ and μ satisfy the algebraic equation

$$A\lambda^2 + B\lambda\mu + C\mu^2 + D\lambda + E\mu + F = 0.$$

Note that when values of λ are selected, appropriate values of μ arise from the quadratic equations in μ that result. The values of μ are, of course, not necessarily real even when λ is real.[†] Similar remarks apply when values of μ are first selected.

[†] The usual rules for differentiating the exponential function in calculus also hold when complex numbers are involved. See the footnote in Sec. 9.

(The possibility of such exponential solutions is suggested by experience with *ordinary* differential equations that have constant coefficients.)

7. Suppose that the functions $u_n = u_n(x, y)$ $(n = 1, 2, \ldots)$ are all solutions of Laplace's equation

$$\frac{\partial^2 u}{\partial x^2} + \frac{\partial^2 u}{\partial y^2} = 0.$$

Show that, for any constants c_n $(n = 1, 2, \ldots, N)$, the linear combination

$$u = \sum_{n=1}^{N} c_n u_n$$

is also a solution. Do this by substituting the sum here into the left-hand side of the differential equation and grouping terms appropriately. [This result is a special case of the *principle of superposition* of solutions, to be developed more fully in Chap. 3 (Sec. 26).]

8. Show that if each of the functions $u_n = u_n(x, t)$ $(n = 1, 2, \ldots)$ satisfies the heat equation

$$\frac{\partial u}{\partial t} = k\frac{\partial^2 u}{\partial x^2},$$

then the same is true of any linear combination

$$u = \sum_{n=1}^{N} c_n u_n.$$

(Compare Problem 7.)

9. Let $u_{mn} = u_{mn}(x, y, z)$ $(m = 1, 2, \ldots;\ n = 1, 2, \ldots)$ denote solutions of Laplace's equation

$$\frac{\partial^2 u}{\partial x^2} + \frac{\partial^2 u}{\partial y^2} + \frac{\partial^2 u}{\partial z^2} = 0.$$

Verify that any linear combination

$$u = \sum_{n=1}^{N} \sum_{m=1}^{M} c_{mn} u_{mn}$$

is also a solution. (Compare Problem 7.)

10. Verify that each of the functions

$$u_0(x, y) = y, \qquad u_n(x, y) = \sinh ny \cos nx \qquad (n = 1, 2, \ldots)$$

satisfies Laplace's equation

$$u_{xx}(x, y) + u_{yy}(x, y) = 0 \qquad (0 < x < \pi,\ 0 < y < 2)$$

and the three boundary conditions

$$u_x(0, y) = u_x(\pi, y) = 0, \qquad u(x, 0) = 0.$$

Then, with the aid of the superposition principle in Problem 7, note that any linear combination

$$u(x, y) = A_0 y + \sum_{n=1}^{N} A_n \sinh ny \cos nx$$

satisfies the same differential equation and boundary conditions. Take $N = 2$ and find values of the coefficients A_0, A_1, A_2 such that a fourth boundary condition

$$u(x,2) = 4 + 3 \cos x - \cos 2x$$

is satisfied. Interpret the result physically.

Answer: $A_0 = 2, \ A_1 = \dfrac{3}{\sinh 2}, \ A_2 = \dfrac{-1}{\sinh 4}.$

11. When the unit of time is chosen so that the diffusivity k in the heat equation is unity (see Problem 9, Sec. 4), the boundary value problem

$$u_t(x,t) = u_{xx}(x,t) \qquad\qquad (0 < x < 1, t > 0),$$

$$u(0,t) = u_x(1,t) = 0, \qquad u(x,0) = f(x)$$

describes temperatures in a slab $0 \leqslant x \leqslant 1$ whose face $x = 0$ is kept at temperature zero, whose face $x = 1$ is insulated, and whose initial temperatures depend only on x.

(*a*) Assuming that the function f is bounded, modify the treatment of the Dirichlet problem in the example in Sec. 9 to discover the following solutions of the above heat equation and the first two boundary conditions:

$$u_n(x,t) = \exp\left[-\frac{(2n-1)^2\pi^2}{4}t\right]\sin\frac{(2n-1)\pi x}{2} \qquad (n = 1,2,\dots).$$

Verify these solutions directly. Then, with the aid of the result in Problem 8, point out how it follows that any linear combination

$$u(x,t) = \sum_{n=1}^{N} B_n \exp\left[-\frac{(2n-1)^2\pi^2}{4}t\right]\sin\frac{(2n-1)\pi x}{2}$$

is also a solution.

(*b*) Use the final result in part (*a*) to obtain the solution

$$u(x,t) = 2\exp\left(-\frac{\pi^2}{4}t\right)\sin\frac{\pi x}{2} - \exp\left(-\frac{25\pi^2}{4}t\right)\sin\frac{5\pi x}{2}$$

of the stated boundary value problem when

$$f(x) = 2\sin\frac{\pi x}{2} - \sin\frac{5\pi x}{2}.$$

12. Verify that each of the products

$$u_{mn}(x, y, z) = \exp\left(-z\sqrt{m^2 + n^2}\right)\cos my \sin nx \qquad (m = 0,1,2,\dots; n = 1,2,\dots)$$

satisfies Laplace's equation

$$u_{xx}(x, y, z) + u_{yy}(x, y, z) + u_{zz}(x, y, z) = 0 \qquad (0 < x < \pi, 0 < y < \pi, z > 0)$$

and the boundary conditions

$$u(0, y, z) = u(\pi, y, z) = 0, \qquad u_y(x, 0, z) = u_y(x, \pi, z) = 0.$$

Then, with the aid of the result in Problem 9, obtain a function $u(x, y, z)$ that satisfies not only Laplace's equation and the stated boundary conditions, but also the condition

$$u_z(x, y, 0) = (-6 + 5\cos 4y)\sin 3x.$$

Interpret that function physically.

   Answer: $u(x, y, z) = (2e^{-3z} - e^{-5z}\cos 4y)\sin 3x.$

13. Let $y(x, t)$ represent transverse displacements in a long stretched string, one end of which is attached to a ring that can slide along the y axis. The other end is so far out on the positive x axis that it may be considered to be infinitely far from the origin. The ring is initially at the origin and is then moved along the y axis (Fig. 13) so that

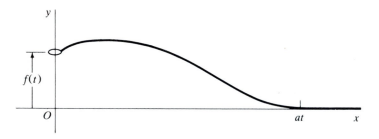

FIGURE 13

$y = f(t)$ when $x = 0$ and $t \geqslant 0$, where f is a prescribed continuous function and $f(0) = 0$. We assume that the string is initially at rest on the x axis; thus $y(x, t) \to 0$ as $x \to \infty$. The boundary value problem for $y(x, t)$ is

$$y_{tt}(x, t) = a^2 y_{xx}(x, t) \qquad\qquad (x > 0, t > 0),$$

$$y(x, 0) = 0, \qquad y_t(x, 0) = 0 \qquad\qquad (x \geqslant 0),$$

$$y(0, t) = f(t) \qquad\qquad (t \geqslant 0).$$

(a) Apply the first two of these boundary conditions to the general solution (Sec. 8)

$$y = \phi(x + at) + \psi(x - at)$$

of the wave equation to show that there is a constant C such that

$$\phi(x) = C \qquad \text{and} \qquad \psi(x) = -C \qquad\qquad \text{when } x \geqslant 0.$$

Then apply the third boundary condition to show that

$$\psi(-x) = f\left(\frac{x}{a}\right) - C \qquad\qquad \text{when } x \geqslant 0,$$

where C is the same constant.

(*b*) With the aid of the results in part (*a*), derive the solution

$$y(x,t) = \begin{cases} 0 & \text{when } x \geqslant at, \\ f\left(t - \dfrac{x}{a}\right) & \text{when } x \leqslant at. \end{cases}$$

Note that the part of the string to the right of the point $x = at$ on the x axis is unaffected by the movement of the ring prior to time t, as shown in Fig. 13.

14. Use the solution obtained in Problem 13 to show that if the ring at the left-hand end of the string in that problem is moved according to the function

$$f(t) = \begin{cases} \sin \pi t & \text{when } 0 \leqslant t \leqslant 1, \\ 0 & \text{when } \quad t \geqslant 1, \end{cases}$$

then

$$y(x,t) = \begin{cases} 0 & \text{when } x \leqslant a(t-1) \text{ or } x \geqslant at, \\ \sin \pi\left(t - \dfrac{x}{a}\right) & \text{when } \quad a(t-1) \leqslant x \leqslant at. \end{cases}$$

Observe that the ring is lifted up 1 unit and then returned to the origin, where it remains after time $t = 1$. The expression for $y(x, t)$ here shows that when $t > 1$, the string coincides with the x axis except on an interval of length a, where it forms one arch of a sine curve (Fig. 14). Furthermore, as t increases, the arch moves to the right with speed a.

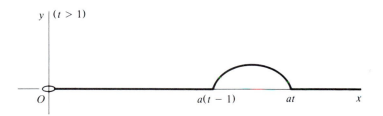

$y \mid (t > 1)$

O $a(t-1)$ at x

FIGURE 14

15. Consider the partial differential equation

$$A y_{xx} + B y_{xt} + C y_{tt} = 0 \qquad (A \neq 0, C \neq 0),$$

where A, B, and C are constants, and assume that it is hyperbolic, so that $B^2 - 4AC > 0$ (Sec. 7).

(*a*) Use the transformation

$$u = x + \alpha t, \qquad v = x + \beta t \qquad (\alpha \neq \beta)$$

to obtain the new differential equation

$$(A + B\alpha + C\alpha^2) y_{uu} + [2A + B(\alpha + \beta) + 2C\alpha\beta] y_{uv}$$
$$+ (A + B\beta + C\beta^2) y_{vv} = 0.$$

(*b*) Show that when α and β have the values

$$\alpha_0 = \frac{-B + \sqrt{B^2 - 4AC}}{2C} \quad \text{and} \quad \beta_0 = \frac{-B - \sqrt{B^2 - 4AC}}{2C},$$

respectively, the differential equation in part (*a*) reduces to $y_{uv} = 0$.

(*c*) Conclude from the result in part (*b*) that the general solution of the original differential equation is

$$y = \phi(x + \alpha_0 t) + \psi(x + \beta_0 t),$$

where ϕ and ψ are arbitrary functions that are twice-differentiable. Then show how the general solution (11), Sec. 8, of the wave equation

$$-a^2 y_{xx} + y_{tt} = 0$$

follows as a special case.

16. Show that under the transformation

$$u = x, \qquad v = \alpha x + \beta t \qquad\qquad (\beta \neq 0),$$

the given differential equation in Problem 15 becomes

$$A y_{uu} + (2A\alpha + B\beta) y_{uv} + (A\alpha^2 + B\alpha\beta + C\beta^2) y_{vv} = 0.$$

Then show that this new equation reduces to

(*a*) Laplace's equation $y_{uu} + y_{vv} = 0$ in two dimensions when the original equation is elliptic ($B^2 - 4AC < 0$) and

$$\alpha = \frac{-B}{\sqrt{4AC - B^2}}, \qquad \beta = \frac{2A}{\sqrt{4AC - B^2}};$$

(*b*) Laplace's equation $y_{uu} = 0$ in one dimension when the original equation is parabolic ($B^2 - 4AC = 0$) and

$$\alpha = -B, \qquad \beta = 2A.$$

CHAPTER

2

FOURIER
SERIES

In this chapter, we shall present the basic theory of Fourier series, which are expansions of arbitrary functions in series of sine and cosine functions. In so doing, we shall introduce the concept of orthonormal sets of functions. That will not only clarify underlying concepts behind the various types of Fourier series treated here but also lay the foundation for finding other types of series expansions that are needed in later chapters.

10. PIECEWISE CONTINUOUS FUNCTIONS

If u_1 and u_2 are functions and c_1 and c_2 are constants, the function $c_1 u_1 + c_2 u_2$ is called a *linear combination* of u_1 and u_2. Note that $u_1 + u_2$ and $c_1 u_1$, as well as the constant function 0, are special cases. A linear space of functions, or *function space*, is a class of functions, all with a common domain of definition, such that each linear combination of any two functions in that class remains in it; that is, if u_1 and u_2 are in the class, then so is $c_1 u_1 + c_2 u_2$. Before developing the theory of Fourier, or trigonometric, series, we need to specify function spaces containing the functions to be represented.

Let a function f be continuous at all points of a bounded open interval $a < x < b$ except possibly for a finite set of points $x_1, x_2, \ldots, x_{n-1}$, where

$$a < x_1 < x_2 < \cdots < x_{n-1} < b.$$

If we write $x_0 = a$ and $x_n = b$, then f is continuous on each of the n open subintervals

$$x_0 < x < x_1, \quad x_1 < x < x_2, \quad \ldots, \quad x_{n-1} < x < x_n.$$

It is not necessarily continuous, or even defined, at their end points. But if, in each of those subintervals, f has finite limits as x approaches the end points from the interior, f is said to be *piecewise continuous* on the interval $a < x < b$. More precisely, the one-sided limits

(1) $$f(x_{k-1}+) = \lim_{\substack{x \to x_{k-1} \\ x > x_{k-1}}} f(x) \qquad \text{and} \qquad f(x_k-) = \lim_{\substack{x \to x_k \\ x < x_k}} f(x)$$

$$(k = 1, 2, \ldots, n)$$

are required to exist.

Note that if the limiting values from the interior of a subinterval are assigned to f at the end points, then f is continuous on the *closed* subinterval. Since any function that is continuous on a closed bounded interval is bounded, it follows that f is bounded on the entire interval $a \leqslant x \leqslant b$. That is, there exists a nonnegative number M such that $|f(x)| \leqslant M$ for all points $x\ (a \leqslant x \leqslant b)$ at which f is defined.

EXAMPLE 1. Consider the function f that has the values

$$f(x) = \begin{cases} x & \text{when } 0 < x < 1, \\ -1 & \text{when } 1 \leqslant x < 2, \\ 1 & \text{when } 2 < x < 3. \end{cases}$$

(See Fig. 15.) Although f is discontinuous at the points $x = 1$ and $x = 2$ in the interval $0 < x < 3$, it is nevertheless piecewise continuous on that interval. This is because the one-sided limits from the interior exist at the end points of each of the three open subintervals on which f is continuous. Note, for instance, that the right-hand limit at $x = 0$ is $f(0+) = 0$ and that the left-hand limit at $x = 1$ is $f(1-) = 1$.

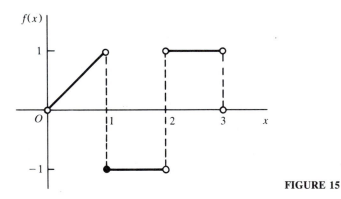

FIGURE 15

A function is piecewise continuous on an interval $a < x < b$ if it is continuous on the *closed* interval $a \leqslant x \leqslant b$. Continuity on the *open* interval $a < x < b$ does not, however, imply piecewise continuity there, as the following example illustrates.

EXAMPLE 2. The function $f(x) = 1/x$ is continuous on the interval $0 < x < 1$, but it is not piecewise continuous there since $f(0 +)$ fails to exist.

When a function f is piecewise continuous on an interval $a < x < b$, the integral of $f(x)$ from $x = a$ to $x = b$ always exists. It is the sum of the integrals of $f(x)$ over the open subintervals on which f is continuous:

$$(2) \qquad \int_a^b f(x)\, dx = \int_a^{x_1} f(x)\, dx + \int_{x_1}^{x_2} f(x)\, dx + \cdots + \int_{x_{n-1}}^b f(x)\, dx.$$

The first integral on the right exists since it is defined as the integral over the interval $a \leqslant x \leqslant x_1$ of the continuous function whose values are $f(x)$ when $a < x < x_1$ and whose values at the end points $x = a$ and $x = x_1$ are $f(a +)$ and $f(x_1 -)$, respectively. The remaining integrals on the right in equation (2) are similarly defined and therefore exist.

EXAMPLE 3. If f is the function in Example 1 and Fig. 15, then

$$\int_0^3 f(x)\, dx = \int_0^1 x\, dx + \int_1^2 -1\, dx + \int_2^3 1\, dx = \frac{1}{2} - 1 + 1 = \frac{1}{2}.$$

Observe that the value of the integral of $f(x)$ over each subinterval is unaffected by the values of f at the end points. The function is, in fact, not even defined at $x = 0, 2$, and 3.

If two functions f_1 and f_2 are each piecewise continuous on an interval $a < x < b$, then there is a finite subdivision of the interval such that both functions are continuous on each closed subinterval when the functions are given their limiting values from the interior at the end points. Hence a linear combination $c_1 f_1 + c_2 f_2$, or the product $f_1 f_2$, has that continuity on each subinterval and is itself piecewise continuous on the interval $a < x < b$. Consequently, the integrals of the functions $c_1 f_1 + c_2 f_2$, $f_1 f_2$, and $[f_1(x)]^2$ all exist on that interval.

Since any linear combination of functions that are piecewise continuous also has that property, we may use the terminology at the beginning of this section and refer to the class of all piecewise continuous functions defined on an interval $a < x < b$ as a function space; we denote it by $C_p(a, b)$. It is analogous to three-dimensional space, where linear combinations of vectors are well-defined vectors in that space. In Sec. 11, we shall extend the analogy by developing the concept of inner products of functions in $C_p(a, b)$.

Other function spaces occur in the theory of Fourier series. An especially important subspace of $C_p(a, b)$ will be introduced in Sec. 17. More advanced texts treat the space of all integrable functions f on an interval $a < x < b$ whose products, including squares $[f(x)]^2$, are integrable. Then a more general type of integral, known as the Lebesgue integral, is often used.

Our treatment of Fourier series involves more elementary concepts in mathematical analysis. Except when otherwise noted, in this book *we shall restrict our attention to functions that are piecewise continuous* on all bounded intervals under consideration. When it is stated that a function is piecewise continuous on an interval, it is to be understood that the interval is bounded; and the notion of piecewise continuity clearly applies regardless of whether the interval is open or closed.

11. INNER PRODUCTS
AND ORTHONORMAL SETS

Let f and g denote any two functions that are continuous on a closed bounded interval $a \leqslant x \leqslant b$. Dividing that interval into N closed subintervals of equal length $\Delta x = (b - a)/N$ and letting x_k denote any point in the kth subinterval, we recall from calculus that when N is large,

$$\int_a^b f(x)g(x)\, dx \doteq \sum_{k=1}^N f(x_k)g(x_k)\, \Delta x,$$

the symbol $\doteq$ here denoting approximate equality. That is,

(1)
$$\int_a^b f(x)g(x)\, dx \doteq \sum_{k=1}^N a_k b_k,$$

where

$$a_k = f(x_k)\sqrt{\Delta x} \qquad \text{and} \qquad b_k = g(x_k)\sqrt{\Delta x}.$$

The left-hand side of expression (1) is, then, approximately equal to the inner product of two vectors in N-dimensional space when N is large. The approximation becomes exact in the limit as N tends to infinity.[†] This suggests defining an *inner product* of the functions f and g:

(2)
$$(f, g) = \int_a^b f(x)g(x)\, dx.$$

The integral here is, of course, also well-defined when f and g are allowed to be piecewise continuous on the *fundamental interval* $a < x < b$. Equation (2) can, therefore, be used to define an inner product of any two functions f and g in the function space $C_p(a, b)$, introduced in Sec. 10.

The function space $C_p(a, b)$, with inner product (2), is analogous to ordinary three-dimensional space. Indeed, the following counterparts of familiar properties of vectors in three-dimensional space hold for any functions f, g,

---

[†] See the book by Lanczos (1966, pp. 210ff), listed in the Bibliography, for an elaboration of this idea.

and h in $C_p(a, b)$:

(3) $$(f, g) = (g, f),$$

(4) $$(f, g + h) = (f, g) + (f, h),$$

(5) $$(cf, g) = c(f, g),$$

where c is any constant, and

(6) $$(f, f) \geqslant 0.$$

The analogy is continued with the introduction of the *norm*

(7) $$\|f\| = (f, f)^{1/2}$$

of a function f in $C_p(a, b)$. It is evident from equation (2) that the norm of f can be written

(8) $$\|f\| = \left\{ \int_a^b [f(x)]^2 \, dx \right\}^{1/2}.$$

The norm of the difference of two functions f and g,

(9) $$\|f - g\| = \left\{ \int_a^b [f(x) - g(x)]^2 \, dx \right\}^{1/2},$$

is a measure of the area of the region between the graphs of $y = f(x)$ and $y = g(x)$ (Fig. 16). To be specific, the quotient $\|f - g\|^2/(b - a)$ is the mean, or average, value of the squares of the vertical distances $|f(x) - g(x)|$ between points on those graphs over the interval $a < x < b$. The quantity $\|f - g\|^2$ is called the *mean square deviation* of one of the functions f and g from the other.

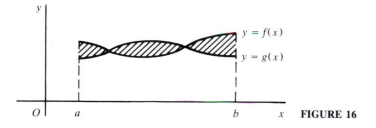

FIGURE 16

Two functions f and g in $C_p(a, b)$ are *orthogonal* when

$$(f, g) = 0,$$

or

(10) $$\int_a^b f(x) g(x) \, dx = 0.$$

Also, if $\|f\| = 1$, the function f is said to be *normalized*. We have carried our analogy too far to preserve the original meaning of the geometric terminology. The orthogonality of two functions f and g signifies nothing about perpendicularity, but instead that the product fg assumes both positive and negative values on the fundamental interval in such a manner that equation (10) holds.

A set of functions $\psi_n(x)$ $(n = 1, 2, \dots)$ is orthogonal on an interval $a < x < b$ if $(\psi_m, \psi_n) = 0$ when $m \neq n$. Assuming that none of the functions ψ_n has zero norm (see Problem 7), one can normalize each of them by dividing it by the positive constant $\|\psi_n\|$. The new set $\{\phi_n(x)\}$ so formed, where

$$(11) \qquad\qquad \phi_n(x) = \frac{\psi_n(x)}{\|\psi_n\|} \qquad\qquad (n = 1, 2, \dots),$$

is *orthonormal* on the fundamental interval; that is,

$$(12) \qquad\qquad (\phi_m, \phi_n) = \delta_{mn} \qquad\qquad (m = 1, 2, \dots; n = 1, 2, \dots),$$

where δ_{mn} is *Kronecker's* δ. Written in full, the characterization (12) of an orthonormal set $\{\phi_n(x)\}$ becomes

$$(13) \qquad\qquad \int_a^b \phi_m(x)\phi_n(x)\,dx = \begin{cases} 0 & \text{when } m \neq n, \\ 1 & \text{when } m = n. \end{cases}$$

EXAMPLE 1.   From the trigonometric identity

$$2 \sin A \sin B = \cos(A - B) - \cos(A + B),$$

we know that

$$\sin mx \sin nx = \frac{1}{2}\cos(m - n)x - \frac{1}{2}\cos(m + n)x,$$

where m and n are positive integers. It is thus easy to verify that

$$(14) \qquad\qquad \int_0^\pi \sin mx \sin nx\,dx = \begin{cases} 0 & \text{when } m \neq n, \\ \dfrac{\pi}{2} & \text{when } m = n. \end{cases}$$

Evidently, then, the set of sine functions

$$(15) \qquad\qquad \psi_n(x) = \sin nx \qquad\qquad (n = 1, 2, \dots)$$

is orthogonal on the interval of $0 < x < \pi$; and the norm $\|\psi_n\|$ of each of these functions is $\sqrt{\pi/2}$. Hence the corresponding orthonormal set $\{\phi_n(x)\}$ consists of the functions

$$(16) \qquad\qquad \phi_n(x) = \sqrt{\frac{2}{\pi}}\,\sin nx \qquad\qquad (n = 1, 2, \dots).$$

It is sometimes more convenient to index an infinite orthogonal or orthonormal set by starting with $n = 0$, rather than $n = 1$. This is the case in the following example. Verification that the given set is orthonormal is left to the problems.

EXAMPLE 2. The functions

$$\phi_0(x) = \frac{1}{\sqrt{\pi}}, \qquad \phi_n(x) = \sqrt{\frac{2}{\pi}} \, \cos nx \qquad (n = 1, 2, \dots)$$

constitute a set $\{\phi_n(x)\}$ $(n = 0, 1, 2, \dots)$ that is orthonormal on the interval $0 < x < \pi$.

PROBLEMS

1. (a) Use the trigonometric identity

$$2 \cos A \cos B = \cos (A - B) + \cos (A + B)$$

to show that if m and n are positive integers,

$$\int_0^\pi \cos mx \cos nx \, dx = \begin{cases} 0 & \text{when } m \neq n, \\ \dfrac{\pi}{2} & \text{when } m = n. \end{cases}$$

(b) With the aid of the integration formula obtained in part (a), verify that the set $\{\phi_n(x)\}$ $(n = 0, 1, 2, \dots)$ in Example 2, Sec. 11, is orthonormal on the interval $0 < x < \pi$.

  Suggestion: Note that, in order to establish orthogonality in part (b), it is necessary to show that $(\phi_0, \phi_n) = 0$ and $(\phi_m, \phi_n) = 0$ $(m \neq n)$ for *positive* integers m and n.

2. (a) Use the fact that the functions (16) in Example 1, Sec. 11, constitute an orthonormal set on the interval $0 < x < \pi$ to show that the functions

$$\phi_n(x) = \frac{1}{\sqrt{\pi}} \sin nx \qquad (n = 1, 2, \dots)$$

form a set that is orthonormal on the interval $-\pi < x < \pi$.

(b) Use the fact that the set in Example 2, Sec. 11, and Problem 1 is orthonormal on the interval $0 < x < \pi$ to show that the functions

$$\phi_0(x) = \frac{1}{\sqrt{2\pi}}, \qquad \phi_n(x) = \frac{1}{\sqrt{\pi}} \cos nx \qquad (n = 1, 2, \dots)$$

form an orthonormal set on the interval $-\pi < x < \pi$. (See the suggestion with Problem 1.)

  Suggestion: Observe that if f is an *even* integrable function, one where $f(-x) = f(x)$, then

$$\int_{-\pi}^\pi f(x) \, dx = 2 \int_0^\pi f(x) \, dx$$

since the graph of $y = f(x)$ is symmetric with respect to the y axis.

3. With the aid of the results obtained in Problem 2, show that the set $\{\phi_n(x)\}$ $(n = 0, 1, 2, \ldots)$ consisting of the functions

$$\phi_0(x) = \frac{1}{\sqrt{2\pi}}, \qquad \phi_{2n-1}(x) = \frac{1}{\sqrt{\pi}} \cos nx, \qquad \phi_{2n}(x) = \frac{1}{\sqrt{\pi}} \sin nx$$

$$(n = 1, 2, \ldots)$$

is orthonormal on the interval $-\pi < x < \pi$. Note that, in view of Problem 2, one need only show that

$$(\phi_0, \phi_{2n}) = 0 \qquad \text{and} \qquad (\phi_{2m-1}, \phi_{2n}) = 0 \quad (m = 1, 2, \ldots; n = 1, 2, \ldots)$$

in order to establish orthogonality.

Suggestion: Observe that if f is an *odd* integrable function, one where $f(-x) = -f(x)$, then

$$\int_{-\pi}^{\pi} f(x)\, dx = 0$$

since the graph of $y = f(x)$ is symmetric with respect to the origin.

4. Show that the functions $\psi_1(x) = 1$ and $\psi_2(x) = x$ are orthogonal on the interval $-1 < x < 1$, and determine constants A and B such that the function

$$\psi_3(x) = 1 + Ax + Bx^2$$

is orthogonal to both ψ_1 and ψ_2 on that interval.
Answer: $A = 0$, $B = -3$.

5. Suppose that two continuous functions $\psi_1(x)$ and $f(x)$, with positive norms, are linearly independent on an interval $a \leqslant x \leqslant b$; that is, one is not a constant times the other. By determining the linear combination $f + A\psi_1$ of those functions that is orthogonal to ψ_1 on the fundamental interval $a < x < b$, obtain an orthogonal pair ψ_1, ψ_2, where

$$\psi_2(x) = f(x) - \frac{(f, \psi_1)}{\|\psi_1\|^2} \psi_1(x).$$

Interpret this expression geometrically when f, ψ_1, and ψ_2 represent vectors in three-dimensional space.

6. In Problem 5, suppose that the fundamental interval is $-\pi < x < \pi$ and that

$$\psi_1(x) = \cos nx \qquad \text{and} \qquad f(x) = \cos nx + \sin nx,$$

where n is a fixed positive integer. Show that the function $\psi_2(x)$ there turns out to be $\sin nx$.

Suggestion: One can avoid evaluating any integrals by using the fact that the set in Problem 3 is orthogonal on the interval $-\pi < x < \pi$.

7. Justify the following two statements, regarding functions f in the space $C_p(a, b)$:
   (a) If $f(x) = 0$ except possibly at a finite number of points in the interval $a < x < b$, then $\|f\| = 0$.
   (b) Conversely, if $\|f\| = 0$, then $f(x) = 0$ except possibly at a finite number of points in the interval $a < x < b$.

Suggestion: In part (*b*), use the fact that a definite integral of a nonnegative continuous function over a closed bounded interval has positive value if the function has a positive value somewhere in that interval.

8. Verify that, for any two functions f and g in the space $C_p(a, b)$,

$$\frac{1}{2} \int_a^b \int_a^b [f(x)g(y) - g(x)f(y)]^2 \, dx \, dy = \|f\|^2 \|g\|^2 - (f, g)^2.$$

Thus establish the *Schwarz inequality*

$$|(f, g)| \leq \|f\| \|g\|,$$

which is also valid when f and g denote vectors in three-dimensional space. In that case, it is known as Cauchy's inequality.

9. Let f and g denote any two functions in the space $C_p(a, b)$. Use the Schwarz inequality (Problem 8) to show that if either of these functions has zero norm, then $(f, g) = 0$.

10. Prove that if f and g are functions in the space $C_p(a, b)$, then

$$\|f + g\| \leq \|f\| + \|g\|.$$

If f and g denote, instead, vectors in three-dimensional space, this is the familiar triangle inequality, which states that the length of one side of a triangle is less than or equal to the sum of the lengths of the other two sides.

Suggestion: Start the proof by showing that

$$\|f + g\|^2 = \|f\|^2 + 2(f, g) + \|g\|^2,$$

and then use the Schwarz inequality (Problem 8).

12. GENERALIZED FOURIER SERIES

Let f be any given function in $C_p(a, b)$, the space of piecewise continuous functions defined on an interval $a < x < b$. When an orthonormal set of functions $\phi_n(x)$ $(n = 1, 2, \ldots)$ in $C_p(a, b)$ is specified, it *may* be possible to represent $f(x)$ by a linear combination of those functions, generalized to an infinite series that converges to $f(x)$ at all but possibly a finite number of points in the fundamental interval $a < x < b$:

$$(1) \qquad\qquad f(x) = \sum_{n=1}^{\infty} c_n \phi_n(x) \qquad\qquad (a < x < b).$$

This is analogous to the expression for any vector in three-dimensional space in terms of three mutually orthogonal vectors of unit length, such as **i**, **j**, and **k**.

In order to discover an expression for the coefficients c_n in representation (1), if such a representation actually exists, we use the index of summation m, rather than n, to write

$$(2) \qquad\qquad f(x) = \sum_{m=1}^{\infty} c_m \phi_m(x) \qquad\qquad (a < x < b).$$

We also assume that after each of the terms here is multiplied by a specific

$\phi_n(x)$, the resulting series is integrable term by term over the interval $a < x < b$. This enables us to write

$$\int_a^b f(x)\phi_n(x)\,dx = \sum_{m=1}^{\infty} c_m \int_a^b \phi_m(x)\phi_n(x)\,dx,$$

or

$$(3) \qquad\qquad (f,\phi_n) = \sum_{m=1}^{\infty} c_m(\phi_m,\phi_n).$$

But $(\phi_m,\phi_n) = 0$ for all values of m here except when $m = n$, in which case $(\phi_m,\phi_n) = \|\phi_n\|^2 = 1$. Hence equation (3) becomes $(f,\phi_n) = c_n$, and c_n is evidently the inner product of f and ϕ_n.

As indicated above, we cannot be certain that representation (1), with coefficients $c_n = (f,\phi_n)$, is actually valid for a specific f and a given orthonormal set $\{\phi_n\}$. Hence we write

$$(4) \qquad\qquad f(x) \sim \sum_{n=1}^{\infty} c_n\phi_n(x) \qquad\qquad (a < x < b),$$

where the tilde symbol $\sim$ merely denotes correspondence when

$$(5) \qquad\qquad c_n = (f,\phi_n) = \int_a^b f(x)\phi_n(x)\,dx \qquad (n = 1,2,\dots).$$

To strengthen the analogy with vectors, we recall that if a vector $\mathbf{A}$ in three-dimensional space is to be written in terms of the orthonormal set $\{\mathbf{i},\mathbf{j},\mathbf{k}\}$ as

$$\mathbf{A} = a_1\mathbf{i} + a_2\mathbf{j} + a_3\mathbf{k},$$

the components can be obtained by taking the inner product of $\mathbf{A}$ with each of the vectors of that set. That is, the inner product of $\mathbf{A}$ with $\mathbf{i}$ is a_1, etc.

The series in correspondence (4) is the *generalized Fourier series*, with respect to the orthonormal set $\{\phi_n\}$, for the function f on the interval $a < x < b$. The coefficients c_n are known as *Fourier constants*.

EXAMPLE. Let f denote any function in the space $C_p(0,\pi)$. We know from Example 2, Sec. 11, that the set $\{\phi_n(x)\}$ $(n = 0,1,2,\dots)$ consisting of the functions

$$(6) \qquad\qquad \phi_0(x) = \frac{1}{\sqrt{\pi}}, \qquad \phi_n(x) = \sqrt{\frac{2}{\pi}}\,\cos nx \qquad (n = 1,2,\dots)$$

is orthonormal on the interval $0 < x < \pi$; and correspondence (4), with the summation starting from $n = 0$, becomes

$$f(x) \sim \frac{c_0}{\sqrt{\pi}} + \sum_{n=1}^{\infty} c_n\sqrt{\frac{2}{\pi}}\,\cos nx \qquad\qquad (0 < x < \pi),$$

where

$$c_0 = \frac{1}{\sqrt{\pi}} \int_0^\pi f(x)\, dx, \qquad c_n = \sqrt{\frac{2}{\pi}} \int_0^\pi f(x) \cos nx\, dx \qquad (n = 1, 2, \dots).$$

By writing

$$a_0 = \frac{2}{\sqrt{\pi}} c_0, \qquad a_n = \sqrt{\frac{2}{\pi}} c_n \qquad (n = 1, 2, \dots),$$

we thus arrive at the *Fourier cosine series* correspondence

(7)
$$f(x) \sim \frac{a_0}{2} + \sum_{n=1}^\infty a_n \cos nx \qquad (0 < x < \pi),$$

where

(8)
$$a_n = \frac{2}{\pi} \int_0^\pi f(x) \cos nx\, dx \qquad (n = 0, 1, 2, \dots).$$

Fourier cosine series will be developed further in the next section.

The generalized Fourier series that we shall encounter will always involve orthonormal sets and functions f in a space of the type $C_p(a, b)$ or subspaces of it, and we say that representation (1) is valid for functions f in a given space if equality holds at all but possibly a finite number of points x in the fundamental interval $a < x < b$. Representation (1) will not, however, always be valid even in very restricted function spaces. We may anticipate this limitation by considering vectors in three-dimensional space. For if only the two vectors $\mathbf{i}$ and $\mathbf{j}$ make up the orthonormal set, any vector $\mathbf{A}$ that is not parallel to the xy plane fails to have a representation of the form $\mathbf{A} = a_1\mathbf{i} + a_2\mathbf{j}$. In particular, the nonzero vector $\mathbf{k}$ is orthogonal to both $\mathbf{i}$ and $\mathbf{j}$, in which case the components $a_1 = \mathbf{k} \cdot \mathbf{i}$ and $a_2 = \mathbf{k} \cdot \mathbf{j}$ would both be zero.

Similarly, an orthonormal set $\{\phi_n(x)\}$ may not be large enough to write a generalized Fourier series. To be specific, if the function $f(x)$ in correspondence (4) is orthogonal to each function in the orthonormal set $\{\phi_n(x)\}$, then the Fourier constants $c_n = (f, \phi_n)$ are all zero. This means, of course, that the sum of the series is the zero function. Consequently, if f has a positive norm, the series is not equal to $f(x)$ at all but possibly a finite number of points in the fundamental interval [see Problem 7(a), Sec. 11].

An orthonormal set is *closed* in $C_p(a, b)$, or a subspace of it, if there is no function in the space, with positive norm, that is orthogonal to each of the functions $\phi_n(x)$. Thus, according to the preceding paragraph, *if an orthonormal set $\{\phi_n(x)\}$ is not closed, then representation (1) cannot be valid for each function f in the space.*

In Sec. 20, we shall identify a certain subspace of $C_p(0, \pi)$ such that series (7) is valid when f is in that subspace. Note that if the function $\phi_0(x)$ is not included with the other functions $\phi_n(x)$ $(n = 1, 2, \dots)$ in the set (6), the resulting set is not closed in the subspace since $\phi_0(x)$ is orthogonal to each of the functions in that smaller set. Hence the term $a_0/2$ is needed, in general, for Fourier cosine series representations to be valid in the subspace.

13.   FOURIER COSINE SERIES

In the example in Sec. 12, we introduced the concept of a Fourier cosine series corresponding to a function $f(x)$ in $C_p(0, \pi)$:

$$(1) \qquad f(x) \sim \frac{a_0}{2} + \sum_{n=1}^{\infty} a_n \cos nx \qquad (0 < x < \pi),$$

where

$$(2) \qquad a_n = \frac{2}{\pi} \int_0^{\pi} f(x) \cos nx \, dx \qquad (n = 0, 1, 2, \dots).$$

The fact that f is piecewise continuous on the interval $0 < x < \pi$ ensures the existence of the integrals in expression (2) for the coefficients a_n. As already noted at the end of the preceding section, we shall, in Sec. 20, establish further conditions on f under which the cosine series actually converges to $f(x)$ when $0 < x < \pi$, in which case correspondence (1) becomes an equality.

Observe that correspondence (1), with coefficients (2), can be written more compactly as

$$(3) \qquad f(x) \sim \frac{1}{\pi} \int_0^{\pi} f(s) \, ds + \frac{2}{\pi} \sum_{n=1}^{\infty} \cos nx \int_0^{\pi} f(s) \cos ns \, ds,$$

where s is used for the variable of integration in order to distinguish it from the free variable x.

If f is defined on the interval $0 \leqslant x \leqslant \pi$ and series (1) converges to $f(x)$ for all x in that interval, the series also converges to the *even periodic extension*, *with period* 2π, of f on the entire x axis. That is, it converges to a function $F(x)$ having the properties

$$(4) \qquad F(x) = f(x) \qquad \text{when } 0 \leqslant x \leqslant \pi$$

and

$$(5) \qquad F(-x) = F(x), \qquad F(x + 2\pi) = F(x) \qquad \text{for all } x.$$

The reason for this is that each term in series (1) is itself even and periodic with period 2π. The graph of the extension $y = F(x)$ is obtained by reflecting the graph of $y = f(x)$ in the y axis, to give a graph for the interval $-\pi \leqslant x \leqslant \pi$, and then repeating that graph on the intervals $\pi \leqslant x \leqslant 3\pi$, $3\pi \leqslant x \leqslant 5\pi$, etc., as well as on the intervals $-3\pi \leqslant x \leqslant -\pi$, $-5\pi \leqslant x \leqslant -3\pi$, etc. It follows from these observations that if one is given a function f that is both even and periodic with period 2π, then the cosine series corresponding to $f(x)$ on the interval $0 < x < \pi$ represents $f(x)$ for all x when that series converges to it on the interval $0 \leqslant x \leqslant \pi$. Clearly, a cosine series cannot represent a function $f(x)$ for all x if $f(x)$ is not both even and periodic with period 2π.

EXAMPLE. Let us find the Fourier cosine series for the function $f(x) = \sin x$ on the interval $0 < x < \pi$. The trigonometric identity

$$2 \sin A \cos B = \sin(A + B) + \sin(A - B)$$

enables us to write

$$a_n = \frac{2}{\pi} \int_0^\pi \sin x \cos nx \, dx$$

$$= \frac{1}{\pi} \int_0^\pi [\sin(1 + n)x + \sin(1 - n)x] \, dx \quad (n = 0, 1, 2, \ldots).$$

Hence, when $n \neq 1$,

$$a_n = \frac{1}{\pi} \left[-\frac{\cos(1 + n)x}{1 + n} - \frac{\cos(1 - n)x}{1 - n} \right]_0^\pi = \frac{2}{\pi} \cdot \frac{1 + (-1)^n}{1 - n^2};$$

and when $n = 1$, the coefficient is

$$a_1 = \frac{1}{\pi} \int_0^\pi \sin 2x \, dx = 0.$$

Correspondence (1) then becomes

$$\sin x \sim \frac{2}{\pi} + \frac{2}{\pi} \sum_{n=2}^\infty \frac{1 + (-1)^n}{1 - n^2} \cos nx \qquad (0 < x < \pi).$$

Observe that $1 + (-1)^n = 0$ when n is odd and that this series can be written more efficiently by summing only the terms that occur when n is even. This is accomplished by replacing n by $2n$ wherever n appears after the summation symbol and starting the summation from $n = 1$. The result is

$$(6) \qquad\qquad \sin x \sim \frac{2}{\pi} - \frac{4}{\pi} \sum_{n=1}^\infty \frac{\cos 2nx}{4n^2 - 1} \qquad (0 < x < \pi).$$

The function $\sin x$ will, in fact, satisfy conditions in Sec. 20 ensuring that the correspondence here is an equality for each value of x in the interval $0 \leqslant x \leqslant \pi$. Thus, at each point on the x axis, the series converges to the even periodic extension, with period 2π, of $\sin x$ $(0 \leqslant x \leqslant \pi)$. That extension, shown in Fig. 17, is the function $y = |\sin x|$.

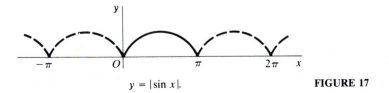

$y = |\sin x|.$ **FIGURE 17**

14.   FOURIER SINE SERIES

We saw in Example 1, Sec. 11, that the sine functions

$$\phi_n(x) = \sqrt{\frac{2}{\pi}} \, \sin nx \qquad\qquad (n = 1, 2, \ldots)$$

constitute an orthonormal set on the interval $0 < x < \pi$. The generalized Fourier series (Sec. 12) corresponding to a function $f(x)$ in $C_p(0, \pi)$ is

$$f(x) \sim \sum_{n=1}^{\infty} c_n \sqrt{\frac{2}{\pi}} \, \sin nx \qquad\qquad (0 < x < \pi),$$

where

$$c_n = \sqrt{\frac{2}{\pi}} \int_0^\pi f(x) \sin nx \, dx \qquad\qquad (n = 1, 2, \ldots).$$

Upon writing

$$b_n = \sqrt{\frac{2}{\pi}} \, c_n \qquad\qquad (n = 1, 2, \ldots),$$

we have the *Fourier sine series* correspondence

(1) $$f(x) \sim \sum_{n=1}^{\infty} b_n \sin nx \qquad\qquad (0 < x < \pi),$$

where

(2) $$b_n = \frac{2}{\pi} \int_0^\pi f(x) \sin nx \, dx \qquad\qquad (n = 1, 2, \ldots).$$

The correspondence can, of course, also be written

(3) $$f(x) \sim \frac{2}{\pi} \sum_{n=1}^{\infty} \sin nx \int_0^\pi f(s) \sin ns \, ds.$$

Suppose that f is defined on the open interval $0 < x < \pi$ and that series (1) converges to $f(x)$ there. Since series (1) clearly converges to zero when $x = 0$ and $x = \pi$, it converges to $f(x)$ for all x in the closed interval $0 \leqslant x \leqslant \pi$ if f is assigned the values $f(0) = 0$ and $f(\pi) = 0$. Remarks similar to ones in Sec. 13, regarding cosine series, show that series (1) then converges to the *odd periodic extension, with period* 2π, of f for all values of x. This time, the extension is the function $F(x)$ defined by the equations

(4) $$\qquad\qquad F(x) = f(x) \qquad\qquad \text{when } 0 \leqslant x \leqslant \pi$$

and

(5) $$\qquad F(-x) = -F(x), \qquad F(x + 2\pi) = F(x) \qquad\qquad \text{for all } x.$$

The extension F is odd and periodic with period 2π since the terms $b_n \sin nx$ in series (1) have those properties. The graph of $y = F(x)$ is symmetric with respect to the origin and can be obtained by first reflecting the graph of $y = f(x)$ in the y axis, then reflecting the result in the x axis, and finally repeating the graph found for the interval $-\pi \leqslant x \leqslant \pi$ every 2π units along the entire x axis. Evidently, a Fourier sine series on the interval $0 < x < \pi$ can also be used to represent a given function that is defined for all x and is both odd and periodic with period 2π, provided the representation is valid when $0 \leqslant x \leqslant \pi$.

EXAMPLE 1. To find the sine series corresponding to the function $f(x) = x$ on the interval $0 < x < \pi$, we refer to expression (2) for the coefficients b_n and use integration by parts to write

$$b_n = \frac{2}{\pi} \int_0^\pi x \sin nx \, dx = \frac{2}{\pi} \left[-\frac{x \cos nx}{n} + \frac{\sin nx}{n^2} \right]_0^\pi = 2 \frac{(-1)^{n+1}}{n}.$$

Thus

$$(6) \qquad\qquad x \sim 2 \sum_{n=1}^{\infty} \frac{(-1)^{n+1}}{n} \sin nx \qquad\qquad (0 < x < \pi).$$

Our theory will show that the series actually converges to $f(x)$ when $0 < x < \pi$. Hence it converges to the odd periodic function $y = F(x)$ that is graphed in Fig. 18. The fact that the series converges to zero at the points $x = 0, \pm\pi, \pm 3\pi, \pm 5\pi, \ldots$ is in agreement with our theory, which will tell us that it must converge to the mean value of the one-sided limits of $F(x)$ at each of those discontinuities.

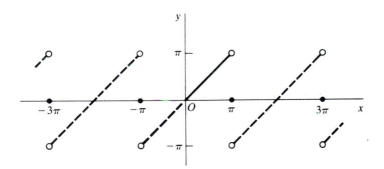

FIGURE 18

In the evaluation of integrals representing Fourier coefficients, it is sometimes necessary to apply integration by parts more than once. We now give an example where this can be accomplished by means of a single formula due to

L. Kronecker (1823–1891). We preface the example with a statement of that formula.[†]

Let $p(x)$ be a polynomial of degree m, and suppose that $f(x)$ is continuous. Then, except for an arbitrary additive constant,

$$(7) \qquad \int p(x) f(x)\, dx = pF_1 - p'F_2 + p''F_3 - \cdots + (-1)^m p^{(m)} F_{m+1},$$

where p is successively differentiated until it becomes zero, where F_1 denotes an indefinite integral of f, F_2 an indefinite integral of F_1, etc., and where alternating signs are affixed to the terms. Note that the differentiation of p begins with the *second* term, whereas the integration of f begins with the *first* term. The formula, which is readily verified by differentiating its right-hand side to obtain $p(x)f(x)$, could even have been used to evaluate the integral in Example 1, where only one integration by parts was needed.

EXAMPLE 2. To illustrate the advantage of formula (7) when successive integration by parts is required, let us find the Fourier sine series for the function $f(x) = x^3$ on the interval $0 < x < \pi$. With the aid of that formula, we may write

$$b_n = \frac{2}{\pi} \int_0^\pi x^3 \sin nx\, dx$$

$$= \frac{2}{\pi} \left[(x^3)\left(-\frac{\cos nx}{n} \right) - (3x^2)\left(-\frac{\sin nx}{n^2} \right) + (6x)\left(\frac{\cos nx}{n^3} \right) - (6)\left(\frac{\sin nx}{n^4} \right) \right]_0^\pi$$

$$= 2(-1)^{n+1} \frac{(n\pi)^2 - 6}{n^3} \qquad\qquad\qquad (n = 1, 2, \dots).$$

Hence

$$(8) \qquad\qquad x^3 \sim 2 \sum_{n=1}^\infty (-1)^{n+1} \frac{(n\pi)^2 - 6}{n^3} \sin nx \qquad (0 < x < \pi).$$

As was the case in Example 1, the series converges to the given function on the interval $0 < x < \pi$. Since x^3 is an odd function whose value is zero when $x = 0$, this series represents x^3 on the interval $-\pi < x < \pi$ also.

We conclude this section by pointing out a computational aid that is useful in finding the coefficients b_n ($n = 1, 2, \dots$) in the Fourier sine series for a linear combination $c_1 f_1(x) + c_2 f_2(x)$ of two functions $f_1(x)$ and $f_2(x)$ whose sine

---

[†] Kronecker actually treated the problem more extensively in papers that originally appeared in the *Berlin Sitzungsberichte* (1885, 1889).

series are already known. Namely, since the expression

$$b_n = \frac{2}{\pi} \int_0^{\pi} [c_1 f_1(x) + c_2 f_2(x)] \sin nx \, dx$$

can be written as

$$b_n = c_1 \frac{2}{\pi} \int_0^{\pi} f_1(x) \sin nx \, dx + c_2 \frac{2}{\pi} \int_0^{\pi} f_2(x) \sin nx \, dx,$$

it is clear that each b_n is simply the same linear combination of the nth coefficients in the sine series for the individual functions $f_1(x)$ and $f_2(x)$. Such an observation applies as well in finding coefficients in cosine and other types of series encountered in the present and later chapters.

EXAMPLE 3. In view of the sine series for x and x^3 found in Examples 1 and 2, respectively, the coefficients b_n in the sine series corresponding to the function

$$f(x) = x(\pi^2 - x^2) = \pi^2 x - x^3 \qquad (0 < x < \pi)$$

are

$$b_n = \pi^2 2 \frac{(-1)^{n+1}}{n} - 2(-1)^{n+1} \frac{(n\pi)^2 - 6}{n^3} = 12 \frac{(-1)^{n+1}}{n^3} \qquad (n = 1, 2, \dots).$$

Thus

$$(9) \qquad\qquad x(\pi^2 - x^2) \sim 12 \sum_{n=1}^{\infty} \frac{(-1)^{n+1}}{n^3} \sin nx \qquad (0 < x < \pi).$$

PROBLEMS

Find (a) the Fourier cosine series and (b) the Fourier sine series on the interval $0 < x < \pi$ that corresponds to each of the functions in Problems 1 through 4.

1. $f(x) = 1 \quad (0 < x < \pi)$.

$$\text{Answers: } (a) \ 1; \quad (b) \ \frac{4}{\pi} \sum_{n=1}^{\infty} \frac{\sin (2n - 1)x}{2n - 1}. \,†$$

2. $f(x) = \pi - x \quad (0 < x < \pi)$.

$$\text{Answers: } (a) \ \frac{\pi}{2} + \frac{4}{\pi} \sum_{n=1}^{\infty} \frac{\cos (2n - 1)x}{(2n - 1)^2}; \quad (b) \ 2 \sum_{n=1}^{\infty} \frac{\sin nx}{n}.$$

---

† In part (b) the coefficients b_n are zero when n is even. The index n in the series can, therefore, be replaced by $2n - 1$ wherever it appears after the summation symbol. (Compare the example in Sec. 13.)

3. $f(x) = \begin{cases} 1 & \text{when } 0 < x < \dfrac{\pi}{2}, \\ 0 & \text{when } \dfrac{\pi}{2} < x < \pi. \end{cases}$

Answers: (a) $\dfrac{1}{2} + \dfrac{2}{\pi} \displaystyle\sum_{n=1}^{\infty} (-1)^{n+1} \dfrac{\cos(2n-1)x}{2n-1}$;

(b) $\dfrac{2}{\pi} \displaystyle\sum_{n=1}^{\infty} \left(1 - \cos\dfrac{n\pi}{2}\right) \dfrac{\sin nx}{n}$.

4. $f(x) = x^2 \quad (0 < x < \pi)$.

Answers: (a) $\dfrac{\pi^2}{3} + 4 \displaystyle\sum_{n=1}^{\infty} \dfrac{(-1)^n}{n^2} \cos nx$;

(b) $2\pi^2 \displaystyle\sum_{n=1}^{\infty} \left[\dfrac{(-1)^{n+1}}{n\pi} - 2\dfrac{1-(-1)^n}{(n\pi)^3} \right] \sin nx$.

5. By referring to the sine series for x in Example 1, Sec. 14, and the one found for x^2 in Problem 4(b), show that

$$x(\pi - x) \sim \frac{8}{\pi} \sum_{n=1}^{\infty} \frac{\sin(2n-1)x}{(2n-1)^3} \qquad (0 < x < \pi).$$

6. What is the Fourier sine series corresponding to the function $f(x) = \sin x$ on the interval $0 < x < \pi$?

Suggestion: To find the coefficients in the series, refer to the integration formula (10), Sec. 11.

Answer: $\sin x$.

7. Find the Fourier cosine series for x on the interval $0 < x < \pi$. Then, given that the correspondence obtained is actually an equality when $0 \leqslant x \leqslant \pi$, point out how it follows that

$$|x| = \frac{\pi}{2} - \frac{4}{\pi} \sum_{n=1}^{\infty} \frac{\cos(2n-1)x}{(2n-1)^2} \qquad (-\pi \leqslant x \leqslant \pi).$$

8. Show that

$$x^4 \sim \frac{\pi^4}{5} + 8 \sum_{n=1}^{\infty} (-1)^n \frac{(n\pi)^2 - 6}{n^4} \cos nx \qquad (0 < x < \pi).$$

Given that this correspondence is actually an equality when $0 \leqslant x \leqslant \pi$, sketch the function represented by the series for all x.

9. Verify Kronecker's formula (7), Sec. 14.

10. Let $\{\psi_n(x)\}$ $(n = 1, 2, \ldots)$ denote an orthogonal, but not necessarily orthonormal, set on a fundamental interval $a < x < b$. Show that the correspondence between a piecewise continuous function $f(x)$ and its generalized Fourier series with respect to the orthonormal set of functions $\phi_n(x) = \psi_n(x)/\|\psi_n\|$ $(n = 1, 2, \ldots)$ can be written

$$f(x) \sim \sum_{n=1}^{\infty} \gamma_n \psi_n(x), \qquad \text{where} \qquad \gamma_n = \frac{(f, \psi_n)}{\|\psi_n\|^2}.$$

11. In the space of *continuous* functions on the interval $a \leqslant x \leqslant b$, prove that if two functions f and g have the same Fourier constants with respect to a *closed* orthonormal set $\{\phi_n(x)\}$, then f and g must be identical. Thus show that f is uniquely determined by its Fourier constants.

  Suggestion: Show that the norm of the difference $f(x) - g(x)$ is zero. Then point out how it follows that $f(x) - g(x) \equiv 0$ (see the suggestion with Problem 7, Sec. 11).

12. Let f be a function in $C_p(-\pi, \pi)$, and write $f(x) = g(x) + h(x)$, where g and h are defined by the equations

$$g(x) = \frac{f(x) + f(-x)}{2} \quad \text{and} \quad h(x) = \frac{f(x) - f(-x)}{2}.$$

The functions g and h are evidently even and odd, respectively, on the interval $-\pi < x < \pi$.

(*a*) Explain why it is reasonable to expect that

$$f(x) = \frac{a_0}{2} + \sum_{n=1}^{\infty} (a_n \cos nx + b_n \sin nx) \quad (-\pi < x < \pi),$$

where a_n $(n = 0, 1, 2, \ldots)$ are the coefficients in the Fourier cosine series for $g(x)$ on the interval $0 < x < \pi$ and b_n $(n = 1, 2, \ldots)$ are the coefficients in the Fourier sine series for $h(x)$ on that same interval.

(*b*) Show that the coefficients a_n and b_n in part (*a*) can be written in the form

$$a_n = \frac{1}{\pi} \int_{-\pi}^{\pi} f(x) \cos nx \, dx \qquad (n = 0, 1, 2, \ldots),$$

$$b_n = \frac{1}{\pi} \int_{-\pi}^{\pi} f(x) \sin nx \, dx \qquad (n = 1, 2, \ldots).$$

(Series of the type introduced here are discussed in the next section.)

  Suggestion: In part (*b*), write

$$a_n = \frac{1}{\pi} \left[\int_0^{\pi} f(x) \cos nx \, dx + \int_0^{\pi} f(-s) \cos ns \, ds \right]$$

and then make the substitution $x = -s$ in the second integral here. The coefficients b_n are to be treated similarly.

15. FOURIER SERIES

In Problem 3, Sec. 11, we found that the functions

$$(1) \quad \phi_0(x) = \frac{1}{\sqrt{2\pi}}, \qquad \phi_{2n-1}(x) = \frac{1}{\sqrt{\pi}} \cos nx, \qquad \phi_{2n}(x) = \frac{1}{\sqrt{\pi}} \sin nx$$

$$(n = 1, 2, \ldots)$$

form an orthonormal set on the fundamental interval $-\pi < x < \pi$. The generalized Fourier series corresponding to a function f in $C_p(-\pi, \pi)$ is, therefore,

$$\sum_{n=0}^{\infty} c_n \phi_n(x) = c_0 \phi_0(x) + \sum_{n=1}^{\infty} [c_{2n-1} \phi_{2n-1}(x) + c_{2n} \phi_{2n}(x)].$$

That is,

$$(2) \qquad f(x) \sim \frac{c_0}{\sqrt{2\pi}} + \sum_{n=1}^{\infty} \left(\frac{c_{2n-1}}{\sqrt{\pi}} \cos nx + \frac{c_{2n}}{\sqrt{\pi}} \sin nx \right) \quad (-\pi < x < \pi),$$

where

$$c_0 = (f, \phi_0) = \frac{1}{\sqrt{2\pi}} \int_{-\pi}^{\pi} f(x) \, dx$$

and

$$c_{2n-1} = (f, \phi_{2n-1}) = \frac{1}{\sqrt{\pi}} \int_{-\pi}^{\pi} f(x) \cos nx \, dx \quad (n = 1, 2, \ldots),$$

$$c_{2n} = (f, \phi_{2n}) = \frac{1}{\sqrt{\pi}} \int_{-\pi}^{\pi} f(x) \sin nx \, dx \qquad (n = 1, 2, \ldots).$$

So if we write

$$(3) \qquad a_0 = \sqrt{\frac{2}{\pi}} \, c_0, \qquad a_n = \frac{c_{2n-1}}{\sqrt{\pi}}, \qquad b_n = \frac{c_{2n}}{\sqrt{\pi}} \qquad (n = 1, 2, \ldots),$$

correspondence (2) becomes

$$(4) \qquad f(x) \sim \frac{a_0}{2} + \sum_{n=1}^{\infty} (a_n \cos nx + b_n \sin nx) \quad (-\pi < x < \pi),$$

where

$$(5) \qquad a_n = \frac{1}{\pi} \int_{-\pi}^{\pi} f(x) \cos nx \, dx \qquad (n = 0, 1, 2, \ldots)$$

and

$$(6) \qquad b_n = \frac{1}{\pi} \int_{-\pi}^{\pi} f(x) \sin nx \, dx \qquad (n = 1, 2, \ldots).$$

Series (4), with coefficients (5) and (6), is the *Fourier series* corresponding to $f(x)$ on the interval $-\pi < x < \pi$. Suppose that the series converges to $f(x)$ when $-\pi < x < \pi$. Then, in view of the periodicity of its terms, it converges to a function $y = F(x)$ that coincides with $y = f(x)$ on that interval and whose graph there is repeated every 2π units along the x axis. The function F is, therefore, the *periodic extension, with period* 2π, of f. If, on the other hand, f is a given periodic function, with period 2π, series (4) represents $f(x)$ everywhere when it converges to $f(x)$ on the interval $-\pi \leqslant x \leqslant \pi$.

EXAMPLE 1. Let us find the Fourier series corresponding to the function $f(x)$ which is defined on the fundamental interval $-\pi < x < \pi$ as follows:

(7)
$$f(x) = \begin{cases} 0 & \text{when } -\pi < x \leqslant 0, \\ x & \text{when } \quad 0 < x < \pi. \end{cases}$$

The graph of $y = f(x)$ is indicated by bold line segments in Fig. 19.

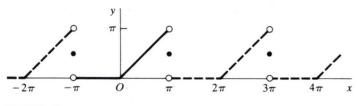

FIGURE 19

According to expression (5),

$$a_n = \frac{1}{\pi}\left(\int_{-\pi}^{0} 0 \cos nx \, dx + \int_{0}^{\pi} x \cos nx \, dx \right)$$

$$= \frac{1}{\pi}\int_{0}^{\pi} x \cos nx \, dx \qquad\qquad (n = 0, 1, 2, \dots).$$

By applying integration by parts, or Kronecker's method (Sec. 14), one can show that

$$a_n = \frac{(-1)^n - 1}{\pi n^2}$$

when $n = 1, 2, \dots$. To avoid division by zero, we must evaluate the integral for a_0 separately:

$$a_0 = \frac{1}{\pi}\int_{0}^{\pi} x \, dx = \frac{\pi}{2}.$$

Expression (6) tells us that

$$b_n = \frac{1}{\pi}\left(\int_{-\pi}^{0} 0 \sin nx \, dx + \int_{0}^{\pi} x \sin nx \, dx \right)$$

$$= \frac{1}{\pi}\int_{0}^{\pi} x \sin nx \, dx = \frac{(-1)^{n+1}}{n}$$

for all positive integers $n = 1, 2, \dots$. Hence, on the interval $-\pi < x < \pi$,

(8)
$$f(x) \sim \frac{\pi}{4} + \sum_{n=1}^{\infty}\left[\frac{(-1)^n - 1}{\pi n^2}\cos nx + \frac{(-1)^{n+1}}{n}\sin nx \right].$$

This series will be shown to converge to $f(x)$ on the fundamental interval, as well as to the periodic extension $F(x)$ that is indicated in Fig. 19, where the graph of $y = F(x)$ is sketched. As in Example 1, Sec. 14, the series must converge to the mean value of the one-sided limits of the periodic extension at each of the discontinuities $x = \pm\pi, \pm 3\pi, \pm 5\pi, \ldots$. Here the mean values are all $\pi/2$.

It may be that the given function f in $C_p(-\pi, \pi)$ is *even* on the interval $-\pi < x < \pi$; that is, $f(-x) = f(x)$ for all such values of x. Then

$$f(-x)\cos(-nx) = f(x)\cos nx \qquad (n = 0, 1, 2, \ldots)$$

and

$$f(-x)\sin(-nx) = -f(x)\sin nx \qquad (n = 1, 2, \ldots)$$

when $-\pi < x < \pi$, and we see that $f(x)\cos nx$ and $f(x)\sin nx$ are even and odd, respectively. Hence expressions (5) and (6) reduce to

$$a_n = \frac{2}{\pi}\int_0^\pi f(x)\cos nx\, dx \qquad (n = 0, 1, 2, \ldots)$$

and $b_n = 0$ $(n = 1, 2, \ldots)$ (see the suggestions with Problems 2 and 3, Sec. 11). Series (4) thus becomes a Fourier cosine series (Sec. 13) for $f(x)$ $(0 < x < \pi)$.

Similarly, if f is *odd* on the interval $-\pi < x < \pi$, it follows from expressions (5) and (6) that $a_n = 0$ $(n = 0, 1, 2, \ldots)$ and

$$b_n = \frac{2}{\pi}\int_0^\pi f(x)\sin nx\, dx \qquad (n = 1, 2, \ldots).$$

In this case, series (4) becomes a Fourier sine series (Sec. 14) for $f(x)$ $(0 < x < \pi)$.

EXAMPLE 2. The function $f(x) = |\sin x|$ $(-\pi < x < \pi)$ is even. Hence the Fourier series corresponding to $f(x)$ on the interval $-\pi < x < \pi$ is actually the Fourier cosine series for the function

$$f(x) = |\sin x| = \sin x \qquad (0 < x < \pi).$$

That series has already been found in the example in Sec. 13; and, rewriting correspondence (6) there, we see that

$$|\sin x| \sim \frac{2}{\pi} - \frac{4}{\pi}\sum_{n=1}^{\infty}\frac{\cos 2nx}{4n^2 - 1} \qquad (-\pi < x < \pi).$$

EXAMPLE 3. Since the function $f(x) = x$ $(-\pi < x < \pi)$ is odd, the Fourier series for f on $-\pi < x < \pi$ is simply the Fourier sine series for that function on $0 < x < \pi$. Hence correspondence (6) in Example 1, Sec. 14, is also a correspondence on the larger interval $-\pi < x < \pi$:

$$x \sim 2\sum_{n=1}^{\infty}\frac{(-1)^{n+1}}{n}\sin nx \qquad (-\pi < x < \pi).$$

Similarly, correspondence (8) in Example 2, Sec. 14, can be written

$$x^3 \sim 2 \sum_{n=1}^{\infty} (-1)^{n+1} \frac{(n\pi)^2 - 6}{n^3} \sin nx \qquad (-\pi < x < \pi).$$

Correspondence (4), when combined with expressions (5) and (6) for the constants a_n and b_n, becomes

$$f(x) \sim \frac{1}{2\pi} \int_{-\pi}^{\pi} f(s)\, ds$$

$$+ \frac{1}{\pi} \sum_{n=1}^{\infty} \left[\cos nx \int_{-\pi}^{\pi} f(s) \cos ns\, ds + \sin nx \int_{-\pi}^{\pi} f(s) \sin ns\, ds \right].$$

The trigonometric identity

$$\cos(A - B) = \cos A \cos B + \sin A \sin B$$

then enables us to write the correspondence in the form

$$(9) \qquad f(x) \sim \frac{1}{2\pi} \int_{-\pi}^{\pi} f(s)\, ds + \frac{1}{\pi} \sum_{n=1}^{\infty} \int_{-\pi}^{\pi} f(s) \cos n(s - x)\, ds.$$

Note that the term

$$\frac{1}{2\pi} \int_{-\pi}^{\pi} f(s)\, ds$$

here, which is the same as the term $a_0/2$ in series (4), is the mean, or average, value of $f(x)$ over the interval $-\pi < x < \pi$.

Form (9) of correspondence (4) will be the starting point of the proof in Sec. 19 of our theorem ensuring the convergence of the Fourier series to $f(x)$ on the interval $-\pi < x < \pi$.

16. BEST APPROXIMATION IN THE MEAN

Let $S_N(x)$ $(N = 1, 2, \ldots)$ denote partial sums of the Fourier series for a function f in $C_p(-\pi, \pi)$:

$$(1) \qquad S_N(x) = \frac{a_0}{2} + \sum_{n=1}^{N} (a_n \cos nx + b_n \sin nx) \qquad (-\pi < x < \pi).$$

We consider here the matter of approximating the function f by these partial sums. While the main result of this section is of interest in itself, it yields a property of the coefficients a_n and b_n that we shall need in our treatment of convergence of Fourier series in the next several sections. Namely,

$$(2) \qquad \lim_{n \to \infty} a_n = 0 \qquad \text{and} \qquad \lim_{n \to \infty} b_n = 0.$$

As is often the case, the discussion is simplified by first treating any orthonormal set $\{\phi_n(x)\}$ $(n = 1, 2, \ldots)$ on a fundamental interval $a < x < b$. We consider the first N functions $\phi_1(x), \phi_2(x), \ldots, \phi_N(x)$ of that set, and we let $\Phi_N(x)$ denote any linear combination of them:

$$(3) \qquad \Phi_N(x) = \gamma_1 \phi_1(x) + \gamma_2 \phi_2(x) + \cdots + \gamma_N \phi_N(x).$$

The norm

$$(4) \qquad \| f - \Phi_N \| = \left\{ \int_a^b [f(x) - \Phi_N(x)]^2 \, dx \right\}^{1/2}$$

is a measure of the deviation of the sum Φ_N from a given function f in $C_p(a, b)$ (see Sec. 11). Let us determine values of the constants γ_n in expression (3) that make $\| f - \Phi_N \|$, or the quantity

$$(5) \qquad E = \| f - \Phi_N \|^2 = \int_a^b [f(x) - \Phi_N(x)]^2 \, dx,$$

as small as possible. The nonnegative number E represents the *mean square error* in the approximation by the function Φ_N to the function f; and we seek the *best approximation in the mean.*[†]

We start with the observation that

$$(f - \Phi_N)^2 = \left(f - \sum_{n=1}^{N} \gamma_n \phi_n \right)^2 = f^2 - 2f \sum_{n=1}^{N} \gamma_n \phi_n + \left(\sum_{n=1}^{N} \gamma_n \phi_n \right)^2.$$

But

$$\left(\sum_{n=1}^{N} \gamma_n \phi_n \right)^2 = \left(\sum_{m=1}^{N} \gamma_m \phi_m \right)\left(\sum_{n=1}^{N} \gamma_n \phi_n \right)$$

$$= \sum_{n=1}^{N} \left(\sum_{m=1}^{N} \gamma_m \phi_m \right) \gamma_n \phi_n$$

$$= \sum_{n=1}^{N} \left(\sum_{m=1}^{N} \gamma_m \gamma_n \phi_m \phi_n \right);$$

[†] The approximation sought here is also called a *least squares approximation.*

and this enables us to write

$$(f - \Phi_N)^2 = f^2 + \sum_{n=1}^{N} \left[\left(\sum_{m=1}^{N} \gamma_m \gamma_n \phi_m \phi_n \right) - 2\gamma_n f \phi_n \right].$$

Integrating each side here over the interval $a < x < b$ and then using the relations $(\phi_m, \phi_n) = \delta_{mn}$ and $(f, \phi_n) = c_n$, where δ_{mn} is Kronecker's δ (Sec. 11) and the c_n are Fourier constants (Sec. 12), we arrive at the following expression for the error E, defined above:

$$E = \|f\|^2 + \sum_{n=1}^{N} \left(\gamma_n^2 - 2\gamma_n c_n \right).$$

That is,

(6) $$E = \|f\|^2 + \sum_{n=1}^{N} \left(\gamma_n - c_n \right)^2 - \sum_{n=1}^{N} c_n^2.$$

In view of the squares in the first summation appearing in equation (6), the smallest possible value of E is, then, obtained when $\gamma_n = c_n$ $(n = 1, 2, \ldots, N)$, that value being

(7) $$E = \|f\|^2 - \sum_{n=1}^{N} c_n^2.$$

We state the result as a theorem.

Theorem. Let c_n $(n = 1, 2, \ldots)$ be the Fourier constants for a function f in $C_p(a, b)$ with respect to an orthonormal set $\{\phi_n(x)\}$ $(n = 1, 2, \ldots)$ in that space. Then, of all possible linear combinations of the functions $\phi_1(x), \phi_2(x), \ldots, \phi_N(x)$, the combination

$$c_1 \phi_1(x) + c_2 \phi_2(x) + \cdots + c_N \phi_N(x)$$

is the best approximation in the mean to $f(x)$ on the fundamental interval $a < x < b$. In that case, the mean square error E is given by equation (7).

This theorem is analogous to, and even suggested by, a corresponding result in three-dimensional space. Namely, suppose that we wish to approximate a vector $\mathbf{A} = a_1 \mathbf{i} + a_2 \mathbf{j} + a_3 \mathbf{k}$ by a linear combination of just the two basis vectors $\mathbf{i}$ and $\mathbf{j}$. If we interpret $\mathbf{A}$ and any linear combination $\alpha_1 \mathbf{i} + \alpha_2 \mathbf{j}$ as radius vectors, it is geometrically evident that the shortest distance d between their tips occurs when $\alpha_1 \mathbf{i} + \alpha_2 \mathbf{j}$ is the vector projection of $\mathbf{A}$ onto the plane of $\mathbf{i}$ and $\mathbf{j}$. That projection is, of course, the vector $a_1 \mathbf{i} + a_2 \mathbf{j}$ (see Fig. 20), the components a_1 and a_2 being the inner products of $\mathbf{A}$ with $\mathbf{i}$ and $\mathbf{j}$, respectively.

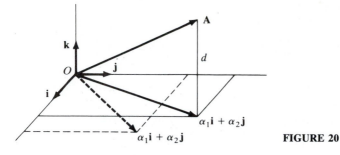

FIGURE 20

Corollary. If c_n $(n = 1, 2, \ldots)$ are the Fourier constants for a function f in $C_p(a, b)$ with respect to an orthonormal set $\{\phi_n(x)\}$ $(n = 1, 2, \ldots)$ in that space, then

$$(8) \qquad \lim_{n \to \infty} c_n = 0.$$

The proof of this corollary is based on *Bessel's inequality*

$$(9) \qquad \sum_{n=1}^{N} c_n^2 \leqslant \|f\|^2 \qquad (N = 1, 2, \ldots),$$

which is an immediate consequence of expression (7) for the mean square error E and the fact that $E \geqslant 0$. We observe that the right-hand side of Bessel's inequality is independent of the positive integer N; and as N increases on the left-hand side, the sums there form a sequence that is bounded and nondecreasing. Since such a sequence must converge and since this particular sequence is the sequence of partial sums of the series whose terms are c_n^2 $(n = 1, 2, \ldots)$, that series must converge. Limit (8) now follows from the fact that the nth term of a convergent series always tends to zero as n tends to infinity.

We recall from Sec. 15 that when the orthonormal set of functions

$$\phi_0(x) = \frac{1}{\sqrt{2\pi}}, \qquad \phi_{2n-1}(x) = \frac{1}{\sqrt{\pi}} \cos nx, \qquad \phi_{2n}(x) = \frac{1}{\sqrt{\pi}} \sin nx$$

$$(n = 1, 2, \ldots)$$

in $C_p(-\pi, \pi)$ is used, the generalized Fourier series

$$(10) \qquad \sum_{n=0}^{\infty} c_n \phi_n(x) = c_0 \phi_0(x) + \sum_{n=1}^{\infty} \left[c_{2n-1} \phi_{2n-1}(x) + c_{2n} \phi_{2n}(x) \right]$$

$$(-\pi < x < \pi)$$

corresponding to a function f in $C_p(-\pi, \pi)$ is the ordinary Fourier series

$$(11) \qquad \frac{a_0}{2} + \sum_{n=1}^{\infty} (a_n \cos nx + b_n \sin nx) \qquad (-\pi < x < \pi),$$

where

$$(12) \qquad a_0 = \sqrt{\frac{2}{\pi}} \, c_0, \qquad a_n = \frac{c_{2n-1}}{\sqrt{\pi}}, \qquad b_n = \frac{c_{2n}}{\sqrt{\pi}} \qquad (n = 1, 2, \dots).$$

The above theorem now tells us that, of all possible linear combinations of the functions

$$\phi_0(x), \quad \phi_1(x), \quad \phi_2(x), \quad \dots, \quad \phi_{2N}(x),$$

the partial sum

$$\sum_{n=0}^{2N} c_n \phi_n(x) = c_0 \phi_0(x) + \sum_{n=1}^{N} \left[c_{2n-1} \phi_{2n-1}(x) + c_{2n} \phi_{2n}(x) \right]$$

of series (10) is the best approximation in the mean to f on the interval $-\pi < x < \pi$. That is, the partial sum (1) is the best approximation of all linear combinations of the functions

$$\frac{1}{2}, \qquad \cos nx, \qquad \sin nx \qquad (n = 1, 2, \dots, N).$$

The corollary, together with relations (12), also yields limits (2). Those limits are, in fact, valid when a_n and b_n are the coefficients in the Fourier cosine and sine series, respectively, for a function f in $C_p(0, \pi)$. To see that the coefficients a_n in the cosine series tend to zero as n tends to infinity, we need only observe that the cosine series is the same as the Fourier series on $-\pi < x < \pi$ for the even extension of f onto the interval $-\pi < x < 0$ (see Sec. 15). Similarly, the sine series can be thought of as the Fourier series for the odd extension of f onto $-\pi < x < 0$. Hence the coefficients b_n in the sine series also tend to zero as n tends to infinity.

PROBLEMS

Find the Fourier series on the interval $-\pi < x < \pi$ that corresponds to each of the functions in Problems 1 through 6.

1. $f(x) = \begin{cases} -\dfrac{\pi}{2} & \text{when } -\pi < x < 0, \\[2mm] \dfrac{\pi}{2} & \text{when } \quad 0 < x < \pi. \end{cases}$

$$\text{Answer: } 2 \sum_{n=1}^{\infty} \frac{\sin(2n-1)x}{2n-1}.^{\dagger}$$

---

† See the footnote with Problem 1(b), Sec. 14.

2. $f(x)$ is the function such that the graph of $y = f(x)$ consists of the two line segments shown in Fig. 21.

$$\text{Answer:} \quad \frac{3}{2} + 2 \sum_{n=1}^{\infty} \left[\frac{1 - (-1)^n}{(n\pi)^2} \cos nx + \frac{(-1)^{n+1}}{n\pi} \sin nx \right].$$

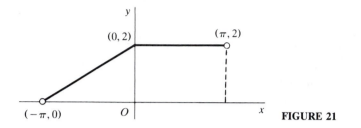

FIGURE 21

3. $f(x) = x + \dfrac{1}{4}x^2 \quad (-\pi < x < \pi)$.

   Suggestion: Use the series for x in Example 3, Sec. 15, and the one for x^2 in Problem 4(a), Sec. 14.

$$\text{Answer:} \quad \frac{\pi^2}{12} + \sum_{n=1}^{\infty} (-1)^n \left(\frac{\cos nx}{n^2} - \frac{2 \sin nx}{n} \right).$$

4. $f(x) = e^{ax} \quad (-\pi < x < \pi)$, where $a \neq 0$.

   Suggestion: Use Euler's formula $e^{i\theta} = \cos \theta + i \sin \theta$, where $i = \sqrt{-1}$, to write

$$a_n + ib_n = \frac{1}{\pi} \int_{-\pi}^{\pi} f(x) e^{inx} \, dx \qquad (n = 1, 2, \ldots).$$

   Then, after evaluating this single integral, equate real and imaginary parts.[†]

$$\text{Answer:} \quad \frac{2 \sinh a\pi}{\pi} \left[\frac{1}{2a} + \sum_{n=1}^{\infty} \frac{(-1)^n}{a^2 + n^2} (a \cos nx - n \sin nx) \right].$$

5. $f(x) = \sinh ax \quad (-\pi < x < \pi)$.

   Suggestion: Use the series found in Problem 4.

$$\text{Answer:} \quad \frac{2 \sinh a\pi}{\pi} \sum_{n=1}^{\infty} (-1)^{n+1} \frac{n}{a^2 + n^2} \sin nx.$$

6. $f(x) = \cos ax \quad (-\pi < x < \pi)$, where $a \neq 0, \pm 1, \pm 2, \ldots$.

   Suggestion: With the aid of Euler's formula, stated in Problem 4, write

$$\cos ax = \frac{e^{iax} + e^{-iax}}{2}.$$

[†] For a justification of Euler's formula and background on complex-variable methods, see the authors' book (1990), listed in the Bibliography.

Then use the series already obtained in that earlier problem.

$$Answer: \quad \frac{2a \sin a\pi}{\pi} \left[\frac{1}{2a^2} + \sum_{n=1}^{\infty} \frac{(-1)^{n+1}}{n^2 - a^2} \cos nx \right].$$

7. Find the Fourier series on the interval $-\pi < x < \pi$ for the function f defined by the equations

$$f(x) = \begin{cases} 0 & \text{when } -\pi \leqslant x \leqslant 0, \\ \sin x & \text{when } 0 < x \leqslant \pi. \end{cases}$$

Then, given that the series converges to $f(x)$ when $-\pi \leqslant x \leqslant \pi$, describe graphically the function that is represented by the series for all x ($-\infty < x < \infty$).

Suggestion: To find the series, write the function in the form

$$f(x) = \frac{\sin x + |\sin x|}{2} \qquad (-\pi \leqslant x \leqslant \pi)$$

and then use the results in Problem 6, Sec. 14, and Example 2, Sec. 15.

$$Answer: \quad \frac{1}{\pi} + \frac{1}{2} \sin x - \frac{2}{\pi} \sum_{n=1}^{\infty} \frac{\cos 2nx}{4n^2 - 1}.$$

8. Let a_n and b_n denote the coefficients in the Fourier cosine and sine series, respectively, corresponding to a function $f(x)$ in $C_p(0, \pi)$.

(a) By referring to the example in Sec. 12, obtain from Bessel's inequality (9), Sec. 16, the inequality

$$\frac{a_0^2}{2} + \sum_{n=1}^{N} a_n^2 \leqslant \frac{2}{\pi} \int_0^{\pi} [f(x)]^2 \, dx \qquad (N = 1, 2, \dots).$$

(b) By referring to Sec. 14, show how it follows from Bessel's inequality (9), Sec. 16, that

$$\sum_{n=1}^{N} b_n^2 \leqslant \frac{2}{\pi} \int_0^{\pi} [f(x)]^2 \, dx \qquad (N = 1, 2, \dots).$$

9. Show that when a_n and b_n are the coefficients in the Fourier series corresponding to a function $f(x)$ in $C_p(-\pi, \pi)$ (Sec. 15), the inequality

$$\frac{a_0^2}{2} + \sum_{n=1}^{N} (a_n^2 + b_n^2) \leqslant \frac{1}{\pi} \int_{-\pi}^{\pi} [f(x)]^2 \, dx \qquad (N = 1, 2, \dots)$$

follows from Bessel's inequality (9), Sec. 16, for Fourier constants.

17. ONE-SIDED DERIVATIVES

In developing sufficient conditions on a function f such that its Fourier series on the interval $-\pi < x < \pi$ converges to $f(x)$ there, we need to generalize the concept of the derivative

$$\text{(1)} \qquad f'(x_0) = \lim_{x \to x_0} \frac{f(x) - f(x_0)}{x - x_0}$$

of f at a point $x = x_0$.

Suppose that the right-hand limit $f(x_0 +)$ exists at x_0 (see Sec. 10). The *right-hand derivative*, or derivative from the right, of f at x_0 is defined as follows:

$$\text{(2)} \qquad f_R'(x_0) = \lim_{\substack{x \to x_0 \\ x > x_0}} \frac{f(x) - f(x_0 +)}{x - x_0},$$

provided the limit here exists. Note that, although $f(x_0)$ need not exist, $f(x_0 +)$ must exist if $f_R'(x_0)$ does. When the ordinary, or two-sided, derivative $f'(x_0)$ exists, f is continuous at x_0; and it is obvious that $f_R'(x_0) = f'(x_0)$.

Similarly, if $f(x_0 -)$ exists, the *left-hand derivative* of f at x_0 is given by the equation

$$\text{(3)} \qquad f_L'(x_0) = \lim_{\substack{x \to x_0 \\ x < x_0}} \frac{f(x) - f(x_0 -)}{x - x_0}$$

when this limit exists; and if $f'(x_0)$ exists, $f_L'(x_0) = f'(x_0)$.

EXAMPLE 1. Let f denote the continuous function defined by the equations

$$f(x) = \begin{cases} x^2 & \text{when } x \leqslant 0, \\ \sin x & \text{when } x > 0. \end{cases}$$

With the aid of l'Hospital's rule, we see that

$$f_R'(0) = \lim_{\substack{x \to 0 \\ x > 0}} \frac{\sin x}{x} = 1;$$

furthermore, $f_L'(0) = 0$. Since these one-sided derivatives have different values, the ordinary derivative $f'(0)$ cannot exist.

The ordinary derivative $f'(x_0)$ can fail to exist even when $f(x_0)$ is defined and $f_R'(x_0)$ and $f_L'(x_0)$ have a common value.

EXAMPLE 2. If f is the step function

$$f(x) = \begin{cases} 0 & \text{when } x < 0, \\ 1 & \text{when } x \geqslant 0, \end{cases}$$

then $f_R'(0) = f_L'(0) = 0$. But the derivative $f'(0)$ does not exist since f is not continuous at $x = 0$.

As is the case with ordinary derivatives, the mere continuity of f at a point x_0 does *not* ensure the existence of either one-sided derivative there.

EXAMPLE 3. The function $f(x) = \sqrt{x}$ $(x \geqslant 0)$ has no right-hand derivative at the point $x = 0$, although it is continuous there.

A number of properties of ordinary derivatives remain valid for one-sided derivatives. If, for example, each of two functions f and g has a right-hand derivative at a point x_0, then so does their product. A direct proof is left to the problems. But a proof can be based on the corresponding property of ordinary derivatives in the following way. We use $f(x_0 +)$ and $g(x_0 +)$ as the values of f and g at x_0, and we define those functions when $x < x_0$ as the linear functions represented by the tangent lines at the points $(x_0, f(x_0 +))$ and $(x_0, g(x_0 +))$ with slopes $f'_R(x_0)$ and $g'_R(x_0)$, respectively (Fig. 22). Those modifications of f and g are differentiable at x_0, with derivatives that are also right-hand derivatives. Thus the derivative of their product exists there, and its value is also the right-hand derivative of $f(x)g(x)$ at x_0.

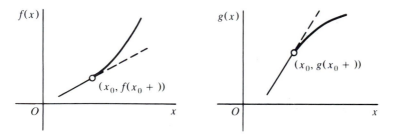

FIGURE 22

Likewise, if $f'_L(x_0)$ and $g'_L(x_0)$ exist, the left-hand derivative of the product $f(x)g(x)$ exists at x_0.

Finally, we turn to a property of one-sided derivatives that is particularly important in the theory of convergence of Fourier series. It concerns the subspace $C'_p(a, b)$ of $C_p(a, b)$ consisting of all piecewise continuous functions f on an interval $a < x < b$ whose derivatives f' are also piecewise continuous on that interval. Such a function is said to be *piecewise smooth* because, over the subintervals on which both f and f' are continuous, any tangents to the graph of $y = f(x)$ that turn do so continuously.

Theorem. If a function f is piecewise smooth on an interval $a < x < b$, then at each point x_0 in the closed interval $a \leqslant x \leqslant b$ the one-sided derivatives of f, from the interior at the end points, exist and are the same as the corresponding one-sided limits of f':

(4) $f'_R(x_0) = f'(x_0 +), \qquad f'_L(x_0) = f'(x_0 -).$

To prove this, we assume for the moment that f and f' are actually continuous on the interval $a < x < b$ and that the one-sided limits of f and f' from the interior exist at the end points $x = a$ and $x = b$. If x_0 is a point in that open interval, $f'(x_0)$ exists. Hence $f_R'(x_0)$ and $f_L'(x_0)$ exist, and both are equal to $f'(x_0)$. Because f' is continuous at x_0, then, equations (4) hold. The following argument shows that it is also true that $f_R'(a)$ exists and is equal to $f'(a+)$. If we let x^* denote any number in the interval $a < x < b$ and define $f(a)$ as $f(a+)$, then f is continuous on the closed interval $a \leqslant x \leqslant x^*$. Since f' exists in the open interval $a < x < x^*$, the mean value theorem for derivatives applies. To be specific, there exists a number c, where $a < c < x^*$, such that

(5)
$$\frac{f(x^*) - f(a+)}{x^* - a} = f'(c).$$

Letting x^*, and therefore c, tend to a in equation (5), we see that since $f'(a+)$ exists, the limit of $f'(c)$ exists and has that value. Consequently, the limit of the difference quotient on the left in equation (5) exists, its value being $f_R'(a)$. Thus $f_R'(a) = f'(a+)$. Similarly, $f_L'(b) = f'(b-)$.

Now any piecewise smooth function f is continuous, along with its derivative f', on a finite number of subintervals at whose end points the one-sided limits of f and f' from the interior exist. If the results of the preceding paragraph are applied to each of those subintervals, the theorem is established.

The following example illustrates the distinction between one-sided derivatives and one-sided limits of derivatives.

EXAMPLE 4. Consider the function f whose values are

$$f(x) = \begin{cases} x^2 \sin \dfrac{1}{x} & \text{when } x \neq 0, \\ 0 & \text{when } x = 0. \end{cases}$$

Since $0 \leqslant |x^2 \sin(1/x)| \leqslant x^2$ when $x \neq 0$, both one-sided limits $f(0+)$ and $f(0-)$ exist and have value zero. Moreover, since $0 \leqslant |x \sin(1/x)| \leqslant |x|$ when $x \neq 0$,

$$f_R'(0) = \lim_{\substack{x \to 0 \\ x > 0}} x \sin \frac{1}{x} = 0 \quad \text{and} \quad f_L'(0) = \lim_{\substack{x \to 0 \\ x < 0}} x \sin \frac{1}{x} = 0.$$

But, from the expression

$$f'(x) = 2x \sin \frac{1}{x} - \cos \frac{1}{x} \qquad (x \neq 0),$$

we see that the one-sided limits $f'(0+)$ and $f'(0-)$ do not exist.

Note that although its one-sided derivatives exist everywhere, the function f is not piecewise smooth on any bounded interval containing the origin. Hence the theorem is not applicable to this function on such an interval.

18.   TWO LEMMAS

We begin our discussion of the convergence of Fourier series with two lemmas, or preliminary theorems. The first is a special case of what is known as the *Riemann-Lebesgue lemma*. That lemma appears later on in Chap. 6 (Sec. 53), where it is needed in full generality.

Lemma 1. *If a function $G(u)$ is piecewise continuous on the interval $0 < u < \pi$, then*

(1)
$$\lim_{N \to \infty} \int_0^\pi G(u) \sin \frac{(2N + 1)u}{2} \, du = 0,$$

where N denotes positive integers.

To prove this, we recall the trigonometric identity

$$\sin(A + B) = \sin A \cos B + \cos A \sin B$$

and write

$$\int_0^\pi G(u) \sin \frac{(2N + 1)u}{2} \, du = \int_0^\pi G(u) \sin\left(\frac{u}{2} + Nu\right) du$$

$$= \int_0^\pi G(u) \sin \frac{u}{2} \cos Nu \, du + \int_0^\pi G(u) \cos \frac{u}{2} \sin Nu \, du.$$

Now, except for a factor of $2/\pi$, the first of these last two integrals is the coefficient a_N in the Fourier cosine series for the piecewise continuous function $G(u) \sin(u/2)$ on the interval $0 < u < \pi$. The other integral is, except for a factor of $2/\pi$, the coefficient b_N in the Fourier sine series for $G(u) \cos(u/2)$ on the same interval. Thus, according to the last paragraph in Sec. 16, the numbers a_N and b_N tend to zero as N tends to infinity; and Lemma 1 is established.

Our second lemma involves the *Dirichlet kernel*

(2)
$$D_N(u) = \frac{1}{2} + \sum_{n=1}^N \cos nu,$$

where N is any positive integer. Note that $D_N(u)$ is continuous, even, and periodic with period 2π. The Dirichlet kernel plays a central role in our theory, and two other properties will also be useful:

(3)
$$\int_0^\pi D_N(u) \, du = \frac{\pi}{2},$$

(4) $$D_N(u) = \frac{\sin\left[(2N + 1)u/2\right]}{2 \sin(u/2)} \qquad (u \neq 0, \pm 2\pi, \pm 4\pi, \dots).$$

Property (3) is obvious upon integrating each side of equation (2). Expression (4) can be derived with the aid of a certain trigonometric identity (Problem 14, Sec. 20).

   Lemma 2. *Suppose that a function $g(u)$ is piecewise continuous on the interval $0 < u < \pi$ and that the right-hand derivative $g'_R(0)$ exists. Then*

(5)
$$\lim_{N \to \infty} \int_0^\pi g(u) D_N(u)\, du = \frac{\pi}{2} g(0+),$$

where $D_N(u)$ is defined by equation (2).

   To start the proof, we write

(6)
$$\int_0^\pi g(u) D_N(u)\, du = I_N + J_N,$$

where

$$I_N = \int_0^\pi [g(u) - g(0+)] D_N(u)\, du \quad \text{and} \quad J_N = \int_0^\pi g(0+) D_N(u)\, du.$$

   In view of expression (4), the first of these two integrals can be put in the form

(7)
$$I_N = \int_0^\pi \frac{g(u) - g(0+)}{2 \sin(u/2)} \sin \frac{(2N+1)u}{2}\, du.$$

Observe that the function

$$G(u) = \frac{g(u) - g(0+)}{2 \sin(u/2)}$$

is a quotient of two functions that are piecewise continuous on the interval $0 < u < \pi$. Although the denominator vanishes at the point $u = 0$, the existence of $g'_R(0)$ ensures the existence of $G(0+)$:

$$\lim_{\substack{u \to 0 \\ u > 0}} G(u) = \lim_{\substack{u \to 0 \\ u > 0}} \frac{g(u) - g(0+)}{u - 0} \lim_{\substack{u \to 0 \\ u > 0}} \frac{(u/2)}{\sin(u/2)} = g'_R(0).$$

Hence $G(u)$ is itself piecewise continuous on the interval $0 < u < \pi$. Applying Lemma 1 to integral (7), we therefore conclude that

(8)
$$\lim_{N \to \infty} I_N = 0.$$

   With property (3) of the Dirichlet kernel, we know that $J_N = (\pi/2)g(0+)$, or

(9)
$$\lim_{N \to \infty} J_N = \frac{\pi}{2} g(0+).$$

The desired result (5) now follows from equation (6) and limits (8) and (9).

19. A FOURIER THEOREM

A theorem that gives conditions under which a Fourier series converges to its function is called a *Fourier theorem*. One such theorem will now be established. Although it is stated for periodic functions of period 2π, it applies as well to functions defined only on the fundamental interval $-\pi < x < \pi$; for we need only consider the periodic extensions of those functions, with period 2π. This will be done in the corollary to follow.

Theorem. *Let f denote a function that is piecewise continuous on the interval $-\pi < x < \pi$ and periodic with period 2π. Its Fourier series*

$$(1) \qquad \frac{a_0}{2} + \sum_{n=1}^{\infty} (a_n \cos nx + b_n \sin nx),$$

where

$$(2) \qquad a_n = \frac{1}{\pi} \int_{-\pi}^{\pi} f(x) \cos nx\, dx \qquad (n = 0, 1, 2, \ldots)$$

and

$$(3) \qquad b_n = \frac{1}{\pi} \int_{-\pi}^{\pi} f(x) \sin nx\, dx \qquad (n = 1, 2, \ldots),$$

converges to the mean value

$$(4) \qquad \frac{f(x+) + f(x-)}{2}$$

of the one-sided limits of f at each point x $(-\infty < x < \infty)$ where both of the one-sided derivatives $f_R'(x)$ and $f_L'(x)$ exist.

Note that if f is actually continuous at x, the quotient (4) becomes $f(x)$. Hence

$$f(x) = \frac{a_0}{2} + \sum_{n=1}^{\infty} (a_n \cos nx + b_n \sin nx)$$

at that point, provided $f_R'(x)$ and $f_L'(x)$ exist.

The integrals in expressions (2) and (3) for the coefficients a_n and b_n always exist since f is piecewise continuous; and we begin our proof of the theorem by writing series (1) as (see Sec. 15)

$$\frac{1}{2\pi} \int_{-\pi}^{\pi} f(s)\, ds + \frac{1}{\pi} \sum_{n=1}^{\infty} \int_{-\pi}^{\pi} f(s) \cos n(s - x)\, ds,$$

with those coefficients incorporated into it. Then, if $S_N(x)$ denotes the partial sum consisting of the sum of the first $N + 1$ ($N \geqslant 1$) terms of the series,

$$(5) \qquad S_N(x) = \frac{1}{2\pi} \int_{-\pi}^{\pi} f(s)\, ds + \frac{1}{\pi} \sum_{n=1}^{N} \int_{-\pi}^{\pi} f(s) \cos n(s - x)\, ds.$$

Using the Dirichlet kernel (Sec. 18)

$$D_N(u) = \frac{1}{2} + \sum_{n=1}^{N} \cos nu,$$

we can put equation (5) in the form

$$S_N(x) = \frac{1}{\pi} \int_{-\pi}^{\pi} f(s) D_N(s - x)\, ds.$$

The periodicity of the integrand here allows us to change the interval of integration to any interval of length 2π without altering the value of the integral (see Problem 13, Sec. 20). Thus

$$(6) \qquad S_N(x) = \frac{1}{\pi} \int_{x-\pi}^{x+\pi} f(s) D_N(s - x)\, ds,$$

where the point x is at the center of the interval we have chosen. It now follows from equation (6) that

$$(7) \qquad S_N(x) = \frac{1}{\pi} [I_N(x) + J_N(x)],$$

where

$$(8) \qquad I_N(x) = \int_{x}^{x+\pi} f(s) D_N(s - x)\, ds$$

and

$$(9) \qquad J_N(x) = \int_{x-\pi}^{x} f(s) D_N(s - x)\, ds.$$

If we replace the variable of integration s in integral (8) by the new variable $u = s - x$, that integral becomes

$$(10) \qquad I_N(x) = \int_{0}^{\pi} f(x + u) D_N(u)\, du.$$

Since f is piecewise continuous on the fundamental interval $-\pi < x < \pi$ and also periodic, it is piecewise continuous on any bounded interval of the x axis. So, for a fixed value of x, the function $g(u) = f(x + u)$ in expression (10) is piecewise continuous on any bounded interval of the u axis and, in particular, on the interval $0 < u < \pi$. Let us assume that the right-hand derivative $f_R'(x)$ exists. After observing that

$$g(0+) = \lim_{\substack{u \to 0 \\ u > 0}} g(u) = \lim_{\substack{u \to 0 \\ u > 0}} f(x + u) = \lim_{\substack{v \to x \\ v > x}} f(v) = f(x+),$$

one can show that the right-hand derivative of g at $u = 0$ exists:

$$g_R'(0) = \lim_{\substack{u \to 0 \\ u > 0}} \frac{g(u) - g(0+)}{u - 0} = \lim_{\substack{u \to 0 \\ u > 0}} \frac{f(x + u) - f(x+)}{u}$$

$$= \lim_{\substack{v \to x \\ v > x}} \frac{f(v) - f(x+)}{v - x} = f_R'(x).$$

According to Lemma 2 in Sec. 18, then,

$$(11) \qquad \lim_{N \to \infty} I_N(x) = \frac{\pi}{2} g(0+) = \frac{\pi}{2} f(x+).$$

If, on the other hand, we make the substitution $u = x - s$ in integral (9) and recall from our discussion in Sec. 18 that $D_N(u)$ is an even function of u, we find that

$$(12) \qquad J_N(x) = \int_0^\pi f(x - u) D_N(u) \, du.$$

This time, we assume that the left-hand derivative $f_L'(x)$ exists; and we note that the function $g(u) = f(x - u)$ in expression (12) is piecewise continuous on the interval $0 < u < \pi$. Furthermore,

$$g(0+) = \lim_{\substack{u \to 0 \\ u > 0}} g(x) = \lim_{\substack{u \to 0 \\ u > 0}} f(x - u) = \lim_{\substack{v \to x \\ v < x}} f(v) = f(x-)$$

and

$$g_R'(0) = \lim_{\substack{u \to 0 \\ u > 0}} \frac{g(u) - g(0+)}{u - 0} = \lim_{\substack{u \to 0 \\ u > 0}} \frac{f(x - u) - f(x-)}{u}$$

$$= - \lim_{\substack{v \to x \\ v < x}} \frac{f(v) - f(x-)}{v - x} = -f_L'(x).$$

So once again by Lemma 2 in Sec. 18,

$$(13) \qquad \lim_{N \to \infty} J_N(x) = \frac{\pi}{2} g(0+) = \frac{\pi}{2} f(x-).$$

Finally, we may conclude from equation (7) and limits (11) and (13) that

$$\lim_{N \to \infty} S_N(x) = \frac{f(x+) + f(x-)}{2};$$

and the theorem is proved.

This theorem is especially suited to functions f that are *piecewise smooth* on the fundamental interval $-\pi < x < \pi$, ones such that both f and f' are piecewise continuous there. For, when f is piecewise smooth on $-\pi < x < \pi$, we know from the theorem in Sec. 17 that its one-sided derivatives, from the interior at the end points $x = \pm\pi$, exist everywhere in the closed interval $-\pi \leqslant x \leqslant \pi$. Hence if F denotes the periodic extension of f, with period 2π, the one-sided derivatives of F exist at each point x $(-\infty < x < \infty)$. According to the theorem just proved, then, the Fourier series for f on $-\pi < x < \pi$ converges everywhere to the mean value of the one-sided limits of F. We state this useful result as follows.

Corollary. *Suppose that a function f is piecewise smooth on the interval $-\pi < x < \pi$, and let F denote its periodic extension, with period 2π. Then, for each x $(-\infty < x < \infty)$, the Fourier series (1), with coefficients (2) and (3), converges to the mean value*

$$(14) \qquad \frac{F(x+) + F(x-)}{2}$$

of the one-sided limits of F at x.

20. DISCUSSION OF THE THEOREM AND ITS COROLLARY

It should be emphasized that the conditions in the theorem in Sec. 19, as well as the corollary there, are only sufficient, and there is no claim that they are *necessary* conditions. More general conditions are given in a number of the references listed in the Bibliography. Indeed, there are functions that even become unbounded at certain points but nevertheless have valid Fourier series representations.[†]

The corollary in Sec. 19 will be adequate for most of the applications in this book, where the functions are generally piecewise smooth. We note that if f and F denote the functions in the corollary, then $F(x+) = f(x+)$ and $F(x-) = f(x-)$ for $-\pi < x < \pi$. Consequently, when $-\pi < x < \pi$, the corollary tells us that the Fourier series

$$(1) \qquad \frac{a_0}{2} + \sum_{n=1}^{\infty} (a_n \cos nx + b_n \sin nx),$$

with coefficients

$$(2) \qquad a_n = \frac{1}{\pi} \int_{-\pi}^{\pi} f(x) \cos nx \, dx \qquad (n = 0, 1, 2, \dots)$$

and

$$(3) \qquad b_n = \frac{1}{\pi} \int_{-\pi}^{\pi} f(x) \sin nx \, dx \qquad (n = 1, 2, \dots),$$

[†] See, for instance, the book by Tolstov (1976, pp. 91–94) that is listed in the Bibliography.

converges to the number

(4)
$$\frac{f(x+) + f(x-)}{2},$$

which becomes $f(x)$ if x is a point of continuity of f. At the end points $x = \pm\pi$, however, the series converges to

(5)
$$\frac{f(-\pi+) + f(\pi-)}{2}.$$

To see that this is so, consider first the point $x = -\pi$. Since

$$F(-\pi+) = f(-\pi+) \quad \text{and} \quad F(-\pi-) = f(\pi-),$$

as is evident from Fig. 23, the quotient

$$\frac{F(x+) + F(x-)}{2}$$

in the corollary becomes the quotient (5) when $x = -\pi$. Because of the periodicity of the series, it also converges to the quotient (5) when $x = \pi$. Observe how it follows that the series converges to $f(-\pi+)$ at $x = -\pi$ and to $f(\pi-)$ at $x = \pi$ if and only if $f(-\pi+) = f(\pi-)$.

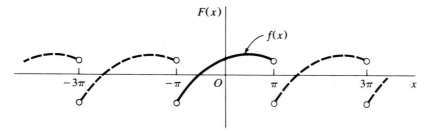

FIGURE 23

EXAMPLE 1. In Example 1, Sec. 15, we obtained the Fourier series

(6)
$$\frac{\pi}{4} + \sum_{n=1}^{\infty} \left[\frac{(-1)^n - 1}{\pi n^2} \cos nx + \frac{(-1)^{n+1}}{n} \sin nx \right]$$

on the interval $-\pi < x < \pi$ for the function f defined by the equations

$$f(x) = \begin{cases} 0 & \text{when } -\pi < x \leqslant 0, \\ x & \text{when } \quad 0 < x < \pi. \end{cases}$$

Since

$$f'(x) = \begin{cases} 0 & \text{when } -\pi < x < 0, \\ 1 & \text{when } \quad 0 < x < \pi, \end{cases}$$

f is clearly piecewise smooth on the fundamental interval $-\pi < x < \pi$. In view of the continuity of f when $-\pi < x < \pi$, the series converges to $f(x)$ at each point x in that open interval. Since $f(-\pi +) = 0$ and $f(\pi -) = \pi$, it converges to $\pi/2$ at the end points $x = \pm\pi$. The series, in fact, converges to $\pi/2$ at each of the points $x = \pm\pi, \pm3\pi, \pm5\pi, \ldots$, as indicated in Fig. 19 (Sec. 15), where the sum of the series for all x is described graphically.

In particular, since series (6) converges to $\pi/2$ when $x = \pi$, we have the identity

$$\frac{\pi}{4} + \sum_{n=1}^{\infty} \frac{(-1)^n - 1}{\pi n^2} (-1)^n = \frac{\pi}{2},$$

which can be written as

$$\sum_{n=1}^{\infty} \frac{1}{(2n - 1)^2} = \frac{\pi^2}{8}.$$

This illustrates how Fourier series can sometimes be used to find the sums of convergent series encountered in calculus. Note that setting $x = 0$ in series (6) also yields this particular summation.

The corollary in Sec. 19 tells us that a function f in the space $C'_p(-\pi, \pi)$ of piecewise smooth functions on the interval $-\pi < x < \pi$ has a valid Fourier series representation on that interval, or one that is equal to $f(x)$ at all but possibly a finite number of points there. It also ensures that a function f in the space $C'_p(0, \pi)$ has valid Fourier cosine and sine series representations on the interval $0 < x < \pi$. For, according to Sec. 15, the cosine series

(7)
$$\frac{a_0}{2} + \sum_{n=1}^{\infty} a_n \cos nx,$$

where

(8)
$$a_n = \frac{2}{\pi} \int_0^{\pi} f(x) \cos nx \, dx \qquad (n = 0, 1, 2, \ldots),$$

is the same as the Fourier series corresponding to the *even* extension of f on the interval $-\pi < x < \pi$; and the sine series

(9)
$$\sum_{n=1}^{\infty} b_n \sin nx,$$

where

(10)
$$b_n = \frac{2}{\pi} \int_0^{\pi} f(x) \sin nx \, dx \qquad (n = 1, 2, \ldots),$$

is the Fourier series for the *odd* extension of f on that same interval.

In view of the even periodic function represented by the series (7), that series converges to $f(0+)$ at the point $x = 0$ and to $f(\pi -)$ at $x = \pi$. The sum of the series (9) is, of course, zero when $x = 0$ and $x = \pi$.

EXAMPLE 2. In the example in Sec. 13, we found the Fourier cosine series corresponding to the function $f(x) = \sin x$ on the interval $0 < x < \pi$:

$$(11) \qquad \sin x \sim \frac{2}{\pi} - \frac{4}{\pi} \sum_{n=1}^{\infty} \frac{\cos 2nx}{4n^2 - 1}.$$

Since $\sin x$ is piecewise smooth on $0 < x < \pi$ and continuous on the closed interval $0 \leqslant x \leqslant \pi$, correspondence (11) is evidently an equality when $0 \leqslant x \leqslant \pi$.

Our final example illustrates how the theorem in Sec. 19 can be useful when the corollary there fails to apply.

EXAMPLE 3. The odd function $x^{1/3}$ is continuous on the closed interval $-\pi \leqslant x \leqslant \pi$ and, therefore, piecewise continuous on the interval $-\pi < x < \pi$. If we let f denote that function, we see that f is not piecewise smooth on $-\pi < x < \pi$ since $f'(0+)$ and $f'(0-)$ do not exist. Hence the corollary does not apply.

If, however, f denotes the periodic extension, with period 2π, of $x^{1/3}$ $(-\pi < x < \pi)$, the theorem can be used. To be precise, since the one-sided derivatives of f exist everywhere in the interval $-\pi < x < \pi$ except at $x = 0$, we find that the Fourier series for $x^{1/3}$ on $-\pi < x < \pi$ converges to $x^{1/3}$ when $-\pi < x < 0$ or $0 < x < \pi$. That series representation is valid even at $x = 0$ since $x^{1/3}$ is odd and the series is actually a Fourier sine series on $0 < x < \pi$, which converges to zero when $x = 0$. We may conclude, then, that the Fourier series representation for $x^{1/3}$ $(-\pi < x < \pi)$ is valid throughout the entire interval $-\pi < x < \pi$.

PROBLEMS

1. For each of the following functions, point out why its Fourier series on the interval $-\pi < x < \pi$ is convergent when $-\pi \leqslant x \leqslant \pi$, and state the sum of the series when $x = \pm\pi$.
   (*a*) The function

$$f(x) = \begin{cases} -\dfrac{\pi}{2} & \text{when } -\pi < x < 0, \\[2mm] \dfrac{\pi}{2} & \text{when } \quad 0 < x < \pi. \end{cases}$$

whose series was found in Problem 1, Sec. 16.
   (*b*) The function $f(x) = e^{ax}$ $(a \neq 0)$, whose series was found in Problem 4, Sec. 16.
   Answers: (*a*) Sum = 0; (*b*) Sum = $\cosh a\pi$.

2. By writing $x = 0$ and $x = \pi/2$ in the representation

$$\sin x = \frac{2}{\pi} - \frac{4}{\pi} \sum_{n=1}^{\infty} \frac{\cos 2nx}{4n^2 - 1} \qquad (0 \leqslant x \leqslant \pi),$$

established in Example 2, Sec. 20, obtain the following summations:

$$\sum_{n=1}^{\infty} \frac{1}{4n^2 - 1} = \frac{1}{2}, \qquad \sum_{n=1}^{\infty} \frac{(-1)^n}{4n^2 - 1} = \frac{1}{2} - \frac{\pi}{4}.$$

3. Point out why the Fourier series in Problem 7, Sec. 16, for the function

$$f(x) = \begin{cases} 0 & \text{when } -\pi \leqslant x \leqslant 0 \\ \sin x & \text{when } 0 < x \leqslant \pi \end{cases}$$

converges to $f(x)$ everywhere in the interval $-\pi \leqslant x \leqslant \pi$.

4. State why the Fourier sine series in Example 1, Sec. 14, for the function $f(x) = x$ $(0 < x < \pi)$ is a valid representation for x on the interval $0 < x < \pi$. Thus verify fully that the series converges for all x $(-\infty < x < \infty)$ to the function whose graph is shown in Fig. 18 (Sec. 14).

5. State why the correspondence (Problem 7, Sec. 14)

$$|x| \sim \frac{\pi}{2} - \frac{4}{\pi} \sum_{n=1}^{\infty} \frac{\cos (2n - 1)x}{(2n - 1)^2} \qquad (-\pi < x < \pi)$$

is actually an equality on the interval $-\pi \leqslant x \leqslant \pi$. Thus show that

$$\sum_{n=1}^{\infty} \frac{1}{(2n - 1)^2} = \frac{\pi^2}{8}.$$

(Compare Example 1, Sec. 20.)

6. (a) Use the correspondence

$$x^2 \sim \frac{\pi^2}{3} + 4 \sum_{n=1}^{\infty} \frac{(-1)^n}{n^2} \cos nx \qquad (0 < x < \pi),$$

found in Problem 4(a), Sec. 14, to show that

$$\sum_{n=1}^{\infty} \frac{(-1)^{n+1}}{n^2} = \frac{\pi^2}{12}, \qquad \sum_{n=1}^{\infty} \frac{1}{n^2} = \frac{\pi^2}{6}.$$

(b) By writing $x = \pi$ in the correspondence (Problem 8, Sec. 14)

$$x^4 \sim \frac{\pi^4}{5} + 8 \sum_{n=1}^{\infty} (-1)^n \frac{(n\pi)^2 - 6}{n^4} \cos nx \qquad (0 < x < \pi)$$

and referring to the second summation obtained in part (a), show that

$$\sum_{n=1}^{\infty} \frac{1}{n^4} = \frac{\pi^4}{90}.$$

7. With the aid of the correspondence (Problem 6, Sec. 16)

$$\cos ax \sim \frac{2a \sin a\pi}{\pi} \left[\frac{1}{2a^2} + \sum_{n=1}^{\infty} \frac{(-1)^{n+1}}{n^2 - a^2} \cos nx \right] \quad (-\pi < x < \pi),$$

where $a \neq 0, \pm 1, \pm 2, \ldots$, show that

$$\frac{a\pi}{\sin a\pi} = 1 + 2a^2 \sum_{n=1}^{\infty} \frac{(-1)^{n+1}}{n^2 - a^2} \quad (a \neq 0, \pm 1, \pm 2, \ldots).$$

8. If we exclude the constant function $\phi_0(x) = 1/\sqrt{\pi}$ from the orthonormal set in the example in Sec. 12, we still have an orthonormal set, consisting of the functions

$$\phi_n(x) = \sqrt{\frac{2}{\pi}} \cos nx \qquad (n = 1, 2, \ldots).$$

State why this set is closed (Sec. 12) in the space of all functions f that are piecewise smooth on the interval $0 < x < \pi$ and satisfy the condition

$$\int_0^{\pi} f(x)\, dx = 0.$$

Suggestion: Refer to the statement in italics near the end of Sec. 12.

9. Without actually finding the Fourier series for $f(x) = x^{2/3}$ on $-\pi < x < \pi$, point out how the theorem in Sec. 19 ensures the convergence of that series to $f(x)$ when $-\pi \leqslant x < 0$ or $0 < x \leqslant \pi$ but not when $x = 0$.

10. With the aid of l'Hospital's rule, find $f(0+)$ and $f_R'(0)$ for the function

$$f(x) = \frac{e^x - 1}{x} \qquad (x \neq 0).$$

Answer: $f(0+) = 1$, $f_R'(0) = \frac{1}{2}$.

11. Show that the function

$$f(x) = \begin{cases} x \sin \dfrac{1}{x} & \text{when } x \neq 0, \\[2mm] 0 & \text{when } x = 0 \end{cases}$$

is continuous at $x = 0$ but that neither $f_R'(0)$ nor $f_L'(0)$ exists. This provides another illustration (see Example 3, Sec. 17) of the fact that the continuity of a function f at a point x_0 is *not* a sufficient condition for the existence of the one-sided derivatives of f at x_0.

12. Given that the right-hand derivatives of two functions f and g exist at a point x_0, prove that the product $f(x)g(x)$ of those functions has a right-hand derivative there by inserting the term $f(x)g(x_0+)$ and its negative in the numerator of the difference quotient

$$\frac{f(x)g(x) - f(x_0+)g(x_0+)}{x - x_0}.$$

13. Let f denote a function that is piecewise continuous on an interval $-c < x < c$ and periodic with period $2c$. Show that, for any number a,

$$\int_{-c}^{c} f(x)\,dx = \int_{a-c}^{a+c} f(x)\,dx.$$

Suggestion: Write

$$\int_{-c}^{c} f(x)\,dx = \int_{-c}^{a+c} f(x)\,dx + \int_{a+c}^{c} f(s)\,ds$$

and then make the substitution $x = s - 2c$ in the second integral on the right-hand side of this equation.

14. Derive the expression

$$D_N(u) = \frac{\sin[(2N+1)u/2]}{2\sin(u/2)} \qquad (u \neq 0,\ \pm 2\pi,\ \pm 4\pi,\ldots)$$

for the Dirichlet kernel (Sec. 18)

$$D_N(u) = \frac{1}{2} + \sum_{n=1}^{N} \cos nu$$

by writing $A = u/2$ and $B = nu$ in the trigonometric identity

$$2\sin A\cos B = \sin(A+B) + \sin(A-B)$$

and then summing each side of the resulting equation from $n = 1$ to $n = N$.
   Suggestion: Note that

$$\sum_{n=1}^{N} \sin\left(\frac{1}{2} - n\right)u = -\sum_{n=0}^{N-1} \sin\left(\frac{1}{2} + n\right)u.$$

21. FOURIER SERIES ON OTHER INTERVALS

Suppose that a function f is piecewise smooth on an interval $-c < x < c$ and periodic with period $2c$, where c is any positive number. For convenience in the discussion below, we assume that $f(x)$ at each point of discontinuity of f is the mean value of the one-sided limits $f(x+)$ and $f(x-)$, as is the case at a point of continuity.
   Let us define the function

(1)
$$g(s) = f\left(\frac{cs}{\pi}\right) \qquad (-\infty < s < \infty),$$

or

(2)
$$g(s) = f(x) \qquad \text{where} \qquad x = \frac{cs}{\pi} \qquad (-\infty < s < \infty),$$

which is periodic with period 2π. The equation $x = cs/\pi$, or $s = \pi x/c$, establishes a one to one correspondence between points on the x axis and points on the s axis; and it is evident from relations (2) that if a specific point x

corresponds to a point s, then

$$g(s+) = f(x+), \qquad g(s-) = f(x-).$$

Since $f(x)$ is always the mean value of $f(x+)$ and $f(x-)$, it follows from these relations between one-sided limits that the number $g(s) = f(x)$ is always the mean value of $g(s+)$ and $g(s-)$. In particular, g is continuous at s when f is continuous at x. Since f is piecewise continuous on the interval $-c < x < c$, then, g is piecewise continuous on the interval $-\pi < s < \pi$. The derivative f' is also piecewise continuous, and a similar argument shows that g' is piecewise continuous. So g is piecewise smooth on the interval $-\pi < s < \pi$; and it is its own periodic extension, with period 2π, on the entire s axis.

An application of the corollary in Sec. 19 now shows that function (1) is represented by its Fourier series everywhere on the s axis. That is, for each s,

$$(3) \qquad f\left(\frac{cs}{\pi}\right) = \frac{a_0}{2} + \sum_{n=1}^{\infty} (a_n \cos ns + b_n \sin ns),$$

where

$$(4) \qquad a_n = \frac{1}{\pi} \int_{-\pi}^{\pi} f\left(\frac{cs}{\pi}\right) \cos ns \, ds \qquad (n = 0, 1, 2, \dots)$$

and

$$(5) \qquad b_n = \frac{1}{\pi} \int_{-\pi}^{\pi} f\left(\frac{cs}{\pi}\right) \sin ns \, ds \qquad (n = 1, 2, \dots).$$

With the substitution $x = cs/\pi$, representation (3) becomes

$$(6) \qquad f(x) = \frac{a_0}{2} + \sum_{n=1}^{\infty} \left(a_n \cos \frac{n\pi x}{c} + b_n \sin \frac{n\pi x}{c}\right);$$

and this is valid for all x. Expressions (4) and (5) can be written as follows, where the new variable of integration $x = cs/\pi$ is used:

$$(7) \qquad a_n = \frac{1}{c} \int_{-c}^{c} f(x) \cos \frac{n\pi x}{c} \, dx \qquad (n = 0, 1, 2, \dots),$$

$$(8) \qquad b_n = \frac{1}{c} \int_{-c}^{c} f(x) \sin \frac{n\pi x}{c} \, dx \qquad (n = 1, 2, \dots).$$

We state the result as a theorem that is sufficient for our applications. It is possible, by appealing directly to the theorem in Sec. 19, to obtain a more general result for periodic functions f that are only piecewise continuous on the fundamental interval $-c < x < c$ but have one-sided derivatives $f_R'(x)$ and $f_L'(x)$ at certain points x.

Theorem. *Let f denote a function that is piecewise smooth on an interval $-c < x < c$ and periodic with period $2c$. If $f(x)$ at each point of discontinuity of f is defined as the mean value of the one-sided limits $f(x+)$ and $f(x-)$, then the*

Fourier series representation (6), *with coefficients* (7) *and* (8), *is valid for all x*
$(-\infty < x < \infty)$.

If the function f is *even*, then so is the function g that is defined by
equation (1); and we know from Sec. 15 that expression (4) for the coefficients
a_n may be written

$$a_n = \frac{2}{\pi} \int_0^\pi f\left(\frac{cs}{\pi}\right) \cos ns \, ds \qquad (n = 0, 1, 2, \ldots).$$

Furthermore, the coefficients b_n are all zero. Hence series (6) reduces to a
cosine series:

$$(9) \qquad\qquad f(x) = \frac{a_0}{2} + \sum_{n=1}^\infty a_n \cos \frac{n\pi x}{c},$$

where

$$(10) \qquad\qquad a_n = \frac{2}{c} \int_0^c f(x) \cos \frac{n\pi x}{c} \, dx \qquad (n = 0, 1, 2, \ldots).$$

Similarly, if f is *odd*, we have a sine series:

$$(11) \qquad\qquad f(x) = \sum_{n=1}^\infty b_n \sin \frac{n\pi x}{c},$$

where

$$(12) \qquad\qquad b_n = \frac{2}{c} \int_0^c f(x) \sin \frac{n\pi x}{c} \, dx \qquad (n = 1, 2, \ldots).$$

It is easy to adapt the above theorem to any piecewise smooth function f
defined only on the interval $-c < x < c$. To do this, we introduce the *periodic
extension*, with period $2c$, of f and denote it by F. The graph of $y = F(x)$ is the
graph of $y = f(x)$ repeated every $2c$ units along the x axis. After defining $F(x)$
at each point of discontinuity of F as the mean value of the one-sided limits
$F(x+)$ and $F(x-)$, one can apply the theorem to that extension. The same
procedure may be used to verify representations (9) and (11) for piecewise
smooth functions defined only on the interval $0 < x < c$, once the even and odd
extensions, respectively, are made on the interval $-c < x < c$. The following
corollary, which summarizes these results, applies to any function f that has the
following properties:

(*a*) f is piecewise smooth on the stated interval;
(*b*) $f(x)$ at each point of discontinuity of f in that interval is the mean value of
the one-sided limits $f(x+)$ and $f(x-)$.

Corollary. *If a function f has properties* (*a*) *and* (*b*) *on an interval*
$-c < x < c$, *then the Fourier series representation* (6), *with coefficients* (7) *and*
(8), *is valid for each x* $(-c < x < c)$. *If f has those properties on an interval*

$0 < x < c$, *then the Fourier cosine series representation* (9), *with coefficients* (10), *is valid for each* x $(0 < x < c)$; *and the same is true of the Fourier sine series representation* (11), *with coefficients* (12).

EXAMPLE The function $f(x) = x^2$ is piecewise smooth on any interval $0 < x < c$, and the sine series representation

$$(13) \qquad x^2 = 2c^2 \sum_{n=1}^{\infty} \left[\frac{(-1)^{n+1}}{n\pi} - 2\frac{1 - (-1)^n}{(n\pi)^3} \right] \sin \frac{n\pi x}{c} \quad (0 < x < c)$$

can be obtained by evaluating the integrals in expression (12) when $f(x) = x^2$. But, since we already know from Problem 4(b), Sec. 14, and the corollary here that

$$(14) \qquad x^2 = 2\pi^2 \sum_{n=1}^{\infty} \left[\frac{(-1)^{n+1}}{n\pi} - 2\frac{1 - (-1)^n}{(n\pi)^3} \right] \sin nx \quad (0 < x < \pi),$$

it is simpler to start with that special case. To be specific,

$$0 < \frac{\pi x}{c} < \pi \qquad \text{when} \qquad 0 < x < c;$$

and so we can replace x by $\pi x/c$ on each side of representation (14) to obtain an equation that is valid when $0 < x < c$. Then, by multiplying through that equation by c^2/π^2, we arrive at representation (13), which is actually valid on the interval $0 \leqslant x < c$.

PROBLEMS

In Problems 1 through 3, use formulas in Sec. 21 to find the coefficients in the Fourier series involved.

1. Show that if

$$f(x) = \begin{cases} 0 & \text{when } -3 < x < 0, \\ 1 & \text{when } \quad 0 < x < 3 \end{cases}$$

and $f(0) = \frac{1}{2}$, then

$$f(x) = \frac{1}{2} + \frac{2}{\pi} \sum_{n=1}^{\infty} \frac{1}{2n-1} \sin \frac{(2n-1)\pi x}{3} \qquad (-3 < x < 3).$$

Describe graphically the function that is represented by this series for all x $(-\infty < x < \infty)$.

2. Let f denote the function whose values are

$$f(x) = \begin{cases} 0 & \text{when } -2 < x < 1, \\ 1 & \text{when } \quad 1 < x < 2 \end{cases}$$

and $f(-2) = f(1) = f(2) = \frac{1}{2}$. Show that, for each x in the closed interval $-2 \leqslant x \leqslant 2$,

$$f(x) = \frac{1}{4} - \frac{1}{\pi} \sum_{n=1}^{\infty} \frac{1}{n} \left[\sin \frac{n\pi}{2} \cos \frac{n\pi x}{2} + \left(\cos n\pi - \cos \frac{n\pi}{2} \right) \sin \frac{n\pi x}{2} \right].$$

3. Obtain the Fourier sine series representation

$$\cos \pi x = \frac{8}{\pi} \sum_{n=1}^{\infty} \frac{n}{4n^2 - 1} \sin 2n\pi x \qquad (0 < x < 1).$$

Suggestion: To evaluate the integrals that arise, recall the trigonometric identity

$$2 \sin A \cos B = \sin (A + B) + \sin (A - B).$$

4. Let f denote the periodic function, of period 2, where

$$f(x) = \begin{cases} \cos \pi x & \text{when } 0 < x < 1 \\ 0 & \text{when } 1 < x < 2 \end{cases}$$

and where $f(0) = \frac{1}{2}$ and $f(1) = -\frac{1}{2}$. With the aid of the series representation found in Problem 3, show that

$$f(x) = \frac{1}{2} \cos \pi x + \frac{4}{\pi} \sum_{n=1}^{\infty} \frac{n}{4n^2 - 1} \sin 2n\pi x \qquad (-\infty < x < \infty).$$

5. (*a*) Use the Fourier sine series found in Example 1, Sec. 14, for $f(x) = x$ $(0 < x < \pi)$ to show that

$$x = \frac{2}{\pi} \sum_{n=1}^{\infty} \frac{(-1)^{n+1}}{n} \sin n\pi x \qquad (-1 < x < 1).$$

(*b*) By referring to the Fourier cosine series in Problem 4(*a*), Sec. 14, for $f(x) = x^2$ $(0 < x < \pi)$, derive the expansion

$$x^2 = \frac{c^2}{3} + \frac{4c^2}{\pi^2} \sum_{n=1}^{\infty} \frac{(-1)^n}{n^2} \cos \frac{n\pi x}{c} \qquad (-c \leqslant x \leqslant c).$$

6. Show how it follows from the expansions obtained in Problem 5 that

$$x + x^2 = \frac{1}{3} + \frac{2}{\pi} \sum_{n=1}^{\infty} (-1)^n \left(\frac{2}{n^2 \pi} \cos n\pi x - \frac{1}{n} \sin n\pi x \right) \qquad (-1 < x < 1).$$

7. (*a*) Use the Fourier sine series in Example 3, Sec. 14, for the function

$$f(x) = x(\pi^2 - x^2) \qquad (0 < x < \pi)$$

to establish the representation

$$x(1 - x^2) = \frac{12}{\pi^3} \sum_{n=1}^{\infty} \frac{(-1)^{n+1}}{n^3} \sin n\pi x \qquad (0 \leqslant x \leqslant 1).$$

(b) Replace x by $1 - x$ on each side of the representation found in part (a) to show that

$$x(x - 1)(x - 2) = \frac{12}{\pi^3} \sum_{n=1}^{\infty} \frac{\sin n\pi x}{n^3} \qquad (0 \leqslant x \leqslant 1).$$

8. Show how it follows from the Fourier sine series for the function

$$f(x) = x(\pi - x) \qquad (0 < x < \pi),$$

found in Problem 5, Sec. 14, that

$$x(2c - x) = \frac{32c^2}{\pi^3} \sum_{n=1}^{\infty} \frac{1}{(2n - 1)^3} \sin \frac{(2n - 1)\pi x}{2c} \qquad (0 \leqslant x \leqslant 2c).$$

9. Let $M(c, t)$ denote the square wave (Fig. 24) defined by the equations

$$M(c, t) = \begin{cases} 1 & \text{when } 0 < t < c, \\ -1 & \text{when } c < t < 2c \end{cases}$$

and $M(c, t + 2c) = M(c, t)$ $(t > 0)$. Show that

$$M(c, t) = \frac{4}{\pi} \sum_{n=1}^{\infty} \frac{1}{2n - 1} \sin \frac{(2n - 1)\pi t}{c} \qquad (t \neq c, 2c, 3c, \dots).$$

Suggestion: The sine series found in Problem 1(b), Sec. 14, for the function $f(x) = 1$ $(0 < x < \pi)$ can be used here.

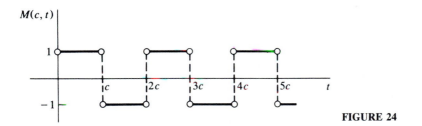

FIGURE 24

10. Let F denote the periodic function, of period c, where

$$F(x) = \begin{cases} \dfrac{c}{4} - x & \text{when } 0 \leqslant x \leqslant \dfrac{c}{2}, \\[2mm] x - \dfrac{3c}{4} & \text{when } \dfrac{c}{2} < x \leqslant c. \end{cases}$$

(a) Describe the function $F(x)$ graphically, and show that it is, in fact, the even periodic extension, with period c, of the function

$$f(x) = \frac{c}{4} - x \qquad \left(0 \leqslant x \leqslant \frac{c}{2}\right).$$

(b) Use the result in part (a) and the Fourier cosine series found in Problem 2(a), Sec. 14, for $f(x) = \pi - x$ $(0 < x < \pi)$ to show that

$$F(x) = \frac{2c}{\pi^2} \sum_{n=1}^{\infty} \frac{1}{(2n-1)^2} \cos \frac{(4n-2)\pi x}{c} \qquad (-\infty < x < \infty).$$

11. Suppose that a function f is piecewise smooth on the interval $0 < x < c$, and let F denote this extension of f to the interval $0 < x < 2c$:

$$F(x) = \begin{cases} f(x) & \text{when } 0 < x < c, \\ f(2c - x) & \text{when } c < x < 2c. \end{cases}$$

[The graph of $y = F(x)$ is evidently symmetric with respect to the line $x = c$.] Show that the coefficients b_n in the Fourier sine series for F on the interval $0 < x < 2c$ can be written

$$b_n = \left[1 - (-1)^n \right] \frac{1}{c} \int_0^c f(x) \sin \frac{n\pi x}{2c}\, dx \qquad (n = 1, 2, \dots).$$

Thus show that

$$f(x) = \sum_{n=1}^{\infty} B_n \sin \frac{(2n-1)\pi x}{2c},$$

where

$$B_n = \frac{2}{c} \int_0^c f(x) \sin \frac{(2n-1)\pi x}{2c}\, dx \qquad (n = 1, 2, \dots),$$

for each point x $(0 < x < c)$ at which f is continuous.

Suggestion: Write

$$b_n = \frac{1}{c} \left[\int_0^c f(x) \sin \frac{n\pi x}{2c}\, dx + \int_c^{2c} f(2c - s) \sin \frac{n\pi s}{2c}\, ds \right],$$

and make the substitution $x = 2c - s$ in the second of these integrals.

12. Use the result in Problem 11 to establish the representation

$$x = \frac{8c}{\pi^2} \sum_{n=1}^{\infty} \frac{(-1)^{n+1}}{(2n-1)^2} \sin \frac{(2n-1)\pi x}{2c} \qquad (-c \leqslant x \leqslant c).$$

13. Show that in Sec. 21 the Fourier series (6), with coefficients (7) and (8), can be written in the compact form

$$\frac{1}{2c} \int_{-c}^{c} f(s)\, ds + \frac{1}{c} \sum_{n=1}^{\infty} \int_{-c}^{c} f(s) \cos \left[\frac{n\pi}{c}(s - x) \right] ds.$$

(See Sec. 15, where this form was obtained when $c = \pi$.)

14. (a) Verify that the set of functions

$$\phi_0(x) = \frac{1}{\sqrt{2c}}, \qquad \phi_{2n-1}(x) = \frac{1}{\sqrt{c}} \cos \frac{n\pi x}{c}, \qquad \phi_{2n}(x) = \frac{1}{\sqrt{c}} \sin \frac{n\pi x}{c}$$

$$(n = 1, 2, \dots),$$

which becomes the set (1), Sec. 15, when $c = \pi$, is orthonormal on the interval $-c < x < c$.

(b) Show that the generalized Fourier series corresponding to a function $f(x)$ in $C_p(-c, c)$ with respect to the orthonormal set in part (a) can be written as the ordinary Fourier series, with coefficients a_n $(n = 0, 1, 2, \ldots)$ and b_n $(n = 1, 2, \ldots)$, for f on the interval $-c < x < c$ (Sec. 21).

(c) Derive Bessel's inequality

$$\frac{a_0^2}{2} + \sum_{n=1}^{N} (a_n^2 + b_n^2) \leq \frac{1}{c} \int_{-c}^{c} [f(x)]^2 \, dx \qquad (N = 1, 2, \ldots)$$

for the coefficients a_n and b_n in part (b) from the general form (9), Sec. 16, of that inequality for Fourier constants. (Compare Problem 9, Sec. 16.)

Suggestion: In part (a), the integrals involved may be transformed by means of the substitution $s = \pi x/c$ into integrals whose values are known, since the set is already known to be orthonormal on $-\pi < x < \pi$ when $c = \pi$.

15. After writing the Fourier series representation (6), Sec. 21, as

$$f(x) = \frac{a_0}{2} + \lim_{N \to \infty} \sum_{n=1}^{N} \left(a_n \cos \frac{n\pi x}{c} + b_n \sin \frac{n\pi x}{c} \right),$$

use the exponential forms[†]

$$\cos \theta = \frac{e^{i\theta} + e^{-i\theta}}{2}, \qquad \sin \theta = \frac{e^{i\theta} - e^{-i\theta}}{2i}$$

of the cosine and sine functions to put that representation in exponential form:

$$f(x) = \lim_{N \to \infty} \sum_{n=-N}^{N} A_n \exp \left(i \frac{n\pi x}{c} \right),$$

where

$$A_0 = \frac{a_0}{2}, \qquad A_n = \frac{a_n - ib_n}{2}, \qquad A_{-n} = \frac{a_n + ib_n}{2} \qquad (n = 1, 2, \ldots).$$

Then use expressions (7) and (8), Sec. 21, for the coefficients a_n and b_n to obtain the single formula

$$A_n = \frac{1}{2c} \int_{-c}^{c} f(x) \exp \left(-i \frac{n\pi x}{c} \right) dx \qquad (n = 0, \pm 1, \pm 2, \ldots).$$

22. UNIFORM CONVERGENCE OF FOURIER SERIES

The reader may at this time pass directly to Chap. 3 without serious disruption. This section, regarding the uniformity of the convergence of Fourier series, and the remaining two sections of the present chapter, dealing with other aspects of

---

[†] For background on these forms and an introduction to series and integrals involving complex-valued functions, see the authors' book (1990), listed in the Bibliography.

the theory of convergence of such series, will be used only occasionally later on. The reader may refer to these sections for results that will be specifically cited as needed.

For convenience, we treat only Fourier series for which the fundamental interval is $-\pi < x < \pi$. Applications of the theorems to series on any fundamental interval $-c < x < c$ can be made by the method used in Sec. 21. We preface our theorem on the uniform convergence of Fourier series with an important lemma.

Lemma. *Suppose that f is continuous on the interval $-\pi \leqslant x \leqslant \pi$, where $f(-\pi) = f(\pi)$, and that its derivative f' is piecewise continuous on the interval $-\pi < x < \pi$. Then if a_n and b_n are the Fourier coefficients*

(1)
$$a_n = \frac{1}{\pi}\int_{-\pi}^{\pi} f(x)\cos nx\,dx, \qquad b_n = \frac{1}{\pi}\int_{-\pi}^{\pi} f(x)\sin nx\,dx,$$

the series

(2)
$$\sum_{n=1}^{\infty} \sqrt{a_n^2 + b_n^2}$$

converges.

The class of functions satisfying the conditions in this theorem is, of course, a subspace of the space of piecewise smooth functions on the interval $-\pi < x < \pi$.

We begin the proof of the lemma with the observation that the Fourier coefficients

(3)
$$\alpha_n = \frac{1}{\pi}\int_{-\pi}^{\pi} f'(x)\cos nx\,dx, \qquad \beta_n = \frac{1}{\pi}\int_{-\pi}^{\pi} f'(x)\sin nx\,dx$$

for f' exist because of the piecewise continuity of f'. Note that

$$\alpha_0 = \frac{1}{\pi}\int_{-\pi}^{\pi} f'(x)\,dx = \frac{1}{\pi}[f(\pi) - f(-\pi)] = 0.$$

Also, since f is continuous and $f(-\pi) = f(\pi)$, integration by parts reveals that when $n = 1, 2, \ldots$,

$$\alpha_n = \frac{1}{\pi}[f(x)\cos nx]_{-\pi}^{\pi} + \frac{n}{\pi}\int_{-\pi}^{\pi} f(x)\sin nx\,dx$$

$$= \frac{(-1)^n}{\pi}[f(\pi) - f(-\pi)] + nb_n = nb_n$$

and

$$\beta_n = \frac{1}{\pi}[f(x)\sin nx]_{-\pi}^{\pi} - \frac{n}{\pi}\int_{-\pi}^{\pi} f(x)\cos nx\,dx = -na_n.$$

That is,

(4)
$$a_n = -\frac{\beta_n}{n}, \qquad b_n = \frac{\alpha_n}{n} \qquad (n = 1, 2, \dots).$$

In view of relations (4), the sum S_N of the first N terms of the infinite series (2) becomes

(5)
$$S_N = \sum_{n=1}^{N} \sqrt{a_n^2 + b_n^2} = \sum_{n=1}^{N} \frac{1}{n} \sqrt{\alpha_n^2 + \beta_n^2}.$$

Cauchy's inequality

$$\left(\sum_{n=1}^{N} p_n q_n \right)^2 \leqslant \left(\sum_{n=1}^{N} p_n^2 \right) \left(\sum_{n=1}^{N} q_n^2 \right),$$

which applies to any two sets of real numbers p_n $(n = 1, 2, \dots, N)$ and q_n $(n = 1, 2, \dots, N)$ (see Problem 6, Sec. 24, for a derivation), can now be used to write

(6)
$$S_N^2 \leqslant \left(\sum_{n=1}^{N} \frac{1}{n^2} \right) \left[\sum_{n=1}^{N} \left(\alpha_n^2 + \beta_n^2 \right) \right] \qquad (N = 1, 2, \dots).$$

The sequence of sums

$$\sum_{n=1}^{N} \frac{1}{n^2} \qquad (N = 1, 2, \dots)$$

here is clearly bounded since each sum is a partial sum of the convergent series whose terms are $1/n^2$ [see Problem 6(a), Sec. 20]. The sequence

$$\sum_{n=1}^{N} \left(\alpha_n^2 + \beta_n^2 \right) \qquad (N = 1, 2, \dots)$$

is also bounded since α_n $(n = 0, 1, 2, \dots)$ and β_n $(n = 1, 2, \dots)$ are the Fourier coefficients for f' on the interval $-\pi < x < \pi$ and must, therefore, satisfy Bessel's inequality:

$$\sum_{n=1}^{N} \left(\alpha_n^2 + \beta_n^2 \right) \leqslant \frac{1}{\pi} \int_{-\pi}^{\pi} [f'(x)]^2 \, dx \qquad (N = 1, 2, \dots).$$

(See Problem 9, Sec. 16.) It now follows from inequality (6) that the sequence S_N^2 $(N = 1, 2, \dots)$ is both bounded and nondecreasing. Hence it converges; and this means that the sequence S_N $(N = 1, 2, \dots)$ converges. Thus series (2) converges.

We turn now to the uniformity of convergence of Fourier series. We begin by recalling some facts about uniformly convergent series of functions.[†]

Let $s(x)$ denote the sum of an infinite series of functions $f_n(x)$, where the series is convergent for all x in some interval $a \leqslant x \leqslant b$. Thus

$$(7) \qquad s(x) = \sum_{n=1}^{\infty} f_n(x) = \lim_{N \to \infty} s_N(x) \qquad (a \leqslant x \leqslant b),$$

where $s_N(x)$ is the partial sum consisting of the sum of the first N terms of the series. The series converges *uniformly* with respect to x if the absolute value of its remainder $r_N(x) = s(x) - s_N(x)$ can be made arbitrarily small for all x in the interval by taking N sufficiently large; that is, for each positive number ε, there exists a positive integer N_ε, *independent of x*, such that

$$(8) \qquad |s(x) - s_N(x)| < \varepsilon \qquad \text{whenever} \qquad N > N_\varepsilon \quad (a \leqslant x \leqslant b).$$

A sufficient condition for uniform convergence is given by the *Weierstrass M-test*. Namely, if there is a convergent series

$$(9) \qquad \sum_{n=1}^{\infty} M_n$$

of positive constants such that

$$(10) \qquad |f_n(x)| \leqslant M_n \qquad (a \leqslant x \leqslant b)$$

for each n, then series (7) is uniformly convergent on the stated interval.

We include here a few properties of uniformly convergent series that are often useful. If the functions f_n are continuous and if series (7) is uniformly convergent, then the sum $s(x)$ of that series is a continuous function. Also, the series can be integrated term by term over the interval $a \leqslant x \leqslant b$ to give the integral of $s(x)$ from $x = a$ to $x = b$. If the functions f_n and their derivatives f_n' are continuous, if series (7) converges, and if the series whose terms are $f_n'(x)$ is uniformly convergent, then $s'(x)$ is found by differentiating series (7) term by term.

Theorem. *When the conditions stated in the above lemma are satisfied, the Fourier series*

$$(11) \qquad \frac{a_0}{2} + \sum_{n=1}^{\infty} (a_n \cos nx + b_n \sin nx),$$

with coefficients (1), converges absolutely and uniformly to $f(x)$ on the interval $-\pi \leqslant x \leqslant \pi$.

[†] See, for instance, the book by Kaplan (1991, chap. 6) or the one by Taylor and Mann (1983, chap. 20), both listed in the Bibliography.

To prove this, we first note that the conditions on f ensure the continuity of the periodic extension of f for all x. Hence it follows from the corollary in Sec. 19 that series (11) converges to $f(x)$ everywhere in the interval $-\pi \leqslant x \leqslant \pi$. Observe that since both $|a_n|$ and $|b_n|$ are less than or equal to $\sqrt{a_n^2 + b_n^2}$,

$$|a_n \cos nx + b_n \sin nx| \leqslant |a_n| + |b_n| \leqslant 2\sqrt{a_n^2 + b_n^2} \quad (n = 1,2,\dots).$$

Since series (2) converges, the comparison test and the Weierstrass M-test thus apply to show that the convergence of series (11) is absolute and uniform on the interval $-\pi \leqslant x \leqslant \pi$, as stated.

In like manner, one can establish the absolute and uniform convergence of the series of cosine or sine terms only. Series (11) is, in fact, the sum of those series:

$$(12) \qquad f(x) = \frac{a_0}{2} + \sum_{n=1}^{\infty} a_n \cos nx + \sum_{n=1}^{\infty} b_n \sin nx \quad (-\pi \leqslant x \leqslant \pi).$$

Modifications of the statements in both the lemma and the theorem are apparent. For instance, it follows from the theorem that the Fourier cosine series on $0 < x < \pi$ for a function f that is continuous on the interval $0 \leqslant x \leqslant \pi$ converges absolutely and uniformly to $f(x)$ if f' is piecewise continuous on the interval $0 < x < \pi$. For the sine series, however, the additional conditions $f(0) = f(\pi) = 0$ are needed.

Since a uniformly convergent series of continuous functions always converges to a continuous function, a Fourier series for a function f cannot converge uniformly on an interval that contains a point at which f is discontinuous. Hence the continuity of f, assumed in the theorem, is necessary in order for the series there to converge uniformly.

Suppose that x_0 is a point at which a piecewise smooth function f is discontinuous. The nature of the deviation near x_0 of the values of the partial sums of a Fourier series for f from the values of f is commonly referred to as the *Gibbs phenomenon* and is illustrated below.[†]

EXAMPLE. Consider the (odd) function f defined by the equations

$$f(x) = \begin{cases} \dfrac{-\pi}{2} & \text{when } -\pi < x < 0, \\ \dfrac{\pi}{2} & \text{when } \quad 0 < x < \pi \end{cases}$$

and $f(0) = 0$. According to Problem 1, Sec. 16, and the corollary in Sec. 21, the

---

[†] For a detailed analysis of this phenomenon, see the book by Carslaw (1952, chap. 9) that is listed in the Bibliography.

Fourier series

$$2 \sum_{n=1}^{\infty} \frac{\sin(2n-1)x}{2n-1} \qquad (-\pi < x < \pi)$$

for f converges to $f(x)$ everywhere in the interval $-\pi < x < \pi$.

Let $s_N(x)$ denote the sum of the first N terms of this series. The sequence $s_N(x)$ $(N = 1, 2, \ldots)$ thus converges to $f(x)$ when $-\pi < x < \pi$. In particular, it converges to the number $\pi/2 = 1.57\ldots$ when $0 < x < \pi$. But, as shown in Problem 7, Sec. 24, there is a fixed number $\sigma = 1.85\ldots$ such that $s_N(\pi/(2N))$ tends to σ. See Fig. 25, which indicates how "spikes" in the graphs of the partial sums $y = s_N(x)$, moving to the left as N increases, are formed, their tips tending to the point σ on the y axis. The behavior of the partial sums is similar on the interval $-\pi < x < 0$.

This illustrates that special care must be taken when a function is approximated by a partial sum of its Fourier series near a point of discontinuity.

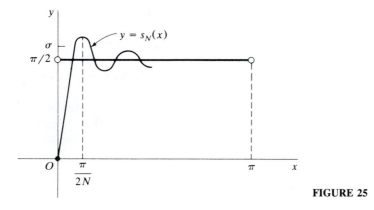

FIGURE 25

23. DIFFERENTIATION AND INTEGRATION OF FOURIER SERIES

According to the corollary in Sec. 21, the Fourier series in Example 3, Sec. 15, for the function $f(x) = x$ $(-\pi < x < \pi)$ converges to $f(x)$ at each point in the interval $-\pi < x < \pi$:

(1) $$x = 2 \sum_{n=1}^{\infty} \frac{(-1)^{n+1}}{n} \sin nx \qquad (-\pi < x < \pi).$$

But the series here is not differentiable. The differentiated series

$$2 \sum_{n=1}^{\infty} (-1)^{n+1} \cos nx$$

does not even converge since its nth term fails to approach zero as n tends to infinity.

The periodic function represented by series (1) for all x has discontinuities at the points $x = \pm\pi, \pm3\pi, \pm5\pi, \ldots$, as shown in Fig. 18 (Sec. 14). Continuity of the periodic extension of the function represented is an important condition for differentiability of a Fourier series on the fundamental interval. Sufficient conditions can be stated as follows.

Theorem 1. *Suppose that the conditions stated in the lemma in Sec. 22 are satisfied. Namely, f is continuous on the interval $-\pi \leqslant x \leqslant \pi$, where $f(-\pi) = f(\pi)$, and f' is piecewise continuous on the interval $-\pi < x < \pi$. Then the Fourier series in the representation*

$$(2) \qquad f(x) = \frac{a_0}{2} + \sum_{n=1}^{\infty} (a_n \cos nx + b_n \sin nx) \qquad (-\pi \leqslant x \leqslant \pi),$$

where

$$a_n = \frac{1}{\pi} \int_{-\pi}^{\pi} f(x) \cos nx \, dx, \qquad b_n = \frac{1}{\pi} \int_{-\pi}^{\pi} f(x) \sin nx \, dx,$$

is differentiable at each point x in the interval $-\pi < x < \pi$ at which $f''(x)$ exists:

$$(3) \qquad f'(x) = \sum_{n=1}^{\infty} n(-a_n \sin nx + b_n \cos nx).$$

Our proof of this theorem is especially brief. To start with, we let x ($-\pi < x < \pi$) be a point at which f'' exists; and we note that f' is, therefore, continuous at x. Hence an application of the Fourier theorem in Sec. 19 to the function f' shows that

$$(4) \qquad f'(x) = \frac{\alpha_0}{2} + \sum_{n=1}^{\infty} (\alpha_n \cos nx + \beta_n \sin nx),$$

where α_n and β_n are the coefficients (3) in Sec. 22. But since f and f' satisfy all the conditions stated in the lemma in Sec. 22, we know from that section that

$$(5) \qquad \alpha_0 = 0, \qquad \alpha_n = nb_n, \qquad \beta_n = -na_n \qquad (n = 1, 2, \ldots).$$

When these substitutions are made, equation (4) takes the form (3); and the proof is complete.

At a point x where $f''(x)$ does not exist, but where f' has one-sided derivatives, differentiation is still valid in the sense that the series in equation (3) converges to the mean of the values $f'(x+)$ and $f'(x-)$. This is also true for the periodic extension of f.

Theorem 1 applies, with obvious changes, to other Fourier series. For instance, if f is continuous when $0 \leqslant x \leqslant \pi$ and f' is piecewise continuous on the interval $0 < x < \pi$, then the Fourier cosine series for f on $0 < x < \pi$ is differentiable at each point x ($0 < x < \pi$) where $f''(x)$ exists.

Integration of a Fourier series is possible under much more general conditions than those for differentiation. This is to be expected because an integration of the series in the correspondence

(6) $$f(x) \sim \frac{a_0}{2} + \sum_{n=1}^{\infty} (a_n \cos nx + b_n \sin nx) \quad (-\pi < x < \pi),$$

where

(7) $$a_n = \frac{1}{\pi} \int_{-\pi}^{\pi} f(x) \cos nx \, dx, \qquad b_n = \frac{1}{\pi} \int_{-\pi}^{\pi} f(x) \sin nx \, dx,$$

introduces a factor n in the denominator of the general term. In the following theorem, it is not even essential that the original series converge in order that the integrated series converge to the integral of the function. The integrated series is not, however, a Fourier series if $a_0 \neq 0$; for it contains a term $(a_0/2)x$.

Theorem 2. *Let f be a function that is piecewise continuous on the interval* $-\pi < x < \pi$. *Regardless of whether series* (6) *converges, the following equation is valid when* $-\pi \leqslant x \leqslant \pi$:

(8) $$\int_{-\pi}^{x} f(s) \, ds = \frac{a_0}{2}(x + \pi) + \sum_{n=1}^{\infty} \frac{1}{n} \left\{ a_n \sin nx - b_n \left[\cos nx + (-1)^{n+1} \right] \right\}.$$

Series (8) is obtained by replacing x by s in series (6) and then integrating term by term from $s = -\pi$ to $s = x$.

Our proof starts with the fact that since f is piecewise continuous, the function

(9) $$F(x) = \int_{-\pi}^{x} f(s) \, ds - \frac{a_0}{2} x \qquad (-\pi \leqslant x \leqslant \pi)$$

is continuous; moreover,

$$F'(x) = f(x) - \frac{a_0}{2} \qquad (-\pi < x < \pi),$$

except at points where f is discontinuous. Hence F' is piecewise continuous on the interval $-\pi < x < \pi$. Since F is piecewise smooth, then, it follows from the corollary in Sec. 21 that

(10) $$F(x) = \frac{A_0}{2} + \sum_{n=1}^{\infty} (A_n \cos nx + B_n \sin nx) \quad (-\pi < x < \pi),$$

where

(11) $$A_n = \frac{1}{\pi} \int_{-\pi}^{\pi} F(x) \cos nx \, dx, \qquad B_n = \frac{1}{\pi} \int_{-\pi}^{\pi} F(x) \sin nx \, dx.$$

We note from expression (9) and the first of expressions (7) that

$$F(\pi) = \int_{-\pi}^{\pi} f(s)\, ds - \frac{a_0}{2}\pi = a_0\pi - \frac{a_0}{2}\pi = \frac{a_0}{2}\pi$$

and $F(-\pi) = (a_0/2)\pi$; thus $F(-\pi) = F(\pi)$. This shows that representation (10) is also valid at the end points of the interval $-\pi < x < \pi$ (see Sec. 20) and, therefore, at each point x in the closed interval $-\pi \leqslant x \leqslant \pi$.

Let us now write the coefficients A_n and B_n in terms of a_n and b_n. When $n \geqslant 1$, we may integrate integrals (11) by parts, using the fact that F is continuous and F' is piecewise continuous. Thus

$$A_n = \frac{1}{n\pi}[F(x)\sin nx]_{-\pi}^{\pi} - \frac{1}{n\pi}\int_{-\pi}^{\pi} F'(x)\sin nx\, dx$$

$$= -\frac{1}{n\pi}\int_{-\pi}^{\pi}\left[f(x) - \frac{a_0}{2}\right]\sin nx\, dx$$

$$= -\frac{1}{n\pi}\int_{-\pi}^{\pi} f(x)\sin nx\, dx + \frac{a_0}{2n\pi}\int_{-\pi}^{\pi}\sin nx\, dx = -\frac{b_n}{n}.$$

Similarly, $B_n = a_n/n$. As for A_0, we know from the preceding paragraph that $F(\pi) = (a_0/2)\pi$ and that representation (10) is also valid when $x = \pi$. So by writing $x = \pi$ in that representation and then solving for A_0, we see that

$$A_0 = a_0\pi - 2\sum_{n=1}^{\infty}(-1)^n A_n = a_0\pi - 2\sum_{n=1}^{\infty}\frac{(-1)^{n+1}}{n}b_n.$$

With these expressions for A_n and B_n, including A_0, representation (10) takes the form

$$F(x) = \frac{a_0}{2}\pi + \sum_{n=1}^{\infty}\frac{1}{n}\left\{a_n\sin nx - b_n\left[\cos nx + (-1)^{n+1}\right]\right\}.$$

Finally, if we use expression (9) to substitute for $F(x)$ here, we arrive at the desired result (8).

The theorem can be written for the integral from x_0 to x, where $-\pi \leqslant x_0 \leqslant \pi$ and $-\pi \leqslant x \leqslant \pi$, by noting that

$$\int_{x_0}^{x} f(s)\, ds = \int_{-\pi}^{x} f(s)\, ds - \int_{-\pi}^{x_0} f(s)\, ds.$$

24. CONVERGENCE IN THE MEAN

A sequence $s_N(x)$ $(N = 1, 2, \ldots)$ of piecewise continuous functions, defined on an interval $a < x < b$, is said to *converge in the mean*, or in the norm, to a function f in $C_p(a, b)$ if the mean square error (Sec. 16) in the approximation by $s_N(x)$ to $f(x)$ tends to zero as N tends to infinity:

(1)
$$\lim_{N \to \infty} \int_a^b [f(x) - s_N(x)]^2 \, dx = 0.$$

That is,

(2)
$$\lim_{N \to \infty} \|f - s_N\| = 0.$$

Sometimes condition (2) is also written

(3)
$$\underset{N \to \infty}{\text{l.i.m.}} \; s_N(x) = f(x),$$

where the abbreviation "l.i.m." stands for *limit in the mean*.

Suppose now that the function s_N are the partial sums of a generalized Fourier series (Sec. 12) corresponding to f on the fundamental interval $a < x < b$:

(4)
$$s_N(x) = \sum_{n=1}^{N} c_n \phi_n(x).$$

This is the linear combination $\Phi_N(x)$ in Sec. 16 when $\gamma_n = c_n$ there. If condition (2) is satisfied by each function f in our function space $C_p(a, b)$, or possibly a subspace containing the orthonormal set $\{\phi_n(x)\}$, we say that $\{\phi_n(x)\}$ is *complete* in that space or subspace. Thus each function f can be approximated arbitrarily closely in the mean by some linear combination of functions $\phi_n(x)$ of a complete set, namely the linear combination (4) when N is large enough.[†]

According to equation (7), Sec. 16, the mean square error in the approximation by $s_N(x)$ to $f(x)$ is

(5)
$$\|f - s_N\|^2 = \|f\|^2 - \sum_{n=1}^{N} c_n^2.$$

Hence, when $\{\phi_n(x)\}$ is complete, it is always true that

(6)
$$\sum_{n=1}^{\infty} c_n^2 = \|f\|^2.$$

Equation (6) is known as *Parseval's equation*. It identifies the sum of the squares of the components of f, with respect to the generalized reference set $\{\phi_n(x)\}$, as the square of the norm of f.

Conversely, if each function f in the space satisfies Parseval's equation, the set $\{\phi_n(x)\}$ is complete in the sense of mean convergence. This is because, in view of equation (5), the limit (2) is merely a restatement of equation (6).

[†] In the mathematical literature, including some earlier editions of this text, the terms *complete* and *closed* are sometimes applied to sets that we have called *closed* (Sec. 12) and *complete*, respectively.

Writing $c_n = (f, \phi_n)$ in equation (6), we now have a theorem that provides an alternative characterization of complete sets.

Theorem 1. *A necessary and sufficient condition for an orthonormal set $\{\phi_n(x)\}$ $(n = 1, 2, \ldots)$ to be complete is that, for each function f in the space considered, Parseval's equation*

$$\text{(7)} \qquad \sum_{n=1}^{\infty} (f, \phi_n)^2 = \|f\|^2$$

be satisfied.

For an application of Theorem 1 to Fourier series on the interval $-\pi < x < \pi$, we recall from Sec. 15 that the functions

$$\text{(8)} \quad \phi_0(x) = \frac{1}{\sqrt{2\pi}}, \qquad \phi_{2n-1}(x) = \frac{1}{\sqrt{\pi}} \cos nx, \qquad \phi_{2n}(x) = \frac{1}{\sqrt{\pi}} \sin nx$$

$$(n = 1, 2, \ldots)$$

form an orthonormal set on the fundamental interval $-\pi < x < \pi$. The Fourier constants c_n $(n = 0, 1, 2, \ldots)$ in the generalized Fourier series for a function f in $C_p(-\pi, \pi)$ with respect to this set were then used to define the constants

$$\text{(9)} \qquad a_0 = \sqrt{\frac{2}{\pi}}\, c_0, \qquad a_n = \frac{c_{2n-1}}{\sqrt{\pi}}, \qquad b_n = \frac{c_{2n}}{\pi} \quad (n = 1, 2, \ldots).$$

This gave rise to the Fourier series correspondence

$$\text{(10)} \qquad f(x) \sim \frac{a_0}{2} + \sum_{n=1}^{\infty} (a_n \cos nx + b_n \sin nx) \qquad (-\pi < x < \pi),$$

where

$$\text{(11)} \qquad a_n = \frac{1}{\pi} \int_{-\pi}^{\pi} f(x) \cos nx \, dx \qquad (n = 0, 1, 2, \ldots)$$

and

$$\text{(12)} \qquad b_n = \frac{1}{\pi} \int_{-\pi}^{\pi} f(x) \sin nx \, dx \qquad (n = 1, 2, \ldots).$$

The following theorem states that the set of functions (8) is complete in a certain subspace of the space $C_p'(-\pi, \pi)$ of piecewise smooth functions (Sec. 17) on the interval $-\pi < x < \pi$.

Theorem 2. *The orthonormal set (8) is complete in the space of functions f that are continuous on the interval $-\pi \leqslant x \leqslant \pi$, where $f(-\pi) = f(\pi)$, and have piecewise continuous derivatives f' on the interval $-\pi < x < \pi$.*

Note that, in view of relations (9), Parseval's equation (6) for the orthonormal set in question can be written

$$(13) \qquad \frac{a_0^2}{2} + \sum_{n=1}^{\infty} (a_n^2 + b_n^2) = \frac{1}{\pi} \int_{-\pi}^{\pi} [f(x)]^2 \, dx.$$

Hence, once we show that the coefficients (11) and (12) actually satisfy equation (13), Theorem 2 is proved.

The fact that Parseval's equation (13) is satisfied is an easy consequence of the theorem in Sec. 22, which tells us that, for functions f in the space considered here, the series in correspondence (10) converges *uniformly* to $f(x)$ on the interval $-\pi \leqslant x \leqslant \pi$:

$$(14) \qquad f(x) = \frac{a_0}{2} + \sum_{n=1}^{\infty} (a_n \cos nx + b_n \sin nx) \qquad (-\pi \leqslant x \leqslant \pi).$$

Now a uniformly convergent series of continuous functions can be integrated term by term (Sec. 22). Hence we may multiply each term in equation (14) by $f(x)$ itself, thus leaving the series still uniformly convergent, and then integrate over the fundamental interval:

$$\int_{-\pi}^{\pi} [f(x)]^2 \, dx$$

$$= \frac{a_0}{2} \int_{-\pi}^{\pi} f(x) \, dx + \sum_{n=1}^{\infty} \left[a_n \int_{-\pi}^{\pi} f(x) \cos nx \, dx + b_n \int_{-\pi}^{\pi} f(x) \sin nx \, dx \right].$$

In view of expressions (11) and (12), the integrals on the right here can be written in terms of a_n and b_n; and we find that

$$\int_{-\pi}^{\pi} [f(x)]^2 \, dx = \pi \left[\frac{a_0^2}{2} + \sum_{n=1}^{\infty} (a_n^2 + b_n^2) \right].$$

Since this is the same as Parseval's equation (13), the proof is finished.

Theorem 2 is readily modified so as to apply to the orthonormal sets leading to Fourier cosine and sine series on the interval $0 < x < \pi$. More specifically, the set of normalized cosine functions in Example 2, Sec. 11, is complete in the space consisting of continuous functions f, on the interval $0 \leqslant x \leqslant \pi$, whose derivatives f' are piecewise continuous. When the normalized sine functions in Example 1, Sec. 11, are used to obtain a sine series, the conditions $f(0) = f(\pi) = 0$ are also needed for the set to be complete.

The function space in the theorem we have just proved is quite restricted. It can be shown that Parseval's equation (13) holds for any function f whose square is integrable over the interval $-\pi < x < \pi$.[†]

[†] See, for instance, the book by Tolstov (1976, pp. 54–57 and 117–120) that is listed in the Bibliography.

This bare introduction to the theory of convergence in the mean will not be developed further here. It should be emphasized, however, that statement (3) is *not* the same as the statement

$$(15) \qquad\qquad \lim_{N \to \infty} s_N(x) = f(x) \qquad\qquad (a < x < b),$$

even if a finite number of points in the interval are excepted.[‡] In fact, neither of statements (3) and (15) implies the other (see Problems 10 and 11).

PROBLEMS

1. Show that the function

$$f(x) = \begin{cases} 0 & \text{when } -\pi \leqslant x \leqslant 0, \\ \sin x & \text{when } \quad 0 < x \leqslant \pi \end{cases}$$

satisfies all the conditions in the lemma and the theorem in Sec. 22. Then, with the aid of the Weierstrass M-test (Sec. 22), verify directly that the Fourier series

$$\frac{1}{\pi} + \frac{1}{2}\sin x - \frac{2}{\pi}\sum_{n=1}^{\infty}\frac{\cos 2nx}{4n^2 - 1} \qquad (-\pi < x < \pi)$$

for f, found in Problem 7, Sec. 16, converges uniformly on the interval $-\pi \leqslant x \leqslant \pi$, as the theorem in Sec. 22 tells us. Also, state why this series is differentiable in the interval $-\pi < x < \pi$, except at the point $x = 0$, and describe graphically the function that is represented by the differentiated series for all x.

2. We know from Problem 7, Sec. 14, that the series

$$\frac{\pi}{2} - \frac{4}{\pi}\sum_{n=1}^{\infty}\frac{\cos(2n-1)x}{(2n-1)^2}$$

is the Fourier cosine series for the function $f(x) = x$ on the interval $0 < x < \pi$. Differentiate this series term by term to obtain a representation for the function $f'(x) = 1$ on that interval. State why the procedure is reliable here.

3. State Theorem 1 in Sec. 23 as it applies to Fourier sine series. Point out, in particular, why the conditions $f(0) = f(\pi) = 0$ are present in this case.

4. Let a_n and b_n denote the Fourier coefficients in the lemma in Sec. 22. Using the fact that the coefficients in the Fourier series for a function in $C_p(-\pi, \pi)$ always tend to zero as n tends to infinity (Sec. 16), show why

$$\lim_{n \to \infty} na_n = 0 \qquad \text{and} \qquad \lim_{n \to \infty} nb_n = 0.$$

5. Integrate from $s = 0$ to $s = x$ $(-\pi \leqslant x \leqslant \pi)$ the Fourier series

$$2\sum_{n=1}^{\infty}\frac{(-1)^{n+1}}{n}\sin ns,$$

[‡] An example of a sequence of functions that converges in the mean to zero but *diverges at each point* of the interval is given in the book by Franklin (1964, p. 408), listed in the Bibliography.

mentioned at the beginning of Sec. 23, and the one

$$2 \sum_{n=1}^{\infty} \frac{\sin (2n - 1)s}{2n - 1},$$

appearing in the example in Sec. 22. In each case, describe graphically the function that is represented by the new series.

6. Let p_n $(n = 1, 2, \ldots, N)$ and q_n $(n = 1, 2, \ldots, N)$ denote real numbers, where at least one of the numbers p_n, say p_m, is nonzero. By writing the quadratic equation

$$x^2 \sum_{n=1}^{N} p_n^2 + 2x \sum_{n=1}^{N} p_n q_n + \sum_{n=1}^{N} q_n^2 = 0$$

in the form

$$\sum_{n=1}^{N} (p_n x + q_n)^2 = 0,$$

show that the number $x_0 = -q_m/p_m$ is the only possible real root. Conclude that, since *there cannot be two distinct real roots*, the discriminant of this quadratic equation is negative or zero. Thus derive Cauchy's inequality (Sec. 22)

$$\left(\sum_{n=1}^{N} p_n q_n \right)^2 \leq \left(\sum_{n=1}^{N} p_n^2 \right) \left(\sum_{n=1}^{N} q_n^2 \right),$$

which is clearly valid even if all the numbers p_n are zero.

7. As in the example in Sec. 22, let $s_N(x)$ denote the partial sum consisting of the sum of the first N terms of the Fourier series

$$2 \sum_{n=1}^{\infty} \frac{\sin (2n - 1)x}{2n - 1} \qquad (-\pi < x < \pi)$$

for the function f there.

(a) By writing $A = x$ and $B = (2n - 1)x$ in the trigonometric identity

$$2 \sin A \cos B = \sin (A + B) + \sin (A - B)$$

and then summing each side of the resulting equation from $n = 1$ to $n = N$, derive the summation formula

$$2 \sum_{n=1}^{N} \cos (2n - 1)x = \frac{\sin 2Nx}{\sin x} \qquad (x \neq 0, \pm \pi, \pm 2\pi, \ldots).$$

Use this formula to write the derivative of $s_N(x)$ on the interval $0 < x < \pi$ as a simple quotient:

$$s_N'(x) = \frac{\sin 2Nx}{\sin x} \qquad (0 < x < \pi).$$

(b) With the aid of the expression for the derivative $s_N'(x)$ in part (a), show that the first extremum of $s_N(x)$ in the interval $0 < x < \pi$ is a relative maximum occurring when $x = \pi/(2N)$.

(c) By integrating each side of the summation formula in part (a) from $x = 0$ to $x = \pi/(2N)$, show that

$$s_N\left(\frac{\pi}{2N}\right) = I_1 + I_2,$$

where

$$I_1 = \int_0^{\pi/(2N)} \frac{x - \sin x}{x \sin x} \sin 2Nx\, dx \quad \text{and} \quad I_2 = \int_0^{\pi/(2N)} \frac{\sin 2Nx}{x}\, dx.$$

Verify that the integrands of these two integrals are piecewise continuous on the interval $0 < x < \pi/(2N)$ and hence that the integrals actually exist.

(d) Using the fact that the integrand of the integral I_1 in part (c) is bounded (see Sec. 10), show that the value of I_1 tends to zero as N tends to infinity. Then conclude that

$$\lim_{N \to \infty} s_N\left(\frac{\pi}{2N}\right) = \int_0^\pi \frac{\sin t}{t}\, dt.$$

The value of this last integral is the number σ in the example in Sec. 22.[†]

8. Use Theorem 1, Sec. 24, to show that *an orthonormal set $\{\phi_n(x)\}$ is closed* (Sec. 12) *in a given function space if it is complete in that space.*

9. Let $\{\phi_n(x)\}$ be an orthonormal set in the space of *continuous* functions on the interval $a \leqslant x \leqslant b$, and suppose that the generalized Fourier series for a function f in that space converges *uniformly* to a sum $s(x)$ on that interval.

(a) Show that s and f have the same Fourier constants with respect to $\{\phi_n(x)\}$.

(b) Use results in part (a) and Problem 11, Sec. 14, to show that if $\{\phi_n(x)\}$ is *closed* (Sec. 12), then $s(x) = f(x)$ on the interval $a \leqslant x \leqslant b$.

   Suggestion: Recall from Sec. 22 that the sum of a uniformly convergent series of continuous functions is continuous and that such a series can be integrated term by term.

10. Consider the sequence of functions $s_N(x)$ ($N = 1, 2, \ldots$) defined on the interval $0 \leqslant x \leqslant 1$ by means of the equations

$$s_N(x) = \begin{cases} 0 & \text{when} \quad 0 \leqslant x \leqslant \dfrac{1}{N}, \\[2mm] \sqrt{N} & \text{when} \quad \dfrac{1}{N} < x < \dfrac{2}{N}, \\[2mm] 0 & \text{when} \quad \dfrac{2}{N} \leqslant x \leqslant 1. \end{cases}$$

Show that this sequence converges pointwise to the function $f(x) = 0$ ($0 \leqslant x \leqslant 1$) but that it does *not* converge in the mean to f in the space $C_p(0, 1)$ or any subspace of $C_p(0, 1)$.

[†] The integral occurs as a particular value of the *sine integral function* $Si(x)$, which is tabulated in, for instance, the handbook edited by Abramowitz and Stegun (1972, p. 244) that is listed in the Bibliography. Approximation methods for evaluating definite integrals can also be used to find σ.

11. Let $s_N(x)$ $(N = 1, 2, \ldots)$ be a sequence of functions defined on the interval $0 \leqslant x \leqslant 1$ by means of the equations

$$
s_N(x) = \begin{cases} 0 & \text{when } x = 1, \dfrac{1}{2}, \ldots, \dfrac{1}{N}, \\[2ex] 1 & \text{when } x \neq 1, \dfrac{1}{2}, \ldots, \dfrac{1}{N}. \end{cases}
$$

Show that this sequence converges in the mean to the function $f(x) = 1$ in $C_p(0, 1)$ but that, for each positive integer p,

$$
\lim_{N \to \infty} s_N\left(\frac{1}{p}\right) = 0.
$$

Suggestion: Observe that $s_N(1/p) = 0$ when $N \geqslant p$.

CHAPTER
3

THE
FOURIER
METHOD

We turn now to a careful presentation of the Fourier method for solving boundary value problems involving partial differential equations, which was touched on in Sec. 9 (Chap. 1). While the example there served to motivate the method, at that time we were seriously limited by our inability to expand functions in Fourier series. Chapter 2 has, of course, addressed that problem.

Once the basics of the Fourier method have been presented, we shall, in Chap. 4, use it to solve a variety of boundary value problems whose solutions entail Fourier series representations. Then, in subsequent chapters, we shall apply the method to problems with solutions involving other, but closely related, types of representations.

25. LINEAR OPERATORS

We recall (Sec. 10) that any two functions u_1 and u_2 in a given function space have the same domain of definition and that each linear combination $c_1 u_1 + c_2 u_2$ is also in the space. A *linear operator* on the space is an operator L that transforms each function u of that space into a function Lu, which need not be in the space, and has the property that, for each pair of functions u_1 and u_2,

(1) $$L(c_1 u_1 + c_2 u_2) = c_1 L u_1 + c_2 L u_2$$

whenever c_1 and c_2 are constants. In particular,

(2) $$L(u_1 + u_2) = L u_1 + L u_2, \qquad L(c_1 u_1) = c_1 L u_1.$$

The function Lu may be a constant function; we note that

$$L(0) = L(0 \cdot 0) = 0L(0) = 0.$$

If u_3 is a third function in the space, then

$$L(c_1u_1 + c_2u_2 + c_3u_3) = L(c_1u_1 + c_2u_2) + L(c_3u_3)$$
$$= c_1Lu_1 + c_2Lu_2 + c_3Lu_3.$$

Proceeding by induction, we find that L transforms linear combinations of N functions in this manner:

(3) $$L\left(\sum_{n=1}^{N} c_n u_n\right) = \sum_{n=1}^{N} c_n Lu_n.$$

EXAMPLE 1. Suppose that both u_1 and u_2 are functions of the independent variables x and y. According to elementary properties of derivatives, a derivative of any linear combination of the two functions can be written as the same linear combination of the individual derivatives. Thus

(4) $$\frac{\partial}{\partial x}(c_1u_1 + c_2u_2) = c_1\frac{\partial u_1}{\partial x} + c_2\frac{\partial u_2}{\partial x},$$

provided $\partial u_1/\partial x$ and $\partial u_2/\partial x$ exist. In view of property (4), the class of functions of x and y that have partial derivatives of the first order with respect to x in the xy plane is a function space. The operator $\partial/\partial x$ is a linear operator on that space. It is naturally classified as a linear *differential operator*.

EXAMPLE 2. Consider a space of functions $u(x, y)$ defined on the xy plane. If $f(x, y)$ is a fixed function, also defined on the xy plane, then the operator L that multiplies each function $u(x, y)$ by $f(x, y)$ is a linear operator, where $Lu = fu$.

If linear operators L and M, distinct or not, are such that M transforms each function u of some function space into a function Mu to which L applies and if u_1 and u_2 are functions in that space, it follows from equation (1) that

(5) $$LM(c_1u_1 + c_2u_2) = L(c_1Mu_1 + c_2Mu_2) = c_1LMu_1 + c_2LMu_2.$$

That is, the *product* LM of linear operators is itself a linear operator.

The *sum* of two linear operators is defined by the equation

(6) $$(L + M)u = Lu + Mu.$$

If we replace u here by $c_1u_1 + c_2u_2$, we can see that the sum $L + M$ is a linear operator and hence that the sum of any finite number of linear operators is linear.

EXAMPLE 3. Let L denote the linear operator $\partial^2/\partial x^2$ defined on the space of functions $u(x, y)$ whose derivatives of the first and second order with respect to x exist in a given domain of the xy plane. The product $M = f\,\partial/\partial x$ of the linear operators in Examples 1 and 2 is linear on the same space, and the sum

$$L + M = \frac{\partial^2}{\partial x^2} + f\frac{\partial}{\partial x}$$

is, therefore, linear.

26. PRINCIPLE OF SUPERPOSITION

Each term of a linear homogeneous differential equation in u (Sec. 1) consists of a product of a function of the independent variables by one of the derivatives of u or by u itself. Hence every linear homogeneous differential equation has the form

(1) $Lu = 0,$

where L is a linear differential operator. For example, if

(2) $L = A\dfrac{\partial^2}{\partial x^2} + B\dfrac{\partial^2}{\partial y\,\partial x} + C\dfrac{\partial^2}{\partial y^2} + D\dfrac{\partial}{\partial x} + E\dfrac{\partial}{\partial y} + F,$

where the letters A through F denote functions of x and y only, equation (1) is the linear homogeneous partial differential equation

(3) $Au_{xx} + Bu_{xy} + Cu_{yy} + Du_x + Eu_y + Fu = 0$

in $u = u(x, y)$.

Linear homogeneous boundary conditions also have the form (1). Then the variables appearing as arguments of u and as arguments of functions that serve as coefficients in the linear operator L are restricted so that they represent points on the boundary of a domain.

Now let u_n $(n = 1, 2, \ldots, N)$ denote functions that satisfy equation (1); that is, $Lu_n = 0$ for each n. It follows from property (3), Sec. 25, of linear operators that each linear combination of those functions also satisfies equation (1). We state that *principle of superposition* of solutions, which is fundamental to the Fourier method for solving linear boundary value problems, as follows.

Theorem 1. *If each of N functions $u_1, u_2, \ldots, u_N$ satisfies a linear homogeneous differential equation $Lu = 0$, then every linear combination*

(4) $u = c_1u_1 + c_2u_2 + \cdots + c_Nu_N,$

where the c's are arbitrary constants, satisfies that differential equation. If each of the N functions satisfies a linear homogeneous boundary condition $Lu = 0$, then every linear combination (4) also satisfies that boundary condition.

The principle of superposition is useful in ordinary differential equations. For example, from the two solutions $y = e^x$ and $y = e^{-x}$ of the linear homogeneous equation $y'' - y = 0$, the general solution $y \doteq c_1 e^x + c_2 e^{-x}$ can be written. In this book, we shall be concerned mainly with applying the principle of superposition to solutions of partial differential equations.

   EXAMPLE. Consider the linear homogeneous heat equation (Sec. 2)

$$(5) \qquad\qquad u_t(x,t) = k u_{xx}(x,t) \qquad\qquad (0 < x < c, t > 0),$$

together with the linear homogeneous boundary conditions

$$(6) \qquad\qquad u_x(0,t) = 0, \qquad u_x(c,t) = 0 \qquad\qquad (t > 0).$$

It is easy to verify that if $L = k\partial^2/\partial x^2 - \partial/\partial t$ and

$$u_0 = \frac{1}{2}, \qquad u_n = \exp\left(-\frac{n^2\pi^2 k}{c^2} t\right) \cos\frac{n\pi x}{c} \qquad (n = 1,2,\dots),$$

then $Lu_n = 0$ $(n = 0,1,2,\dots)$. Hence it follows from Theorem 1 that $Lu = 0$ for every linear combination

$$u = a_0 u_0 + a_1 u_1 + a_2 u_2 + \cdots + a_N u_N.$$

That is, when $N \geqslant 1$, the function

$$(7) \qquad u(x,t) = \frac{a_0}{2} + \sum_{n=1}^{N} a_n \exp\left(-\frac{n^2\pi^2 k}{c^2} t\right) \cos\frac{n\pi x}{c}$$

satisfies the heat equation (5). Although it seems more natural to write $u_0 = 1$, with a_0 instead of $a_0/2$ in expression (7), our choice of $u_0 = \frac{1}{2}$ is simply for convenience in notation later on (Sec. 27).
   As for boundary conditions (6), we write $L = \partial/\partial x$ and observe that Lu_n $(n = 0,1,2,\dots)$ has value zero when $x = 0$ and $x = c$. So, again by Theorem 1, Lu has value zero when $x = 0$ and $x = c$. This shows that the linear combination (7) also satisfies boundary conditions (6).

   In order to state a general principle of superposition, similar to Theorem 1, that applies to an *infinite* set of functions $u_1, u_2, \dots$, one must deal with the convergence and differentiability of infinite series involving those functions. This is indicated below.
   Suppose that the functions u_n and the constants c_n are such that the infinite series whose terms are $c_n u_n$ converges throughout some domain of the independent variables. The sum of that series is a function

$$(8) \qquad\qquad u = \sum_{n=1}^{\infty} c_n u_n.$$

Let x represent one of the independent variables. The series is *differentiable*, or termwise differentiable, with respect to x if the derivatives $\partial u_n/\partial x$ and $\partial u/\partial x$

exist and if the series of functions $c_n \, \partial u_n / \partial x$ converges to $\partial u / \partial x$:

(9)
$$\frac{\partial u}{\partial x} = \sum_{n=1}^{\infty} c_n \frac{\partial u_n}{\partial x}.$$

Note that a series must be convergent if it is to be differentiable. Sufficient conditions for differentiability were noted in Sec. 22. If, in addition, series (9) is differentiable with respect to x, then series (8) is differentiable twice with respect to x.

Let L be a linear operator where Lu is a product of a function of the independent variables by u or by a derivative of u, or where Lu is a sum of a finite number of such terms. We now show that if series (8) is differentiable for all the derivatives involved in L and if each of the functions u_n in series (8) satisfies the linear homogeneous differential equation $Lu_n = 0$, then so does u; that is, $Lu = 0$.

To accomplish this, we first note that, according to the definition of the sum of an infinite series,

$$f \frac{\partial u}{\partial x} = f \lim_{N \to \infty} \sum_{n=1}^{N} c_n \frac{\partial u_n}{\partial x}$$

when series (8) is differentiable with respect to x. Thus

(10)
$$f \frac{\partial u}{\partial x} = \lim_{N \to \infty} f \frac{\partial}{\partial x} \sum_{n=1}^{N} c_n u_n.$$

Here the operator $\partial / \partial x$ can be replaced by other derivatives if the series is so differentiable. Then, by adding corresponding sides of equations similar to equation (10), including one that may not have any derivative at all, we find that

(11)
$$Lu = \lim_{N \to \infty} L \left(\sum_{n=1}^{N} c_n u_n \right).$$

The sum on the right-hand side of equation (11) is a linear combination of the functions $u_1, u_2, \ldots, u_N$; and if $Lu_n = 0 \ (n = 1, 2, \ldots)$, Theorem 1 allows us to write

$$L \left(\sum_{n=1}^{N} c_n u_n \right) = 0$$

for every N. Hence, from equation (11), we have the desired result $Lu = 0$.

A linear homogeneous boundary condition is also represented by an equation $Lu = 0$. In that case, we may require the function Lu to satisfy a condition of continuity at points on the boundary so that its values there will represent limiting values as those points are approached from the interior of the domain.

The following generalization of Theorem 1 is now established.

Theorem 2. *Suppose that each function of an infinite set $u_1, u_2, \ldots$ satisfies a linear homogeneous differential equation or boundary condition $Lu = 0$. Then the infinite series*

$$(12) \qquad\qquad u = \sum_{n=1}^{\infty} c_n u_n,$$

where the c_n are constants, also satisfies $Lu = 0$, provided the series converges and is differentiable for all derivatives involved in L and provided any required continuity condition at the boundary is satisfied by Lu when $Lu = 0$ is a boundary condition.

With Theorem 2, we are now ready to begin our illustration of the Fourier method for solving boundary value problems.

27. A TEMPERATURE PROBLEM

The linear boundary value problem

$$(1) \qquad\qquad u_t(x, t) = k u_{xx}(x, t) \qquad\qquad (0 < x < c, t > 0),$$

$$(2) \qquad\qquad u_x(0, t) = 0, \qquad u_x(c, t) = 0 \qquad\qquad (t > 0),$$

$$(3) \qquad\qquad u(x, 0) = f(x) \qquad\qquad (0 < x < c)$$

is a problem for the temperatures $u(x, t)$ in an infinite slab of material, bounded by the planes $x = 0$ and $x = c$, if its faces are insulated and the initial temperature distribution is a prescribed function $f(x)$ of the distance from the face $x = 0$. (See Fig. 26). It is also a problem of determining temperatures in a

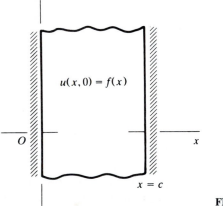

FIGURE 26

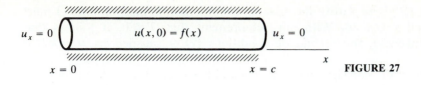

$u_x = 0$ $u(x,0) = f(x)$ $u_x = 0$

$x = 0$ $x = c$ x **FIGURE 27**

bar of uniform cross section, such as one in the shape of a right circular cylinder (Fig. 27), when its bases in the planes $x = 0$ and $x = c$ and its lateral surface, parallel to the x axis, are insulated and its initial temperatures are $f(x)$ ($0 < x < c$). We assume that the thermal diffusivity k of the material is constant throughout the slab, or bar, and that no heat is generated within it.

We saw in the example in Sec. 26 that the functions

$$(4) \qquad u_0 = \frac{1}{2}, \qquad u_n = \exp\left(-\frac{n^2\pi^2 k}{c^2}t\right)\cos\frac{n\pi x}{c} \qquad (n = 1, 2, \dots)$$

all satisfy the homogeneous conditions (1) and (2) in the stated temperature problem. In this section, we describe the method to be used in subsequent chapters for finding such functions. We shall also take into account the nonhomogeneous condition (3) and complete the solution of the boundary value problem. A number of the steps to be taken are *formal*, or manipulative. The validity of the solution obtained will be established in Sec. 28.

To determine nontrivial ($u \neq 0$) solutions of the homogeneous conditions (1) and (2), we seek separated functions (Sec. 9), or functions of the form

$$(5) \qquad u = X(x)T(t),$$

that satisfy those conditions. Note that X is a function of x alone and T a function of t alone. Note, too, that X and T must be nontrivial ($X \neq 0$, $T \neq 0$).

If $u = XT$ satisfies equation (1), then

$$X(x)T'(t) = kX''(x)T(t);$$

and, for values of x and t such that $X(x)T(t)$ is nonzero, we can divide by $kX(x)T(t)$ to separate the variables:

$$(6) \qquad \frac{T'(t)}{kT(t)} = \frac{X''(x)}{X(x)}.$$

Since the left-hand side here is a function of t alone, it does not vary with x. However, it is equal to a function of x alone, and so it cannot vary with t. Hence the two sides must have some constant value $-\lambda$ in common. That is,

$$(7) \qquad \frac{T'(t)}{kT(t)} = -\lambda, \qquad \frac{X''(x)}{X(x)} = -\lambda.$$

Our choice of $-\lambda$, rather than λ, for the *separation constant* is, of course, a minor matter of notation. It is only for convenience later on (Chap. 5) that we have written $-\lambda$.

If $u = XT$ is to satisfy the first of conditions (2), then $X'(0)T(t)$ must vanish for all t ($t > 0$). With our requirement that $T \neq 0$, it follows that $X'(0) = 0$. Likewise, the second of conditions (2) is satisfied by $u = XT$ if $X'(c) = 0$.

Thus $u = XT$ satisfies conditions (1) and (2) when X and T satisfy these two homogeneous problems:

$$(8) \qquad X''(x) + \lambda X(x) = 0, \qquad X'(0) = 0, \qquad X'(c) = 0,$$

$$(9) \qquad T'(t) + \lambda k T(t) = 0,$$

where the parameter λ has the *same* value in both problems. To find nontrivial solutions of this pair of problems, we first note that problem (9) has no boundary conditions. Hence it has nontrivial solutions for all values of λ. Since problem (8) has two boundary conditions, it may have nontrivial solutions for only particular values of λ. Problem (8) is called a *Sturm-Liouville problem*. The general theory of such problems is developed in Chap. 5, where it is shown that λ *must be real-valued* in order for there to be nontrivial solutions.

If $\lambda = 0$, the differential equation in problem (8) becomes $X''(x) = 0$. Its general solution is $X(x) = Ax + B$, where A and B are constants. Since $X'(x) = A$, the boundary conditions $X'(0) = 0$ and $X'(c) = 0$ are both satisfied when $A = 0$. So $X(x) = B$; and, except for a constant factor, problem (8) has the solution $X(x) = \frac{1}{2}$ if $\lambda = 0$. Note that any nonzero value of B might have been selected here.

If $\lambda > 0$, we may write $\lambda = \alpha^2$ ($\alpha > 0$). The differential equation in problem (8) then takes the form $X''(x) + \alpha^2 X(x) = 0$, its general solution being

$$X(x) = C_1 \cos \alpha x + C_2 \sin \alpha x.$$

Writing

$$X'(x) = -C_1 \alpha \sin \alpha x + C_2 \alpha \cos \alpha x$$

and keeping in mind that α is positive and, in particular, nonzero, we see that the condition $X'(0) = 0$ implies that $C_2 = 0$. Also, from the condition $X'(c) = 0$, it follows that $C_1 \alpha \sin \alpha c = 0$. Now if $X(x)$ is to be a nontrivial solution of problem (8), $C_1 \neq 0$. Hence α must be a positive root of the equation $\sin \alpha c = 0$. That is,

$$(10) \qquad \alpha = \frac{n\pi}{c} \qquad (n = 1, 2, \dots).$$

So, except for the constant factor C_1,

$$(11) \qquad X(x) = \cos \frac{n\pi x}{c} \qquad (n = 1, 2, \dots).$$

If $\lambda < 0$, we write $\lambda = -\alpha^2$ ($\alpha > 0$). This time, the differential equation in problem (8) has the general solution

$$X(x) = C_1 e^{\alpha x} + C_2 e^{-\alpha x}.$$

Since

$$X'(x) = C_1 \alpha e^{\alpha x} - C_2 \alpha e^{-\alpha x},$$

the condition $X'(0) = 0$ implies that $C_2 = C_1$. Hence

$$X(x) = C_1(e^{\alpha x} + e^{-\alpha x}),$$

or

$$X(x) = 2C_1 \cosh \alpha x.$$

But the condition $X'(c) = 0$ requires that $C_1 \sinh \alpha c = 0$; and, since $\sinh \alpha c \neq 0$, it follows that $C_1 = 0$. So problem (8) has only the trivial solution $X(x) \equiv 0$ if $\lambda < 0$.

The values $\lambda_0 = 0$, $\lambda_n = (n\pi/c)^2$ $(n = 1, 2, \dots)$ of λ for which problem (8) has nontrivial solutions are called *eigenvalues* of that problem, and the solutions $X_0(x) = \frac{1}{2}$, $X_n(x) = \cos(n\pi x/c)$ $(n = 1, 2, \dots)$ are the corresponding *eigenfunctions*.

We turn now to the differential equation (9) and determine its solution when λ is an eigenvalue. The solution when $\lambda = \lambda_0 = 0$ is, except for a constant factor, $T_0(t) = 1$. When $\lambda = \lambda_n = (n\pi/c)^2$ $(n = 1, 2, \dots)$, any solution of equation (9) is evidently a constant multiple of $T_n(t) = \exp(-n^2\pi^2 kt/c^2)$. Hence each of the products

$$(12) \qquad\qquad u_0 = X_0(x)T_0(t) = \frac{1}{2}$$

and

$$(13) \qquad u_n = X_n(x)T_n(t) = \exp\left(-\frac{n^2\pi^2 k}{c^2}t\right)\cos\frac{n\pi x}{c} \qquad (n = 1, 2, \dots)$$

satisfies the homogeneous conditions (1) and (2). These are the solutions (4). The procedure just used to obtain them is called the *method of separation of variables*.

Assuming that the conditions in Theorem 2 of Sec. 26 are satisfied, we may now use that theorem to see that the generalized linear combination

$$(14) \qquad u(x,t) = \frac{a_0}{2} + \sum_{n=1}^{\infty} a_n \exp\left(-\frac{n^2\pi^2 k}{c^2}t\right)\cos\frac{n\pi x}{c}$$

of the functions (12) and (13) also satisfies the homogeneous conditions (1) and (2). The remaining (nonhomogeneous) condition $u(x,0) = f(x)$ requires that

$$f(x) = \frac{a_0}{2} + \sum_{n=1}^{\infty} a_n \cos\frac{n\pi x}{c} \qquad (0 < x < c),$$

or that the constants a_n be, in fact, the coefficients

$$(15) \qquad a_n = \frac{2}{c} \int_0^c f(x) \cos \frac{n\pi x}{c} \, dx \qquad (n = 0, 1, 2, \dots)$$

in the Fourier cosine series for f on the interval $0 < x < c$ (Sec. 21).

Our formal solution of the temperature problem (1)–(3) is now complete. It consists of series (14) together with expression (15) for the coefficients a_n. The method used, involving separation of variables, superposition, and Fourier series, is the Fourier method.

Note that the steady-state temperatures, occurring when t tends to infinity, are $a_0/2$. That constant temperature is evidently the mean, or average, value of the initial temperatures $f(x)$ over the interval $0 < x < c$.

EXAMPLE. Suppose that the thickness c of the slab is unity and that the initial temperatures are $f(x) = x$ $(0 \leqslant x \leqslant 1)$. Here

$$a_0 = 2\int_0^1 x \, dx = 1.$$

Using integration by parts and observing that $\sin n\pi = 0$ and $\cos n\pi = (-1)^n$ when n is an integer, we find that

$$a_n = 2\int_0^1 x \cos n\pi x \, dx = 2\left[\frac{x \sin n\pi x}{n\pi} + \frac{\cos n\pi x}{n^2\pi^2} \right]_0^1 = 2\frac{(-1)^n - 1}{n^2\pi^2}$$

$$(n = 1, 2, \dots).$$

We have evaluated a_0 separately in order to avoid dividing by zero.

When $c = 1$ and these values for a_n $(n = 0, 1, 2, \dots)$ are used, expression (14) becomes

$$u(x,t) = \frac{1}{2} + \frac{2}{\pi^2} \sum_{n=1}^{\infty} \frac{(-1)^n - 1}{n^2} \exp\left(-n^2\pi^2 kt\right) \cos n\pi x,$$

or [see the footnote with Problem 1(b), Sec. 14]

$$(16) \quad u(x,t) = \frac{1}{2} - \frac{4}{\pi^2} \sum_{n=1}^{\infty} \frac{\exp\left[-(2n-1)^2\pi^2 kt\right]}{(2n-1)^2} \cos(2n-1)\pi x.$$

28. VERIFICATION OF SOLUTION

We turn now to the full verification of the solution of the boundary value problem

$$(1) \qquad\qquad u_t(x,t) = ku_{xx}(x,t) \qquad\qquad (0 < x < c,\ t > 0),$$

$$(2) \qquad\qquad u_x(0,t) = 0, \qquad u_x(c,t) = 0 \qquad\qquad (t > 0),$$

$$(3) \qquad\qquad u(x,0) = f(x) \qquad\qquad (0 < x < c)$$

that was obtained in Sec. 27. We recall that the continuous functions

(4) $$u_0 = \frac{1}{2}, \quad u_n = \exp\left(-\frac{n^2\pi^2 k}{c^2}t\right)\cos\frac{n\pi x}{c} \quad (n = 1, 2, \dots)$$

were found to satisfy the homogeneous conditions (1) and (2). As already pointed out in the example in Sec. 26, Theorem 1 in that section ensures that any linear combination

$$u = \sum_{n=0}^{N} a_n u_n$$

of those functions also satisfies conditions (1) and (2). The generalization

(5) $$u = \sum_{n=0}^{\infty} a_n u_n$$

of that linear combination to an infinite series is, of course, the solution

(6) $$u(x,t) = \frac{a_0}{2} + \sum_{n=1}^{\infty} a_n \exp\left(-\frac{n^2\pi^2 k}{c^2}t\right)\cos\frac{n\pi x}{c}$$

in Sec. 27 when the coefficients a_n are assigned the values

(7) $$a_n = \frac{2}{c}\int_0^c f(x)\cos\frac{n\pi x}{c}\,dx \quad (n = 0, 1, 2, \dots).$$

We assume that f is piecewise smooth (Sec. 17) on the interval $0 < x < c$. Also, at a point of discontinuity of f in that interval, we define $f(x)$ as the mean value of the one-sided limits $f(x+)$ and $f(x-)$. Note how it follows from expression (7) that

$$|a_n| \leqslant \frac{2}{c}\int_0^c |f(x)|\,dx \quad (n = 0, 1, 2, \dots)$$

and hence that there is a positive constant M, independent of n, such that

(8) $$|a_n| \leqslant M \quad (n = 0, 1, 2, \dots).$$

We begin our verification by showing that series (5), with coefficients (7), actually converges in the region $0 \leqslant x \leqslant c, t > 0$ of the xt plane and that it satisfies the homogeneous conditions (1) and (2). To accomplish this, we first note from expressions (4) and inequalities (8) that, if t_0 is a fixed positive number,

(9) $$|a_n u_n| \leqslant M \exp\left(-\frac{n^2\pi^2 k}{c^2}t_0\right) \quad (n = 0, 1, 2, \dots)$$

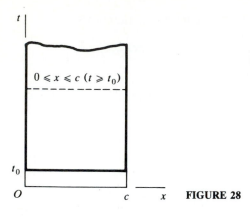

FIGURE 28

whenever $0 \leqslant x \leqslant c$ and $t \geqslant t_0$ (Fig. 28). An application of the ratio test shows that the series

$$(10) \qquad \sum_{n=0}^{\infty} n^i \exp\left(-\frac{n^2 \pi^2 k}{c^2} t_0\right)$$

of constants converges when i is any nonnegative integer and, in particular, when $i = 0$. So we know from the comparison and absolute convergence tests that the series (5) converges when $0 \leqslant x \leqslant c$, $t \geqslant t_0$. One can use series (10) and the Weierstrass M-test (Sec. 22) to show that the series

$$(11) \qquad \sum_{n=0}^{\infty} (a_n u_n)_x, \qquad \sum_{n=0}^{\infty} (a_n u_n)_{xx}$$

of derivatives converge uniformly on the interval $0 \leqslant x \leqslant c$ for any fixed t $(t \geqslant t_0)$. Likewise, the series

$$(12) \qquad \sum_{n=0}^{\infty} (a_n u_n)_t$$

converges uniformly on the semi-infinite interval $t \geqslant t_0$ for any fixed x $(0 \leqslant x \leqslant c)$.

The uniformity of the convergence of these series ensures that the series (5) is differentiable twice with respect to x and once with respect to t when $0 \leqslant x \leqslant c$, $t \geqslant t_0$. Consequently, if we write $L = k\partial^2/\partial x^2 - \partial/\partial t$ and recall from the example in Sec. 26 that $Lu_n = 0$ $(n = 0, 1, 2, \ldots)$, we know from Theorem 2 there that $Lu = 0$ when $0 \leqslant x \leqslant c$, $t \geqslant t_0$. Thus series (5) converges and satisfies the heat equation (1) in the domain $0 < x < c$, $t > 0$ since the positive number t_0 can be chosen arbitrarily small.

Writing $L = \partial/\partial x$ and again using Theorem 2 in Sec. 26, we see that series (5) also satisfies boundary conditions (2). Observe that since the first of series (11) is uniformly convergent on the interval $0 \leqslant x \leqslant c$ for any fixed t $(t \geqslant t_0)$, the derivative $u_x(x, t)$ of series (5) is *continuous* in x on that interval.

(See Fig. 28.) Hence the one-sided limits

$$u_x(0+,t) = \lim_{\substack{x \to 0 \\ x > 0}} u_x(x,t), \qquad u_x(c-,t) = \lim_{\substack{x \to c \\ x < c}} u_x(x,t)$$

at the end points of the interval $0 \leqslant x \leqslant c$ $(t \geqslant t_0)$ exist and have the values $u_x(0,t)$ and $u_x(c,t)$, respectively. Since conditions (2) are satisfied and since t_0 can be chosen arbitrarily small, then,

$$(13) \qquad\qquad u_x(0+,t) = 0, \qquad u_x(c-,t) = 0 \qquad\qquad (t > 0).$$

In seeking solutions of boundary value problems, we shall tacitly require that those solutions satisfy such continuity conditions at boundary points. Thus, when conditions (2) are part of a boundary value problem, it is understood that conditions (13) must also be satisfied. As we have just seen, series (5) has that property.

The nonhomogeneous condition (3) is clearly satisfied by our solution since series (6) reduces to the Fourier cosine series

$$(14) \qquad\qquad \frac{a_0}{2} + \sum_{n=1}^{\infty} a_n \cos \frac{n\pi x}{c} \qquad\qquad (0 < x < c)$$

for f when $t = 0$; and the corollary in Sec. 21 ensures that series (14) converges to $f(x)$ when $0 < x < c$.

It remains to show that

$$(15) \qquad\qquad u(x,0+) = f(x) \qquad\qquad (0 < x < c).$$

This is a continuity requirement that must be satisfied when $t = 0$, just as conditions (13) must hold when $x = 0$ and $x = c$. One can show that solution (6) has this property by appealing to a convergence theorem, due to Abel,[†] that is to be proved in Chap. 9 (Sec. 79). According to that theorem, the series formed by multiplying the terms of a convergent series of constants, such as series (14) with x fixed, by corresponding terms of a bounded sequence of functions of t whose values never increase with n, such as $\exp(-n^2\pi^2 kt/c^2)$ $(n = 0,1,2,\dots)$, is uniformly convergent with respect to t. So, for any fixed x $(0 < x < c)$, the series in expression (6) converges uniformly with respect to t when $t \geqslant 0$ and thus represents a function that is continuous in t $(t \geqslant 0)$. This shows that our solution $u(x,t)$ is continuous in t when $t \geqslant 0$, in particular when $t = 0$. That is,

$$\lim_{\substack{t \to 0 \\ t > 0}} u(x,t) = u(x,0),$$

or $u(x,0+) = u(x,0)$, for each fixed x $(0 < x < c)$. Property (15) now follows from the fact that $u(x,0) = f(x)$ $(0 < x < c)$. This completes the verification that the function (6) is a solution of the boundary value problem (1)–(3).

[†] Niels Henrik Abel, Norwegian, 1802–1829.

PROBLEMS

1. Show that the solution of the temperature problem in Sec. 27 reduces to

$$u(x,t) = \frac{\pi^2}{3} + 4 \sum_{n=1}^{\infty} \frac{(-1)^n}{n^2} e^{-n^2 kt} \cos nx$$

when $c = \pi$ and $f(x) = x^2 \, (-\pi < x < \pi)$.

   Suggestion: Refer to the Fourier cosine series for x^2 that was found in Problem 4(a), Sec. 14.

2. In Problem 10, Sec. 9, the functions

$$u_0 = y, \qquad u_n = \sinh ny \cos nx \qquad\qquad (n = 1, 2, \dots)$$

were shown to satisfy Laplace's equation

$$u_{xx}(x,y) + u_{yy}(x,y) = 0 \qquad (0 < x < \pi, 0 < y < 2)$$

and the homogeneous boundary conditions

$$u_x(0,y) = u_x(\pi,y) = 0, \qquad u(x,0) = 0.$$

After writing $u = X(x)Y(y)$ and separating variables, use the solutions of the Sturm-Liouville problem in Sec. 27 to show how these functions can be discovered. Then, by proceeding formally, derive the following solution of the boundary value problem resulting when the nonhomogeneous condition $u(x,2) = f(x)$ is included:

$$u(x,y) = A_0 y + \sum_{n=1}^{\infty} A_n \sinh ny \cos nx,$$

where

$$A_0 = \frac{1}{2\pi} \int_0^{\pi} f(x)\, dx, \qquad A_n = \frac{2}{\pi \sinh 2n} \int_0^{\pi} f(x) \cos nx\, dx \qquad (n = 1, 2, \dots).$$

(The final result in Problem 10, Sec. 9, is a special case of this.)

3. For each of the following partial differential equations in $u = u(x,t)$, determine if it is possible to write $u = X(x)T(t)$ and separate variables to obtain two ordinary differential equations in X and T. If it can be done, find those ordinary differential equations.

   (a) $u_{xx} - xtu_{tt} = 0$; (b) $(x+t)u_{xx} - u_t = 0$;
   (c) $xu_{xx} + u_{xt} + tu_{tt} = 0$; (d) $u_{xx} - u_{tt} - u_t = 0$.

4. Suppose that equation (6), Sec. 27, had been written in the form

$$\frac{T'(t)}{T(t)} = k \frac{X''(x)}{X(x)}.$$

Set each side here equal to $-\lambda$ and show how the functions (12) and (13) in Sec. 27 still follow. (This illustrates how it is generally simpler to keep the physical constant in the heat equation out of the Sturm-Liouville problem, as we did in Sec. 27.)

5. Show that if an operator L has the two properties

$$L(u_1 + u_2) = Lu_1 + Lu_2, \qquad L(c_1 u_1) = c_1 Lu_1$$

for all functions u_1, u_2 in some space and for every constant c_1, then L is linear; that is, show that it has property (1), Sec. 25.

6. Use special cases of linear operators, such as $L = x$ and $M = \partial/\partial x$, to illustrate that products LM and ML are not always the same.

7. Let u and v denote functions of x and t that satisfy the one-dimensional heat equation:

$$u_t = ku_{xx} \qquad \text{and} \qquad v_t = kv_{xx}.$$

Multiply each side of these two equations by constants c_1 and c_2, respectively, and add to show that the linear combination $c_1u + c_2v$ also satisfies the heat equation. This illustrates a variation in the proof of Theorem 1 in Sec. 26.

8. Show that each of the functions $y_1 = 1/x$ and $y_2 = 1/(1 + x)$ satisfies the *nonlinear* differential equation $y' + y^2 = 0$. Then show that the sum $y_1 + y_2$ fails to satisfy that equation. Also show that if c is a constant, where $c \neq 0$ and $c \neq 1$, neither cy_1 nor cy_2 satisfies the equation.

9. Let u_1 and u_2 satisfy a linear *nonhomogeneous* differential equation $Lu = f$, where f is a function of the independent variables only. Prove that the linear combination $c_1u_1 + c_2u_2$ fails to satisfy that equation when $c_1 + c_2 \neq 1$.

10. Let L denote a linear differential operator, and suppose that f is a function of the independent variables. Show that the solutions u of the equation $Lu = f$ are of the form $u = u_1 + u_2$, where the u_1 are the solutions of the equation $Lu_1 = 0$ and u_2 is any particular solution of $Lu_2 = f$. (This is a principle of superposition of solutions for *nonhomogeneous* differential equations.)

11. Assuming that b_n is a bounded sequence of constants, prove that the series

$$u(x, y) = \sum_{n=1}^{\infty} b_n e^{-ny} \sin nx$$

converges and is twice-differentiable with respect to x and y when $y \geq y_0$, where y_0 is any positive constant. Then show that u satisfies Laplace's equation $u_{xx} + u_{yy} = 0$ in the half plane $y > 0$.

12. Prove that if $n^4|b_n| \leq M$ $(n = 1, 2, \ldots)$, where M is a positive constant, then the series

$$y(x, t) = \sum_{n=1}^{\infty} b_n \sin nx \cos nt$$

converges and satisfies the wave equation $y_{tt} = y_{xx}$ for all x and t.

29.   A VIBRATING STRING PROBLEM

To illustrate further the Fourier method, we now consider a boundary value problem for displacements in a vibrating string. This time, the nonhomogeneous condition will require us to expand a function $f(x)$ into a Fourier sine series, rather than a cosine series.

Let us find an expression for the transverse displacements $y(x, t)$ in a string, stretched between the points $x = 0$ and $x = c$ on the x axis and with no external forces acting along it, if the string is initially displaced into a position $y = f(x)$ and released at rest from that position. The function $y(x, t)$ must satisfy the wave equation (Sec. 5)

$$(1) \qquad y_{tt}(x,t) = a^2 y_{xx}(x,t) \qquad (0 < x < c, t > 0).$$

It must also satisfy the boundary conditions

$$(2) \qquad y(0,t) = 0, \qquad y(c,t) = 0, \qquad y_t(x,0) = 0,$$

$$(3) \qquad y(x,0) = f(x) \qquad (0 \leqslant x \leqslant c),$$

where the prescribed displacement function f is continuous on the interval $0 \leqslant x \leqslant c$ and $f(0) = f(c) = 0$.

We assume a product solution

$$(4) \qquad y = X(x)T(t)$$

of the homogeneous conditions (1) and (2) and substitute it into those conditions. This leads to the two homogeneous problems

$$(5) \qquad X''(x) + \lambda X(x) = 0, \qquad X(0) = 0, \qquad X(c) = 0,$$

$$(6) \qquad T''(t) + \lambda a^2 T(t) = 0, \qquad T'(0) = 0.$$

Problem (5) is another instance of a Sturm-Liouville problem. The method of solution that was used to solve the one in Sec. 27 can be applied here. It turns out (Problem 5, Sec. 30) that the eigenvalues are $\lambda_n = (n\pi/c)^2$ ($n = 1, 2, \ldots$) and that the corresponding eigenfunctions are $X_n(x) = \sin(n\pi x/c)$. When $\lambda = \lambda_n$, problem (6) becomes

$$T''(t) - \left(\frac{n\pi a}{c}\right)^2 T(t) = 0, \qquad T'(0) = 0;$$

and it follows that, except for a constant factor, the solution is $T_n(t) = \cos(n\pi a t/c)$. Consequently, each of the products

$$(7) \qquad y_n = X_n(x)T_n(t) = \sin\frac{n\pi x}{c} \cos\frac{n\pi a t}{c} \qquad (n = 1, 2, \ldots)$$

satisfies the homogeneous conditions (1) and (2).

According to Theorem 2 in Sec. 26, the generalized linear combination

$$(8) \qquad y(x,t) = \sum_{n=1}^{\infty} b_n \sin\frac{n\pi x}{c} \cos\frac{n\pi a t}{c}$$

also satisfies the homogeneous conditions (1) and (2), provided the constants b_n can be restricted so that the infinite series is suitably convergent and differentiable. That series will satisfy the nonhomogeneous condition (3) if the b_n are such that

$$(9) \qquad f(x) = \sum_{n=1}^{\infty} b_n \sin\frac{n\pi x}{c} \qquad (0 < x < c).$$

Note that this series converges to zero at the end points $x = 0$ and $x = c$. Hence if representation (9) is valid, it also holds on the closed interval $0 \leqslant x \leqslant c$.

The constants b_n in expression (9) are evidently the coefficients in the Fourier sine series for f on the interval $0 < x < c$ (Sec. 21):

(10) $$b_n = \frac{2}{c} \int_0^c f(x) \sin \frac{n\pi x}{c} \, dx \qquad (n = 1, 2, \dots).$$

The formal solution of our boundary value problem for the displacements in a vibrating string is, therefore, series (8) with coefficients (10).

EXAMPLE. Suppose that the string has length $c = 2$ and that its midpoint is initially raised to a height h above the horizontal axis. The rest position from which the string is released thus consists of two line segments (Fig. 29).

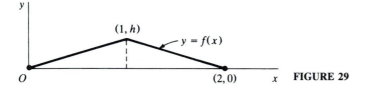

FIGURE 29

The function f, which describes the initial position of this plucked string, is given by the equations

(11) $$f(x) = \begin{cases} hx & \text{when } 0 \leqslant x \leqslant 1, \\ -h(x - 2) & \text{when } 1 < x \leqslant 2; \end{cases}$$

and the coefficients b_n in the Fourier sine series for that function on the interval $0 < x < 2$ can be written

$$b_n = \int_0^2 f(x) \sin \frac{n\pi x}{2} \, dx = h \int_0^1 x \sin \frac{n\pi x}{2} \, dx - h \int_1^2 (x - 2) \sin \frac{n\pi x}{2} \, dx.$$

After integrating by parts and simplifying, we find that

$$b_n = \frac{8h}{n^2 \pi^2} \sin \frac{n\pi}{2} \qquad (n = 1, 2, \dots).$$

Series (8) then becomes

(12) $$y(x, t) = \frac{8h}{\pi^2} \sum_{n=1}^{\infty} \frac{1}{n^2} \sin \frac{n\pi}{2} \sin \frac{n\pi x}{2} \cos \frac{n\pi a t}{2}.$$

Since $\sin(n\pi/2) = 0$ when n is even and since

$$\sin \frac{(2n - 1)\pi}{2} = \sin\left(n\pi - \frac{\pi}{2}\right) = -\cos n\pi = (-1)^{n+1} \qquad (n = 1, 2, \dots),$$

expression (12) for the displacements of points on the string in question can also be written

$$(13) \quad y(x,t) = \frac{8h}{\pi^2} \sum_{n=1}^{\infty} \frac{(-1)^{n+1}}{(2n-1)^2} \sin \frac{(2n-1)\pi x}{2} \cos \frac{(2n-1)\pi at}{2}.$$

Before verifying our solution of the boundary value problem (1)–(3), we comment briefly on its physical interpretation. From expression (8), we can see that, for each fixed x, the displacement $y(x,t)$ is a period function of time t, with period

$$(14) \qquad\qquad\qquad\qquad T_0 = \frac{2c}{a}.$$

The period is independent of the initial displacement $f(x)$. Since $a^2 = H/\delta$, where H is the magnitude of the x component of the tensile force and δ is the mass per unit length of the string (Sec. 5), the period varies directly with c and $\sqrt{\delta}$ and inversely with $\sqrt{H}$.

It is also evident from expression (8) that, for a given length c and initial displacement $f(x)$, the displacement y depends on only the value of x and the value of the product at. That is, $y = \phi(x, at)$ where the function ϕ is the same function regardless of the value of the constant a. Let a_1 and a_2 denote different values of that constant, and let $y_1(x, t)$ and $y_2(x, t)$ be the corresponding displacements. Then

$$(15) \qquad\qquad y_1(x, t_1) = y_2(x, t_2) \quad \text{if} \quad a_1 t_1 = a_2 t_2 \qquad (0 \leqslant x \leqslant c).$$

In particular, suppose that only the constant H has different values, H_1 and H_2. The same set of instantaneous positions is taken by the string when $H = H_1$ and when $H = H_2$. But the times t_1 and t_2 required to reach any one position have the ratio

$$(16) \qquad\qquad\qquad\qquad \frac{t_1}{t_2} = \sqrt{\frac{H_2}{H_1}}.$$

Except for the nonhomogeneous condition (3), our boundary value problem is satisfied by any partial sum

$$(17) \qquad\qquad y_N(x, t) = \sum_{n=1}^{N} b_n \sin \frac{n\pi x}{c} \cos \frac{n\pi at}{c}$$

of the series solution (8). Instead of meeting requirement (3), however, it satisfies the condition

$$(18) \qquad\qquad y(x, 0) = \sum_{n=1}^{N} b_n \sin \frac{n\pi x}{c}.$$

The sum on the right-hand side of equation (18) is, of course, the partial sum consisting of the sum of the first N terms of the Fourier sine series for f on the interval $0 < x < c$. Since the odd periodic extension of f is clearly continuous and f' is piecewise continuous, that series converges uniformly to $f(x)$ on the interval $0 \leqslant x \leqslant c$ (Sec. 22). Hence, if N is taken sufficiently large, the sum $y_N(x, 0)$ can be made to approximate $f(x)$ arbitrarily closely for all values of x in that interval.

The function $y_N(x, t)$, which is everywhere continuous together with all its partial derivatives, is therefore established as a solution of the approximating problem obtained by replacing condition (3) in the original problem by condition (18).

Corresponding approximations can be made to other problems. But a remarkable feature in the present case is that $y_N(x, t)$ never deviates from the actual displacement $y(x, t)$ by more than the greatest deviation of $y_N(x, 0)$ from $f(x)$. To see this, we need only recall the trigonometric identity

$$2 \sin A \cos B = \sin (A + B) + \sin (A - B)$$

and write

$$(19) \qquad 2 \sin \frac{n \pi x}{c} \cos \frac{n \pi a t}{c} = \sin \frac{n \pi (x + a t)}{c} + \sin \frac{n \pi (x - a t)}{c}.$$

Expression (17) then becomes

$$y_N(x, t) = \frac{1}{2} \left[\sum_{n=1}^{N} b_n \sin \frac{n \pi (x + a t)}{c} + \sum_{n=1}^{N} b_n \sin \frac{n \pi (x - a t)}{c} \right];$$

and the two sums here are those of the first N terms of the sine series for the odd periodic extension F of the function f, with arguments $x + at$ and $x - at$. But the greatest deviation of the first sum from $F(x + at)$, or of the second from $F(x - at)$, is the same as the greatest deviation of $y_N(x, 0)$ from $f(x)$.

30. VERIFICATION OF SOLUTION

In this section, we shall verify the formal solution that we found in Sec. 29 for the boundary value problem

$$(1) \qquad\qquad y_{tt}(x, t) = a^2 y_{xx}(x, t) \qquad\qquad (0 < x < c, t > 0),$$

$$(2) \qquad\qquad y(0, t) = 0, \qquad y(c, t) = 0,$$

$$(3) \qquad\qquad y(x, 0) = f(x), \qquad y_t(x, 0) = 0.$$

The given function f was assumed to be continuous on the interval $0 \leqslant x \leqslant c$; also, $f(0) = f(c) = 0$. Assuming further that f' is at least piecewise continuous, we know (Sec. 21) that $f(x)$ is represented by its Fourier sine series when $0 \leqslant x \leqslant c$. The coefficients

$$(4) \qquad\qquad b_n = \frac{2}{c} \int_0^c f(x) \sin \frac{n \pi x}{c} \, dx \qquad\qquad (n = 1, 2, \dots)$$

in that series are the ones in the series solution

(5) $$y(x,t) = \sum_{n=1}^{\infty} b_n \sin \frac{n\pi x}{c} \cos \frac{n\pi at}{c}$$

that we obtained. Hence, when $t = 0$, the series in expression (5) converges to $f(x)$; that is, $y(x,0) = f(x)$ when $0 \leqslant x \leqslant c$.

The nature of the problem calls for a solution $y(x, t)$ that is continuous in x and t when $0 \leqslant x \leqslant c$ and $t \geqslant 0$ and is such that $y_t(x, t)$ is continuous in t at $t = 0$. Hence the prescribed boundary values in conditions (2) and (3) are also limiting values on the boundary of the domain $0 < x < c$, $t > 0$:

$$y(0+,t) = 0, \qquad y(c-,t) = 0 \qquad\qquad (t \geqslant 0).$$
$$y(x,0+) = f(x), \qquad y_t(x,0+) = 0 \qquad\qquad (0 \leqslant x \leqslant c).$$

To verify that expression (5) represents a solution, we must prove that the series there converges to a continuous function $y(x, t)$ which satisfies the wave equation (1) and all the boundary conditions. But series (5), with coefficients (4), can fail to be twice-differentiable with respect to x and t even when it has a sum that satisfies the wave equation. This was, in fact, the case with the solution in the example in Sec. 29, where the coefficients b_n were

$$b_n = \frac{8h}{n^2\pi^2} \sin \frac{n\pi}{2} \qquad\qquad (n = 1,2,\dots).$$

After series (5) is differentiated twice with respect to x or t when those values of b_n are used, it is apparent that the resulting series cannot converge since its nth term does not tend to zero. It is possible, however, to write series (5) in a closed form, which does not involve infinite series. That will enable us to verify our solution.

To do this, we first refer to identity (19) in Sec. 29 and write series (5) as

(6) $$y(x,t) = \frac{1}{2}\left[\sum_{n=1}^{\infty} b_n \sin \frac{n\pi(x + at)}{c} + \sum_{n=1}^{\infty} b_n \sin \frac{n\pi(x - at)}{c} \right].$$

Now the odd periodic extension F of f, with the properties

(7) $$F(x) = f(x) \qquad\qquad \text{when } 0 \leqslant x \leqslant c$$

and

(8) $$F(-x) = -F(x), \qquad F(x + 2c) = F(x) \qquad\qquad \text{for all } x,$$

is represented for *all* x by the sine series for f:

(9) $$F(x) = \sum_{n=1}^{\infty} b_n \sin \frac{n\pi x}{c} \qquad\qquad (-\infty < x < \infty).$$

Consequently, expression (6) can be written

(10) $$y(x,t) = \frac{1}{2}[F(x + at) + F(x - at)].$$

Note that the convergence of series (5) and (6) is ensured for all x and t by the convergence of series (9) for all x.

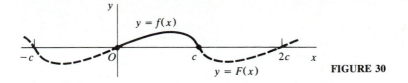

FIGURE 30

We turn now to the verification of our solution in the form (10). From our assumption that f is continuous when $0 \leqslant x \leqslant c$ and that $f(0) = f(c) = 0$, we see that the odd periodic extension F is continuous for all x (Fig. 30). Let us also *assume that f' and f'' are continuous when $0 \leqslant x \leqslant c$ and that $f''(0) = f''(c) = 0$.* It is then easy to show that the derivatives F' and F'' are continuous for all x. For, by recalling that $F(x) = -F(-x)$ and then applying the chain rule, we can write

$$F'(x) = -\frac{d}{dx}F(-x) = F'(-x),$$

where $F'(-x)$ denotes the derivative of F evaluated at $-x$. Thus F' is an *even* periodic function; likewise, F'' is an *odd* periodic function. Consequently, F' and F'' are continuous, as indicated in Fig. 31.

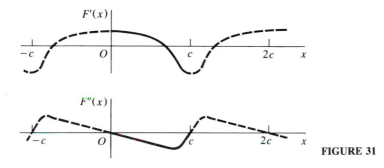

FIGURE 31

To show that the function (10) satisfies the wave equation, we write it as

$$y = \frac{1}{2}F(u) + \frac{1}{2}F(v),$$

where $u = x + at$ and $v = x - at$. The chain rule for differentiating composite functions reveals that

$$\frac{\partial y}{\partial t} = \frac{\partial y}{\partial u}\frac{\partial u}{\partial t} + \frac{\partial y}{\partial v}\frac{\partial v}{\partial t},$$

or

$$\frac{\partial y}{\partial t} = \frac{a}{2}F'(u) - \frac{a}{2}F'(v);$$

and, by letting $\partial y / \partial t$ play the role of y in this last expression, we find that

$$(11) \qquad \frac{\partial^2 y}{\partial t^2} = \frac{\partial}{\partial t} \left(\frac{\partial y}{\partial t} \right) = \frac{a^2}{2} F''(u) + \frac{a^2}{2} F''(v).$$

Similarly,

$$(12) \qquad \frac{\partial^2 y}{\partial x^2} = \frac{1}{2} F''(u) + \frac{1}{2} F''(v).$$

In view of expressions (11) and (12), the function (10) satisfies the wave equation (1). Furthermore, because F is continuous for all x, the function (10) is continuous for all x and t, in particular when $0 \leqslant x \leqslant c$ and $t \geqslant 0$.

While it is evident from series (5) that our solution $y(x,t)$ satisfies the conditions $y(0, t) = y(c, t) = 0$ and $y(x, 0) = f(x)$, expression (10) can also be used to verify this. For example, when $x = c$ in expression (10), one can write

$$F(c - at) = -F(-c + at) = -F(-c + at + 2c) = -F(c + at).$$

Therefore,

$$y(c, t) = \frac{1}{2} [F(c + at) - F(c + at)] = 0.$$

As for the final boundary condition $y_t(x, 0) = 0$, we observe that

$$y_t(x, t) = \frac{a}{2} [F'(x + at) - F'(x - at)].$$

Hence $y_t(x, 0) = 0$, and the continuity of F' ensures that $y_t(x, t)$ is continuous. The function (10) is now fully verified as a solution of the boundary value problem (1)–(3). In Chap. 9 (Sec. 82), we shall show why it is the only possible solution which, together with its derivatives of the first and second order, is continuous throughout the region $0 \leqslant x \leqslant c,\ t \geqslant 0$ of the xt plane.

If the conditions on f' and f'' are relaxed by merely requiring those two functions to be *piecewise continuous*, we find that at each instant t there may be a finite number of points x $(0 \leqslant x \leqslant c)$ where the partial derivatives of y fail to exist. Except at those points, our function satisfies the wave equation and the condition $y_t(x, 0) = 0$. The other boundary conditions are satisfied as before, but we have a solution of our boundary value problem in a broader sense.

PROBLEMS

1. A string is stretched between the fixed points 0 and 1 on the x axis and released at rest from the position $y = A \sin \pi x$, where A is a constant. Obtain from expression (10), Sec. 30, the subsequent displacements $y(x, t)$, and verify the result fully. Sketch the position of the string at several instants of time.

   Answer: $y(x, t) = A \sin \pi x \cos \pi a t.$

2. Solve Problem 1 when the initial displacement there is changed to $y = B \sin 2\pi x$, where B is a constant.

   Answer: $y(x, t) = B \sin 2\pi x \cos 2\pi a t.$

3. Show why the sum of the two functions $y(x, t)$ found in Problems 1 and 2 represents the displacements after the string is released at rest from the position

$$y = A \sin \pi x + B \sin 2\pi x.$$

4. By assuming a product solution $y = X(x)T(t)$, obtain conditions (5) and (6) on X and T in Sec. 29 from the homogeneous conditions (1) and (2) of the string problem there.

5. Derive the eigenvalues and eigenfunctions, stated in Sec. 29, of the Sturm-Liouville problem

$$X''(x) + \lambda X(x) = 0, \qquad X(0) = 0, \qquad X(c) = 0.$$

6. For the initially displaced string of length c considered in Secs. 29 and 30, show why the *frequency* ν of the vibration, in cycles per unit time, has the value

$$\nu = \frac{a}{2c} = \frac{1}{2c}\sqrt{\frac{H}{\delta}}.$$

Show that if $H = 200$ lb, the weight per foot is 0.01 lb ($g\delta = 0.01$, $g = 32$), and the length is 2 ft, then $\nu = 200$ cycles/s.

7. In Secs. 29 and 30, the position of the string at each instant can be shown graphically by moving the graph of the periodic function $\frac{1}{2}F(x)$ to the right with velocity a and an identical curve to the left at the same rate and then adding ordinates, on the interval $0 \leqslant x \leqslant c$, of the two curves so obtained at the instant t. Show how this follows from expression (10), Sec. 30.

8. Plot some positions of the plucked string considered in the example in Sec. 29 by the method described in Problem 7 to verify that the string assumes such positions as those indicated by the bold line segments in Fig. 32.

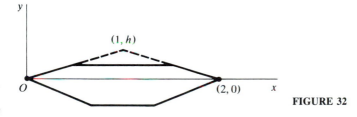

FIGURE 32

9. Write the boundary value problem (1)–(3), Sec. 29, in terms of the two independent variables x and $\tau = at$ to show that the problem in y as a function of x and τ does not involve the constant a (see Sec. 5). Thus, without solving the problem, deduce that the solution has the form $y = \phi(x, \tau) = \phi(x, at)$ and hence that relation (15), Sec. 29, is true.

31.   HISTORICAL DEVELOPMENT

Mathematical sciences experienced a burst of activity following the invention of calculus by Newton (1642–1727) and Leibnitz (1646–1716). Among topics in mathematical physics that attracted the attention of great scientists during that period were boundary value problems in vibrations of strings, elastic bars, and columns of air, all associated with mathematical theories of musical vibrations. Early contributors to the theory of vibrating strings included the English mathematician Brook Taylor (1685–1731), the Swiss mathematicians Daniel Bernoulli (1700–1782) and Leonhard Euler (1707–1783), and Jean d'Alembert (1717–1783) in France.

By the 1750s d'Alembert, Bernoulli, and Euler had advanced the theory of vibrating strings to the stage where the partial differential equation $y_{tt} = a^2 y_{xx}$ was known and a solution of a boundary value problem for strings had been found from the general solution of that equation. Also, the concept of fundamental modes of vibration led those men to the notion of superposition of solutions, to a solution of the form (8), Sec. 29, where a series of trigonometric functions appears, and thus to the matter of representing arbitrary functions by trigonometric series. Euler later found expressions for the coefficients in those series. But the general concept of a function had not been clarified, and a lengthy controversy took place over the question of representing arbitrary functions on a bounded interval by such series. The question of representation was finally settled by the German mathematician Peter Gustav Lejeune Dirichlet (1805–1859) about 70 years later.

The French mathematical physicist Jean Baptiste Joseph Fourier (1768–1830) presented many instructive examples of expansions in trigonometric series in connection with boundary value problems in the conduction of heat. His book *Théorie analytique de la chaleur*, published in 1822, is a classic in the theory of heat conduction. It was actually the third version of a monograph that he originally submitted to the Institut de France on December 21, 1807.[†] He effectively illustrated the basic procedures of separation of variables and superposition, and his work did much toward arousing interest in trigonometric series representations.

But Fourier's contributions to the representation problem did not include conditions of validity; he was interested in applications and methods. As noted above, Dirichlet was the first to give such conditions. In 1829 he firmly established general conditions on a function sufficient to ensure that it can be represented by a series of sine and cosine functions.[‡]

Representation theory has been refined and greatly extended since Dirichlet's time. It is still growing.

[†] A. Freeman's early translation of Fourier's book into English was reprinted by Dover, New York, in 1955. The original 1807 monograph itself remained unpublished until 1972, when the critical edition by Grattan-Guinness that is listed in the Bibliography appeared.

[‡] For supplementary reading on the history of these series, see the articles by Langer (1947) and Van Vleck (1914) that are listed in the Bibliography.

CHAPTER
4

BOUNDARY
VALUE
PROBLEMS

This chapter is devoted to the application of Fourier series in solving various types of boundary value problems that are mathematical formulations of problems in physics. The basic method has already been described in Chap. 3. Except for the final section of this chapter (Sec. 40), we shall limit our attention to problems whose solutions follow from the solutions of the two Sturm-Liouville problems encountered in Secs. 27 and 29 of Chap. 3. To be specific, we saw there that the Sturm-Liouville problem

$$(1) \qquad X''(x) + \lambda X(x) = 0, \qquad X'(0) = 0, \qquad X'(c) = 0,$$

on the interval $0 \leqslant x \leqslant c$, has nontrivial solutions only when λ is one of the eigenvalues

$$\lambda_0 = 0, \qquad \lambda_n = \left(\frac{n\pi}{c}\right)^2 \qquad (n = 1, 2, \dots)$$

and that the corresponding solutions, or eigenfunctions, are

$$X_0(x) = \frac{1}{2}, \qquad X_n(x) = \cos\frac{n\pi x}{c} \qquad (n = 1, 2, \dots).$$

For the Sturm-Liouville problem

$$(2) \qquad X''(x) + \lambda X(x) = 0, \qquad X(0) = 0, \qquad X(c) = 0,$$

on the same interval $0 \leqslant x \leqslant c$,

$$\lambda_n = \left(\frac{n\pi}{c}\right)^2 \qquad\qquad (n = 1, 2, \ldots)$$

and

$$X_n(x) = \sin\frac{n\pi x}{c} \qquad\qquad (n = 1, 2, \ldots).$$

As illustrated in Chap. 3, the solutions of problems (1) and (2) lead to Fourier cosine and sine series representations, respectively. A third Sturm-Liouville problem, to be solved in Sec. 40, leads to Fourier series with both cosines and sines. Boundary value problems whose solutions involve terms other than $\cos(n\pi x/c)$ and $\sin(n\pi x/c)$ are taken up in Chap. 5, where the general theory of Sturm-Liouville problems is developed, and in subsequent chapters.

In Chap. 3, we indicated ways of proving that a solution found for a given boundary value problem truly satisfies the partial differential equation and all the boundary conditions and continuity requirements. When that is done, the solution is rigorously established. But, even for many of the simpler problems, the verification of solutions may be lengthy or difficult. The boundary value problems in this chapter will be solved only *formally* in the sense that we shall not always explicitly mention needed conditions on functions whose Fourier series are used and we shall not verify the solutions.

We shall also ignore questions of uniqueness, but the physics of a given boundary value problem that is well posed generally suggests that there should be only one solution of that problem. In Chap. 9 we shall give some attention to uniqueness of solutions.

32. A SLAB WITH VARIOUS BOUNDARY CONDITIONS

We consider here the problem of finding temperatures in the same slab (or bar) as in Sec. 27 when its boundary surfaces are subjected to other simple thermal conditions. For convenience, however, we take the thickness of the slab as π units, so that $c = \pi$. Eigenvalues $\lambda_n = (n\pi/c)^2$ $(n = 1, 2, \ldots)$ then become simply $\lambda_n = n^2$. As illustrated in Problem 4 of this section, temperature formulas for a slab of arbitrary thickness c follow readily once they are found when $c = \pi$. In each of the three examples below, the temperature function $u = u(x, t)$ is to satisfy the one-dimensional heat equation

$$(1) \qquad\qquad u_t(x, t) = k u_{xx}(x, t) \qquad\qquad (0 < x < \pi, t > 0).$$

EXAMPLE 1. If both faces of the slab are kept at temperature zero and the initial temperatures are $f(x)$ (Fig. 33), then

$$(2) \qquad u(0, t) = 0, \qquad u(\pi, t) = 0, \qquad \text{and} \qquad u(x, 0) = f(x).$$

Conditions (1) and (2) make up the boundary value problem; and, by separation

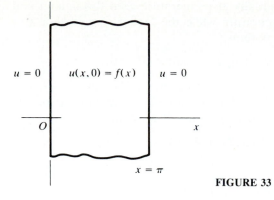

FIGURE 33

of variables, we find that a function $u = X(x)T(t)$ satisfies the homogeneous conditions if

$$(3) \qquad X''(x) + \lambda X(x) = 0, \qquad X(0) = 0, \qquad X(\pi) = 0$$

and

$$(4) \qquad T'(t) + \lambda k T(t) = 0.$$

According to Sec. 29, the Sturm-Liouville problem (3) has eigenvalues $\lambda_n = n^2$ and eigenfunctions $X_n(x) = \sin nx$ $(n = 1, 2, \dots)$. The corresponding functions of t arising from equation (4) are, except for constant factors, $T_n(t) = \exp(-n^2 kt)$. Formally, then, the function

$$(5) \qquad u(x, t) = \sum_{n=1}^{\infty} b_n e^{-n^2 kt} \sin nx$$

satisfies all the conditions in the boundary value problem, including the nonhomogeneous condition $u(x, 0) = f(x)$, if

$$(6) \qquad f(x) = \sum_{n=1}^{\infty} b_n \sin nx \qquad\qquad (0 < x < \pi).$$

Let us assume that f is piecewise smooth on the interval $0 < x < \pi$. Then $f(x)$ is represented by its Fourier sine series (6), where

$$(7) \qquad b_n = \frac{2}{\pi} \int_0^{\pi} f(x) \sin nx \, dx \qquad\qquad (n = 1, 2, \dots).$$

The function (5), with coefficients (7), is our formal solution of the boundary value problem (1)–(2). It can be expressed more compactly in the form

$$u(x, t) = \frac{2}{\pi} \sum_{n=1}^{\infty} e^{-n^2 kt} \sin nx \int_0^{\pi} f(s) \sin ns \, ds,$$

where the variable of integration s is used to avoid confusion with the free variable x.

EXAMPLE 2. If the slab is initially at temperature zero throughout and the face $x = 0$ is kept at that temperature while the face $x = \pi$ is kept at a constant temperature u_0 when $t > 0$, then

$$(8) \qquad u(0, t) = 0, \qquad u(\pi, t) = u_0, \qquad u(x, 0) = 0.$$

The boundary value problem consisting of equations (1) and (8) is not in proper form for the method of separation of variables because one of the two-point boundary conditions is nonhomogeneous. If we write

$$(9) \qquad u(x, t) = U(x, t) + \Phi(x),$$

however, those equations become

$$U_t(x, t) = k[U_{xx}(x, t) + \Phi''(x)]$$

and

$$U(0, t) + \Phi(0) = 0, \qquad U(\pi, t) + \Phi(\pi) = u_0, \qquad U(x, 0) + \Phi(x) = 0.$$

Suppose now that

$$(10) \qquad \Phi'' = 0 \qquad \text{and} \qquad \Phi(0) = 0, \qquad \Phi(\pi) = u_0.$$

Then U satisfies the conditions

$$(11) \qquad U_t = kU_{xx}, \qquad U(0, t) = U(\pi, t) = 0, \qquad U(x, 0) = -\Phi(x).$$

Conditions (10) tell us that $\Phi(x) = (u_0/\pi)x$. Hence problem (11) is a special case of the one in Example 1, where $f(x) = (-u_0/\pi)x$. When $f(x)$ is this particular function, the coefficients b_n in solution (5) can be found by evaluating the integrals in expression (7). But since we already know from Example 1, Sec. 14, that

$$(12) \qquad x = \sum_{n=1}^{\infty} 2\frac{(-1)^{n+1}}{n} \sin nx \qquad (0 < x < \pi)$$

and since the numbers b_n are the coefficients in the Fourier sine series for the function $f(x) = (-u_0/\pi)x$ on the interval $0 < x < \pi$, we can see at once that

$$b_n = -\frac{u_0}{\pi} 2\frac{(-1)^{n+1}}{n} = \frac{u_0}{\pi} 2\frac{(-1)^n}{n} \qquad (n = 1, 2, \ldots).$$

Consequently,

$$(13) \qquad u(x, t) = \frac{u_0}{\pi}\left[x + 2\sum_{n=1}^{\infty} \frac{(-1)^n}{n} e^{-n^2 kt} \sin nx \right].$$

By letting t tend to infinity in solution (13), we see that the function $\Phi(x) = (u_0/\pi)x$ represents the *steady-state* temperatures in the slab. In fact, conditions (10) consist of Laplace's equation in one dimension together with the conditions that the temperature be 0 and u_0 at $x = 0$ and $x = \pi$, respectively.

Expression (9), in the form

$$U(x,t) = u(x,t) - \Phi(x),$$

reveals that $U(x,t)$ is merely the desired solution minus the steady-state temperatures.

Finally, note that one can replace the term x in solution (13) by its representation (12) and write that solution as

(14) $$u(x,t) = \frac{2u_0}{\pi} \sum_{n=1}^{\infty} \frac{(-1)^{n+1}}{n}(1 - e^{-n^2 kt}) \sin nx.$$

This alternative form can be more useful in approximating $u(x,t)$ by a few terms of the series when t is small. For the factors $1 - \exp(-n^2 kt)$ are then small compared to the factors $\exp(-n^2 kt)$ in expression (13). Hence the terms that are discarded are smaller. The terms in series (13) are, of course, smaller when t is large.

EXAMPLE 3. Suppose that the face $x = 0$ is kept at temperature zero and that the face $x = \pi$ is insulated. Then

(15) $u(0,t) = 0$ and $u_x(\pi,t) = 0$ $(t > 0).$

Also, let the initial temperatures be

(16) $u(x,0) = f(x)$ $(0 < x < \pi),$

where f is piecewise smooth. By writing $u = X(x)T(t)$ and separating variables, we find that

$$X''(x) + \lambda X(x) = 0, \qquad X(0) = 0, \qquad X'(\pi) = 0.$$

Although this problem in X can be treated by methods to be developed in Chap. 5, we are not fully prepared to handle it at this time. The stated temperature problem can, however, be solved here by considering a related problem in a larger slab $0 \leqslant x \leqslant 2\pi$ (Fig. 34).

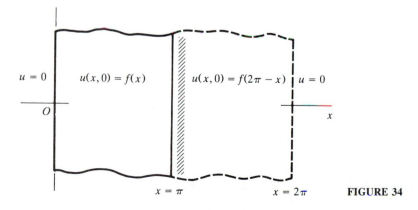

$x = \pi$ $x = 2\pi$ **FIGURE 34**

Let the two faces $x = 0$ and $x = 2\pi$ of that larger slab be kept at temperature zero; and let the initial temperatures be

$$(17) \qquad\qquad u(x,0) = F(x) \qquad\qquad (0 < x < 2\pi),$$

where

$$(18) \qquad\qquad F(x) = \begin{cases} f(x) & \text{when} \quad 0 < x < \pi, \\ f(2\pi - x) & \text{when} \quad \pi < x < 2\pi. \end{cases}$$

The function F is a piecewise smooth extension of the function f on the interval $0 < x < 2\pi$, and the graph of $y = F(x)$ is symmetric with respect to the line $x = \pi$. This procedure is suggested by the fact that, with the initial condition (17), no heat will flow across the midsection $x = \pi$ of the larger slab. So, when the variable x is restricted to the interval $0 < x < \pi$, the temperature function for the larger slab will be the desired one for the original slab.

According to Problem 4(b), which gives the solution of the boundary value problem in Example 1 for a slab of arbitrary thickness, the temperature function for the larger slab is

$$(19) \qquad\qquad u(x,t) = \sum_{n=1}^{\infty} b_n \exp\left(-\frac{n^2 k}{4} t\right) \sin \frac{nx}{2},$$

where the b_n are the coefficients in the Fourier sine series for the function F on the interval $0 < x < 2\pi$:

$$b_n = \frac{1}{\pi} \int_0^{2\pi} F(x) \sin \frac{nx}{2}\, dx \qquad\qquad (n = 1, 2, \dots).$$

This expression can be written in terms of the original function $f(x)$ by simply referring to Problem 11, Sec. 21, which tells us that

$$b_n = \left[1 - (-1)^n\right] \frac{1}{\pi} \int_0^{\pi} f(x) \sin \frac{nx}{2}\, dx \qquad\qquad (n = 1, 2, \dots);$$

that is, $b_{2n} = 0$ and

$$(20) \qquad\qquad b_{2n-1} = \frac{2}{\pi} \int_0^{\pi} f(x) \sin \frac{(2n-1)x}{2}\, dx \qquad\qquad (n = 1, 2, \dots).$$

Solution (19) then becomes

$$(21) \qquad\qquad u(x,t) = \sum_{n=1}^{\infty} b_{2n-1} \exp\left[-\frac{(2n-1)^2 k}{4} t\right] \sin \frac{(2n-1)x}{2},$$

with coefficients (20).

PROBLEMS†

1. Let the initial temperature distribution be uniform over the slab in Example 1, Sec. 32, so that $f(x) = u_0$. Find $u(x, t)$ and the flux $- Ku_x(x_0, t)$ across a plane $x = x_0$ $(0 \leqslant x_0 \leqslant \pi)$ when $t > 0$. Show that no heat flows across the center plane $x = \pi/2$.

2. Suppose that $f(x) = \sin x$ in Example 1, Sec. 32. Find $u(x, t)$ and verify the result fully.

> *Suggestion:* Use the integration formula (10), Sec. 11.
> *Answer:* $u(x, t) = e^{-kt} \sin x$.

3. Let $v(x, t)$ and $w(x, t)$ denote the solutions found in Examples 1 and 2 in Sec. 32. Assuming that those solutions are valid, show that the sum $u = v + w$ gives a temperature formula for a slab $0 \leqslant x \leqslant \pi$ whose faces $x = 0$ and $x = \pi$ are kept at temperatures 0 and u_0, respectively, and whose initial temperature distribution is $f(x)$.

4. The faces $x = 0$ and $x = c$ of a slab $0 \leqslant x \leqslant c$, which is initially at temperatures $f(x)$, are kept at temperature zero. Use the following method to derive an expression for the temperatures $u = u(x, t)$ throughout the slab when $t > 0$.

   (a) After writing the boundary value problem for the temperatures, make the substitution $s = \pi x/c$ to show that $u_t = (k\pi^2/c^2)u_{ss}$, $u = 0$ when $s = 0$ and $s = \pi$, and $u = f(cs/\pi)$ when $t = 0$.

   (b) By referring to the solution (5), with coefficients (7), of the problem in Example 1, Sec. 32, write an expression for u in terms of s and t. Then, with the aid of the relation $s = \pi x/c$ that was used in part (a), show that

$$u(x, t) = \sum_{n=1}^{\infty} b_n \exp\left(-\frac{n^2\pi^2 k}{c^2}t\right) \sin\frac{n\pi x}{c},$$

where

$$b_n = \frac{2}{c}\int_0^c f(x) \sin\frac{n\pi x}{c}\, dx \qquad (n = 1, 2, \dots).$$

5. (a) Show that if A is a constant and

$$f(x) = \begin{cases} A & \text{when } 0 < x < \dfrac{c}{2}, \\ 0 & \text{when } \dfrac{c}{2} < x < c, \end{cases}$$

the temperature formula in Problem 4(b) becomes

$$u(x, t) = \frac{4A}{\pi}\sum_{n=1}^{\infty}\frac{\sin^2(n\pi/4)}{n}\exp\left(-\frac{n^2\pi^2 k}{c^2}t\right)\sin\frac{n\pi x}{c}.$$

   (b) Two slabs of iron ($k = 0.15$ cgs unit), each 20 cm thick, are such that one is at $100°C$ and the other at $0°C$ throughout. They are placed face to face in perfect contact, and their outer faces are kept at $0°C$. Use the result in part (a) here to

---

† Only formal solutions of the boundary value problems here and in the sets of problems to follow are expected, unless the problem specifically states that the solution is to be fully verified. Partial verification is often easy and helpful.

show that the temperature at the common face 10 min after contact has been made is approximately 36°C. Then show that if the slabs are made of concrete ($k = 0.005$ cgs unit), it takes approximately 5 h for the common face to reach that temperature of 36°C. [Note that $u(x, t)$ depends on the product kt.]

6. Let $u(r, t)$ denote *temperatures in a solid sphere* $r \leqslant a$, where r is the spherical coordinate (Sec. 4), when that solid is initially at temperatures $f(r)$ and its surface $r = a$ is kept at temperature zero (Fig. 35). The function $u = u(r, t)$ satisfies the conditions

$$\frac{\partial u}{\partial t} = \frac{k}{r}\frac{\partial^2}{\partial r^2}(ru), \qquad u(a, t) = 0, \qquad u(r, 0) = f(r).$$

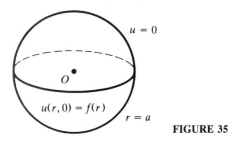

FIGURE 35

Introduce the new function $v(r, t) = ru(r, t)$, and note that $v(0, t) = 0$ because u is continuous at the center $r = 0$. Set up a new boundary value problem in v; and, with the aid of the solution in Problem 4(*b*), derive the expression

$$u(r, t) = \frac{2}{ar}\sum_{n=1}^{\infty} \exp\left(-\frac{n^2\pi^2 k}{a^2}t\right)\sin\frac{n\pi r}{a}\int_0^a sf(s)\sin\frac{n\pi s}{a}\,ds.$$

7. A solid spherical body 40 cm in diameter, initially at 100°C throughout, is cooled by keeping its surface at 0°C. Use the temperature formula in Problem 6, and also the fact that $(\sin\theta)/\theta$ tends to unity as θ tends to zero, to show formally that

$$u(0 + , t) = 200\sum_{n=1}^{\infty}(-1)^{n+1}\exp\left(-\frac{n^2\pi^2 k}{400}t\right).$$

Thus find the approximate temperature at the center of the sphere 10 min after cooling begins when the material is (*a*) iron, for which $k = 0.15$ cgs unit; (*b*) concrete, for which $k = 0.005$ cgs unit.

    Answers: (*a*) 22°C; (*b*) 100°C.

8. The initial temperature of a slab $0 \leqslant x \leqslant \pi$ is zero throughout, and the face $x = 0$ is kept at that temperature. Heat is supplied through the face $x = \pi$ at a constant rate A ($A > 0$) per unit area, so that $Ku_x(\pi, t) = A$ (see Sec. 3). Use the solution of the problem in Example 3, Sec. 32, to derive the expression

$$u(x, t) = \frac{A}{K}\left\{x + \frac{8}{\pi}\sum_{n=1}^{\infty}\frac{(-1)^n}{(2n-1)^2}\exp\left[-\frac{(2n-1)^2 k}{4}t\right]\sin\frac{(2n-1)x}{2}\right\}$$

for the temperatures in this slab.

9. Let $v(x, t)$ denote temperatures in a slender wire lying along the x axis. Variations of the temperature over each cross section are to be neglected. At the lateral surface, the linear law of surface heat transfer between the wire and its surroundings is assumed to apply (see Problem 7, Sec. 4). Let the surroundings be at temperature zero; then

$$v_t(x, t) = kv_{xx}(x, t) - bv(x, t),$$

where b is a positive constant. The ends $x = 0$ and $x = c$ of the wire are insulated (Fig. 36), and the initial temperature distribution is $f(x)$. Solve the boundary value problem for v by separation of variables. Then show that

$$v(x, t) = u(x, t)e^{-bt},$$

where u is the temperature function found in Sec. 27.

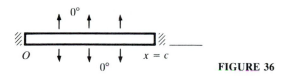

FIGURE 36

10. Use the substitution $v(x, t) = u(x, t) \exp(-bt)$ to reduce the boundary value problem in Problem 9 to the one in Sec. 27.

11. Assuming that the ends of the wire in Problem 9 are not insulated, but kept at temperature zero instead, find the temperature function.

12. Solve the boundary value problem consisting of the differential equation

$$u_t(x, t) = u_{xx}(x, t) - bu(x, t) \qquad (0 < x < \pi, t > 0),$$

where b is a positive constant, and the boundary conditions

$$u(0, t) = 0, \qquad u(\pi, t) = 1, \qquad u(x, 0) = 0.$$

Also, give a physical interpretation of this problem (see Problem 9).
    Suggestion: The Fourier series for $\sinh ax$ in Problem 5, Sec. 16, is useful here.

$$\textit{Answer: } u(x, t) = \frac{\sinh x\sqrt{b}}{\sinh \pi\sqrt{b}} + \frac{2}{\pi}e^{-bt}\sum_{n=1}^{\infty}(-1)^n\frac{n}{n^2 + b}e^{-n^2 t}\sin nx.$$

33. THE SLAB WITH INTERNALLY GENERATED HEAT

We consider here the same infinite slab $0 \leqslant x \leqslant \pi$ as in Sec. 32, but we assume that there is a source that generates heat at a rate per unit volume which depends on time. The slab is initially at temperatures $f(x)$, and both faces are maintained at temperature zero. According to Sec. 2, the temperatures $u(x, t)$ in the slab must satisfy the modified form

$$(1) \qquad u_t(x, t) = ku_{xx}(x, t) + q(t) \qquad (0 < x < \pi, t > 0)$$

of the one-dimensional heat equation, where $q(t)$ is assumed to be a continuous function of t. The conditions

$$(2) \qquad u(0,t) = 0, \qquad u(\pi,t) = 0, \qquad \text{and} \qquad u(x,0) = f(x)$$

complete the statement of this boundary value problem.

Since the differential equation (1) is nonhomogeneous, the method of separation of variables cannot be applied directly. We shall use here, instead, a method known as the *method of variation of parameters*. Also called the method of eigenfunction expansions, it is often useful when the differential equation is nonhomogeneous, especially when the term making it so is time-dependent. To be specific, we seek a solution of the boundary value problem in the form

$$(3) \qquad u(x,t) = \sum_{n=1}^{\infty} B_n(t) \sin nx$$

of a Fourier sine series whose coefficients $B_n(t)$ are differentiable functions of t. The form (3) is suggested by Example 1, Sec. 32, where the problem is the same as this one when $q(t) \equiv 0$ in equation (1). We anticipate that the function $q(t)$ in equation (1) will cause the coefficients b_n in solution (5), Sec. 32, of the homogeneous part of that earlier problem to depend on t. Instead of writing $b_n(t)\exp(-n^2kt)$, we combine $b_n(t)$ with the exponential function and denote the product by $B_n(t)$. So our approach here is, in fact, to start with a generalized linear combination, with coefficients depending on t, of the eigenfunctions $\sin nx$ $(n = 1,2,\ldots)$ of the Sturm-Liouville problem arising in Example 1, Sec. 32. The reader will note that the method of finding a solution of the form (3) is similar in spirit to the method of variation of parameters which is used in solving linear ordinary differential equations that are nonhomogeneous.

We assume that series (3) can be differentiated term by term. Then, by substituting it into equation (1) and recalling [Problem 1(b), Sec. 14] that

$$1 = \sum_{n=1}^{\infty} \frac{2[1-(-1)^n]}{n\pi} \sin nx \qquad (0 < x < \pi),$$

we may write

$$\sum_{n=1}^{\infty} B_n'(t) \sin nx = k \sum_{n=1}^{\infty} [-n^2 B_n(t)] \sin nx + q(t) \sum_{n=1}^{\infty} \frac{2[1-(-1)^n]}{n\pi} \sin nx,$$

or

$$\sum_{n=1}^{\infty} [B_n'(t) + n^2 k B_n(t)] \sin nx = \sum_{n=1}^{\infty} \frac{2[1-(-1)^n]}{n\pi} q(t) \sin nx.$$

By identifying the coefficients in the sine series on each side of this last equation, we now see that

$$(4) \qquad B_n'(t) + n^2 k B_n(t) = \frac{2[1-(-1)^n]}{n\pi} q(t) \qquad (n = 1,2,\ldots).$$

Moreover, according to the third of conditions (2),

$$\sum_{n=1}^{\infty} B_n(0) \sin nx = f(x) \qquad (0 < x < \pi);$$

and this means that

(5) $$B_n(0) = b_n \qquad (n = 1, 2, \dots),$$

where b_n are the coefficients

(6) $$b_n = \frac{2}{\pi} \int_0^{\pi} f(x) \sin nx \, dx \qquad (n = 1, 2, \dots)$$

in the Fourier sine series for $f(x)$ on the interval $0 < x < \pi$.

For each value of n, equations (4) and (5) make up an initial value problem in ordinary differential equations. To solve the linear differential equation (4), we observe that an integrating factor is[†]

$$\exp \int n^2 k \, dt = \exp n^2 kt.$$

Multiplication through equation (4) by this integrating factor puts it in the form

$$\frac{d}{dt}\left[e^{n^2 kt} B_n(t) \right] = \frac{2\left[1 - (-1)^n\right]}{n\pi} e^{n^2 kt} q(t),$$

where the left-hand side is an exact derivative. If we replace the variable t here by τ and integrate each side from $\tau = 0$ to $\tau = t$, we find that

$$\left[e^{n^2 k\tau} B_n(\tau) \right]_0^t = \frac{2\left[1 - (-1)^n\right]}{n\pi} \int_0^t e^{n^2 k\tau} q(\tau) \, d\tau.$$

In view of condition (5), then,

(7) $$B_n(t) = b_n e^{-n^2 kt} + \frac{2\left[1 - (-1)^n\right]}{n\pi} \int_0^t e^{-n^2 k(t-\tau)} q(\tau) \, d\tau.$$

Finally, by substituting this expression for $B_n(t)$ into series (3), we arrive at the formal solution of our boundary value problem:

(8) $$u(x, t) = \sum_{n=1}^{\infty} b_n e^{-n^2 kt} \sin nx$$

$$+ \frac{4}{\pi} \sum_{n=1}^{\infty} \frac{\sin(2n-1)x}{2n-1} \int_0^t e^{-(2n-1)^2 k(t-\tau)} q(\tau) \, d\tau.$$

[†] The reader will recall that any linear first-order equation $y' + p(t)y = g(t)$ has an integrating factor of the form $\exp \int p(t) \, dt$. See, for instance, the book by Rainville and Bedient (1989, chap. 2) that is listed in the Bibliography.

Observe that the first of these series represents the solution of the boundary value problem in Example 1, Sec. 32, where $q(t) \equiv 0$.

To illustrate how interesting special cases of solution (8) are readily obtained, suppose now that $f(x) \equiv 0$ in the third of conditions (2) and that $q(t)$ is the constant function $q(t) = q_0$. Since $b_n = 0$ $(n = 1, 2, \ldots)$ and

$$\int_0^t e^{-(2n-1)^2 k(t-\tau)} q_0 \, d\tau = \frac{q_0}{k} \cdot \frac{1 - \exp\left[-(2n-1)^2 kt\right]}{(2n-1)^2},$$

solution (8) reduces to[†]

$$(9) \qquad u(x,t) = \frac{4q_0}{\pi k} \sum_{n=1}^{\infty} \frac{1 - \exp\left[-(2n-1)^2 kt\right]}{(2n-1)^3} \sin(2n-1)x.$$

In view of the Fourier sine series representation (Problem 5, Sec. 14)

$$x(\pi - x) = \frac{8}{\pi} \sum_{n=1}^{\infty} \frac{\sin(2n-1)x}{(2n-1)^3} \qquad (0 < x < \pi),$$

solution (9) can also be written

$$(10) \quad u(x,t) = \frac{q_0}{2k} x(\pi - x) - \frac{4q_0}{\pi k} \sum_{n=1}^{\infty} \frac{\exp\left[-(2n-1)^2 kt\right]}{(2n-1)^3} \sin(2n-1)x.$$

(See remarks at the end of Example 2, Sec. 32.)

PROBLEMS

1. The boundary value problem

$$u_t(x,t) = u_{xx}(x,t) + xp(t) \qquad (0 < x < 1, t > 0),$$
$$u(0,t) = 0, \qquad u(1,t) = 0, \qquad u(x,0) = 0$$

describes temperatures in an internally heated slab, where the units for t are chosen so that the thermal conductivity k of the material can be taken as unity (compare Problem 10, Sec. 4). Solve this problem by recalling [Problem 5(a), Sec. 21] the expansion

$$x = \frac{2}{\pi} \sum_{n=1}^{\infty} \frac{(-1)^{n+1}}{n} \sin n\pi x \qquad (0 < x < 1)$$

and using the method of variation of parameters.

$$\textit{Answer: } u(x,t) = \frac{2}{\pi} \sum_{n=1}^{\infty} \frac{(-1)^{n+1}}{n} \sin n\pi x \int_0^t e^{-n^2\pi^2(t-\tau)} p(\tau) \, d\tau.$$

[†] This result occurs, for example, in the theory of gluing wood with the aid of radio-frequency heating. See G. H. Brown, *Proc. Inst. Radio Engrs.*, vol. 31, no. 10, pp. 537–548, 1943, where operational methods are used.

2. Show that when the function $p(t)$ in Problem 1 is the constant function $p(t) = a$, the solution obtained there reduces to

$$u(x,t) = \frac{2a}{\pi^3} \sum_{n=1}^{\infty} (-1)^{n+1} \frac{1 - \exp(-n^2\pi^2 t)}{n^3} \sin n\pi x.$$

3. Let $u(x, t)$ denote temperatures in a slab $0 \leqslant x \leqslant 1$ that is initially at temperature zero throughout and whose faces are at temperatures

$$u(0, t) = 0 \quad \text{and} \quad u(1, t) = F(t),$$

where $F(t)$ and $F'(t)$ are continuous when $t \geqslant 0$ and where $F(0) = 0$. The unit of time is chosen so that the one-dimensional heat equation has the form $u_t(x, t) = u_{xx}(x, t)$. Write

$$u(x,t) = U(x,t) + xF(t),$$

and observe how it follows from the stated conditions on the faces of the slab that

$$U(0, t) = 0 \quad \text{and} \quad U(1, t) = 0.$$

Transform the remaining conditions on $u(x, t)$ into conditions on $U(x, t)$, and then refer to the solution found in Problem 1 to show that

$$u(x,t) = xF(t) + \frac{2}{\pi} \sum_{n=1}^{\infty} \frac{(-1)^n}{n} \sin n\pi x \int_0^t e^{-n^2\pi^2(t-\tau)} F'(\tau)\, d\tau.$$

4. Show that when $F(t) = At$, where A is a constant, the expression for $u(x, t)$ derived in Problem 3 becomes

$$u(x,t) = A\left[xt + \frac{2}{\pi^3} \sum_{n=1}^{\infty} (-1)^n \frac{1 - \exp(-n^2\pi^2 t)}{n^3} \sin n\pi x \right].$$

5. By writing

$$u(x,t) = \frac{A_0(t)}{2} + \sum_{n=1}^{\infty} A_n(t) \cos \frac{n\pi x}{c}$$

and recalling [Problem 5(b), Sec. 21] that

$$x^2 = \frac{c^2}{3} + \frac{4c^2}{\pi^2} \sum_{n=1}^{\infty} \frac{(-1)^n}{n^2} \cos \frac{n\pi x}{c} \qquad (0 < x < c),$$

solve the following temperature problem for a slab $0 \leqslant x \leqslant c$ with insulated faces:

$$u_t(x,t) = k u_{xx}(x,t) + ax^2 \qquad (0 < x < c, t > 0),$$

$$u_x(0, t) = 0, \qquad u_x(c, t) = 0, \qquad u(x,0) = 0,$$

where a is a constant. Thus show that

$$u(x,t) = ac^2\left\{ \frac{t}{3} + \frac{4c^2}{\pi^4 k} \sum_{n=1}^{\infty} \frac{(-1)^n}{n^4} \left[1 - \exp\left(-\frac{n^2\pi^2 k}{c^2} t \right) \right] \cos \frac{n\pi x}{c} \right\}.$$

6. A bar, with its lateral surface insulated, is initially at temperature zero, and its ends $x = 0$ and $x = c$ are kept at that temperature. Because of internally generated heat, the temperatures in the bar satisfy the differential equation

$$u_t(x,t) = k u_{xx}(x,t) + q(x,t) \qquad (0 < x < c, t > 0).$$

Use the method of variation of parameters to derive the temperature formula

$$u(x,t) = \frac{2}{c} \sum_{n=1}^{\infty} I_n(t) \sin \frac{n\pi x}{c},$$

where $I_n(t)$ denotes the iterated integrals

$$I_n(t) = \int_0^t \exp\left[-\frac{n^2\pi^2 k}{c^2}(t - \tau) \right] \int_0^c q(x,\tau) \sin \frac{n\pi x}{c} \, dx \, d\tau \qquad (n = 1, 2, \ldots).$$

Suggestion: Write

$$q(x,t) = \sum_{n=1}^{\infty} b_n(t) \sin \frac{n\pi x}{c}, \qquad \text{where} \qquad b_n(t) = \frac{2}{c} \int_0^c q(x,t) \sin \frac{n\pi x}{c} \, dx.$$

7. Use the method of variation of parameters to solve the temperature problem

$$u_t(x,t) = u_{xx}(x,t) - b(t)u(x,t) + q_0 \qquad (0 < x < \pi, t > 0),$$

$$u(0,t) = 0, \qquad u(\pi,t) = 0, \qquad u(x,0) = 0,$$

where q_0 is a constant.[†] (See Problem 7, Sec. 4.)

$$\text{Answer: } u(x,t) = \frac{4q_0}{\pi a(t)} \sum_{n=1}^{\infty} \frac{\sin(2n-1)x}{2n-1} \int_0^t e^{-(2n-1)^2(t-\tau)} a(\tau) \, d\tau,$$

where

$$a(t) = \exp \int_0^t b(\sigma) \, d\sigma.$$

8. When the term making the heat equation nonhomogeneous is a constant or a function of x only, the substitution

$$u(x,t) = U(x,t) + \Phi(x),$$

used in Example 2, Sec. 32, is often a convenient alternative to the method of variation of parameters. Use that substitution and the solution of the problem in Example 1, Sec. 32, to derive the following solution of the boundary value problem (1)–(2) in Sec. 33 when $q(t) = q_0$ there:

$$u(x,t) = \frac{q_0}{2k}x(\pi - x) + \sum_{n=1}^{\infty} b_n e^{-n^2 kt} \sin nx,$$

where

$$b_n = \frac{2}{\pi} \int_0^\pi \left[f(x) - \frac{q_0}{2k}x(\pi - x) \right] \sin nx \, dx \qquad (n = 1, 2, \ldots).$$

9. Show that when $f(x) \equiv 0$, the solution obtained in Problem 8 can be put in the form (9), Sec. 33.

10. A solid sphere $r \leqslant 1$ is initially at temperature zero, and its surface is kept at that temperature. Heat is generated at a constant uniform rate per unit volume throughout the interior of the sphere, so that the temperature function $u = u(r, t)$ satisfies

---

[†] In finding an integrating factor for the ordinary differential equation that arises, it is useful to note that $\int_0^t b(\sigma) \, d\sigma$ is an antiderivative of $b(t)$.

the nonhomogeneous heat equation

$$\frac{\partial u}{\partial t} = \frac{k}{r}\frac{\partial^2}{\partial r^2}(ru) + q_0 \qquad (0 < r < 1, t > 0),$$

where q_0 is a positive constant. Make the substitution

$$u(r,t) = U(r,t) + \Phi(r)$$

in the temperature problem for this sphere, where U and Φ are to be continuous when $r = 0$. [Note that this continuity condition implies that $r\Phi(r)$ tends to zero as r tends to zero.] Then refer to the solution derived in Problem 6, Sec. 32, to write the solution of a new boundary value problem for $U(r,t)$ and thus show that

$$u(r,t) = \frac{q_0}{kr}\left[\frac{1}{6}r(1-r^2) + \frac{2}{\pi^3}\sum_{n=1}^{\infty}\frac{(-1)^n}{n^3}e^{-n^2\pi^2kt}\sin n\pi r\right].$$

Suggestion: It is useful to note that, in view of the formula for the coefficients in a Fourier sine series, the values of certain integrals that arise are, except for a constant factor, the coefficients in the series [Problem 7(a), Sec. 21]

$$x(1-x^2) = \frac{12}{\pi^3}\sum_{n=1}^{\infty}\frac{(-1)^{n+1}}{n^3}\sin n\pi x \qquad (0 < x < 1).$$

34.   DIRICHLET PROBLEMS

As already noted in Sec. 7, a boundary value problem in u is said to be a Dirichlet problem when it consists of Laplace's equation $\nabla^2 u = 0$, which states that u is harmonic in a given domain, together with prescribed values of u on the boundary of that domain. We now illustrate the use of the Fourier method in solving such problems for certain domains in the plane.

EXAMPLE 1. Let u be harmonic in the interior of a rectangular region $0 \leqslant x \leqslant a, 0 \leqslant y \leqslant b$, so that

(1) $u_{xx}(x,y) + u_{yy}(x,y) = 0$ $(0 < x < a, 0 < y < b)$.

These values are prescribed on the boundary (Fig. 37):

(2) $u(0,y) = 0,$ $u(a,y) = 0$ $(0 < y < b),$

(3) $u(x,0) = f(x),$ $u(x,b) = 0$ $(0 < x < a).$

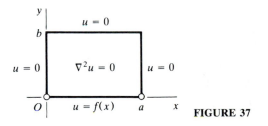

FIGURE 37

Separation of variables, with $u = X(x)Y(y)$, leads to the Sturm-Liouville problem

(4) $X''(x) + \lambda X(x) = 0,$ $X(0) = 0,$ $X(a) = 0,$

whose eigenvalues and eigenfunctions are (Sec. 29)

$$\lambda_n = \left(\frac{n\pi}{a}\right)^2, \qquad X_n(x) = \sin\frac{n\pi x}{a} \qquad (n = 1, 2, \dots),$$

and to the conditions

(5) $Y''(y) - \lambda Y(y) = 0,$ $Y(b) = 0.$

When λ is a particular eigenvalue λ_n of the Sturm-Liouville problem (4), the function $Y_n(y)$ satisfying conditions (5) is found to be

$$Y_n(y) = C_1\left[\exp\frac{n\pi y}{a} - \exp\frac{n\pi(2b - y)}{a}\right],$$

where C_1 denotes an arbitrary nonzero constant. Instead of setting $C_1 = 1$, as we have always done in such cases, let us write

$$C_1 = -\frac{1}{2}\exp\left(-\frac{n\pi b}{a}\right).$$

Then $Y_n(y)$ takes the compact form

$$Y_n(y) = \sinh\frac{n\pi(b - y)}{a}.$$

Thus the function

(6) $$u(x, y) = \sum_{n=1}^{\infty} B_n \sinh\frac{n\pi(b - y)}{a} \sin\frac{n\pi x}{a}$$

formally satisfies all the conditions (1) through (3), provided that

(7) $$f(x) = \sum_{n=1}^{\infty} B_n \sinh\frac{n\pi b}{a} \sin\frac{n\pi x}{a} \qquad (0 < x < a).$$

We assume that f is piecewise smooth. Then series (7) is the Fourier sine series representation of $f(x)$ on the interval $0 < x < a$ if $B_n \sinh(n\pi b/a) = b_n$, where

$$b_n = \frac{2}{a}\int_0^a f(x)\sin\frac{n\pi x}{a}\,dx \qquad (n = 1, 2, \dots).$$

The function defined by equation (6), with coefficients

(8) $$B_n = \frac{2}{a\sinh(n\pi b/a)}\int_0^a f(x)\sin\frac{n\pi x}{a}\,dx \qquad (n = 1, 2, \dots),$$

is, therefore, our formal solution.

If y is replaced by the new variable $b - y$ in the problem above, as well as its solution, and if $f(x)$ is replaced by $g(x)$, the nonhomogeneous condition satisfied by u becomes $u(x, b) = g(x)$. An interchange of x and y then places nonhómogeneous conditions on the edge $x = 0$ or $x = a$. Superposition of the four solutions thus gives the harmonic function whose values are prescribed as functions of position along the entire boundary of the rectangular domain, except for the corners.

From equations (1) through (3), we note that $u(x, y)$ represents the *steady-state temperatures* in a rectangular plate, with insulated faces, when $u = f(x)$ on the edge $y = 0$ and $u = 0$ on the other three edges. The function u also represents the *electrostatic potential* in a space formed by the planes $x = 0$, $x = a$, $y = 0$, and $y = b$ when the space is free of charges and the planar surfaces are kept at potentials given by conditions (2) and (3).

EXAMPLE 2. Let $u(\rho, \phi)$ denote a function of the cylindrical, or polar, coordinates ρ and ϕ that is harmonic in the domain $1 < \rho < b$, $0 < \phi < \pi$ of the plane $z = 0$ (Fig. 38). Thus (Sec. 4)

$$(9) \quad \rho^2 u_{\rho\rho}(\rho, \phi) + \rho u_\rho(\rho, \phi) + u_{\phi\phi}(\rho, \phi) = 0 \qquad (1 < \rho < b, 0 < \phi < \pi).$$

Suppose further that

$$(10) \qquad\qquad u(\rho, 0) = 0, \qquad u(\rho, \pi) = 0 \qquad\qquad (1 < \rho < b),$$

$$(11) \qquad\qquad u(1, \phi) = 0, \qquad u(b, \phi) = u_0 \qquad\qquad (0 < \phi < \pi),$$

where u_0 is a constant.

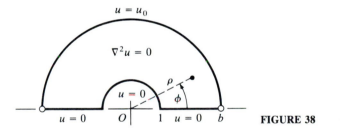

FIGURE 38

Substituting $u = R(\rho)\Phi(\phi)$ into the homogeneous conditions and separating variables, we find that

$$(12) \qquad\qquad \rho^2 R''(\rho) + \rho R'(\rho) - \lambda R(\rho) = 0, \qquad R(1) = 0$$

and

$$(13) \qquad\qquad \Phi''(\phi) + \lambda\Phi(\phi) = 0, \qquad \Phi(0) = 0, \qquad \Phi(\pi) = 0.$$

Except for notation, the problem in Φ is the Sturm-Liouville problem in Sec. 29, with $c = \pi$, whose eigenvalues and eigenfunctions are

$$\lambda_n = n^2, \qquad \Phi_n(\phi) = \sin n\phi \qquad\qquad (n = 1, 2, \ldots).$$

The corresponding functions $R_n(\rho)$ are determined by solving the differential equation

$$\rho^2 R''(\rho) + \rho R'(\rho) - n^2 R(\rho) = 0 \qquad (1 < \rho < b),$$

where $R(1) = 0$. This is a Cauchy-Euler equation (see Problem 3, Sec. 35), and the substitution $\rho = \exp s$ transforms it into the differential equation

$$\frac{d^2 R}{ds^2} - n^2 R = 0.$$

Hence

$$R = C_1 e^{ns} + C_2 e^{-ns} = C_1 \rho^n + C_2 \rho^{-n}.$$

Because $R(1) = 0$, it follows that, except for constant factors, the desired functions of ρ are

$$R_n(\rho) = \rho^n - \rho^{-n} \qquad (n = 1, 2, \dots).$$

Thus, formally,

$$u(\rho, \phi) = \sum_{n=1}^{\infty} B_n(\rho^n - \rho^{-n}) \sin n\phi,$$

where, according to the second of conditions (11), the constants B_n are such that

$$u_0 = \sum_{n=1}^{\infty} B_n(b^n - b^{-n}) \sin n\phi \qquad (0 < \phi < \pi).$$

Since this is in the form of a Fourier sine series representation for the constant function u_0 on the interval $0 < \phi < \pi$,

$$B_n(b^n - b^{-n}) = \frac{2}{\pi} \int_0^\pi u_0 \sin n\phi \, d\phi = \frac{2u_0}{\pi} \cdot \frac{1 - (-1)^n}{n} \qquad (n = 1, 2, \dots).$$

The complete solution of our Dirichlet problem is, therefore,

$$u(\rho, \phi) = \frac{2u_0}{\pi} \sum_{n=1}^{\infty} \frac{\rho^n - \rho^{-n}}{b^n - b^{-n}} \cdot \frac{1 - (-1)^n}{n} \sin n\phi,$$

or

$$(14) \qquad u(\rho, \phi) = \frac{4u_0}{\pi} \sum_{n=1}^{\infty} \frac{\rho^{2n-1} - \rho^{-(2n-1)}}{b^{2n-1} - b^{-(2n-1)}} \cdot \frac{\sin (2n-1)\phi}{2n-1}.$$

35. OTHER TYPES OF BOUNDARY CONDITIONS

Boundary value problems consisting of Laplace's equation $\nabla^2 u = 0$ and boundary conditions not all of which are of the Dirichlet type are also important in applications. In the following example, values of a derivative of the function u, rather than values of u itself, are prescribed along a portion of the boundary of the domain in which u is harmonic.

EXAMPLE. Using cylindrical coordinates, let us derive an expression for the steady temperatures $u = u(\rho, \phi)$ in a long rod, with uniform semicircular cross section and occupying the region $0 \leqslant \rho \leqslant a$, $0 \leqslant \phi \leqslant \pi$, which is insulated on its planar surface and maintained at temperatures $f(\phi)$ on the semicircular part (Fig. 39).

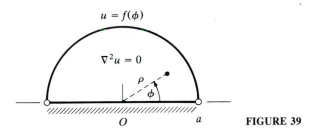

$$u = f(\phi)$$

$$\nabla^2 u = 0$$

$$\rho$$

$$\phi$$

$$O \qquad\qquad a \qquad\qquad \textbf{FIGURE 39}$$

As in Example 2, Sec. 34, $u(\rho, \phi)$ satisfies Laplace's equation

$$(1) \qquad \rho^2 u_{\rho\rho}(\rho, \phi) + \rho u_\rho(\rho, \phi) + u_{\phi\phi}(\rho, \phi) = 0,$$

but now in the domain $0 < \rho < a$, $0 < \phi < \pi$. It also satisfies the homogeneous conditions [see Problem 12(b), Sec. 4]

$$(2) \qquad u_\phi(\rho, 0) = 0, \qquad u_\phi(\rho, \pi) = 0 \qquad\qquad (0 < \rho < a),$$

as well as the nonhomogeneous one

$$(3) \qquad u(a, \phi) = f(\phi) \qquad\qquad (0 < \phi < \pi).$$

The function f is understood to be piecewise smooth and, therefore, bounded. We assume further that $|u(\rho, \phi)| \leqslant M$, where M denotes some positive constant. The need for such a boundedness condition is physically evident and has been only tacitly assumed in earlier problems. Here it serves as a condition at the origin, which may be thought of as the limiting case of a smaller semicircle (compare Fig. 38) as its radius tends to zero.

The substitution $u = R(\rho)\Phi(\phi)$ in the homogeneous conditions (1) and (2) leads to the condition

$$(4) \qquad \rho^2 R''(\rho) + \rho R'(\rho) - \lambda R(\rho) = 0 \qquad\qquad (0 < \rho < a)$$

on $R(\rho)$ and to the Sturm-Liouville problem

$$(5) \qquad \Phi''(\phi) + \lambda\Phi(\phi) = 0, \qquad \Phi'(0) = 0, \qquad \Phi'(\pi) = 0,$$

whose eigenvalues and eigenfunctions are

$$\lambda_0 = 0, \qquad \lambda_n = n^2 \qquad\qquad (n = 1, 2, \dots)$$

and

$$\Phi_0(\phi) = \frac{1}{2}, \qquad \Phi_n(\phi) = \cos n\phi \qquad (n = 1, 2, \dots),$$

according to Sec. 27. The function $R_0(\rho)$, corresponding to $\Phi_0(\phi)$, satisfies the Cauchy-Euler equation

$$\rho^2 R''(\rho) + \rho R'(\rho) - 0R(\rho) = 0;$$

and the general solution is readily found to be $R = A \ln \rho + B$, where A and B are constants. But, since the product $R(\rho)\Phi(\phi)$ is expected to be bounded in the domain $0 < \rho < a$, $0 < \phi < \pi$ and since $\ln \rho$ tends to $-\infty$ as ρ tends to zero through positive values, the constant A must be zero. So, except for a constant factor, $R_0(\rho) = 1$. Similarly, when $\lambda = n^2$, where n is any fixed positive integer, our boundedness condition requires that the constant C_2 in the general solution $R = C_1\rho^n + C_2\rho^{-n}$ of equation (4) be zero. Hence we may write $R_n(\rho) = \rho^n$ $(n = 1, 2, \dots)$, and the homogeneous conditions (1) and (2) are formally satisfied by the function

$$(6) \qquad u(\rho, \phi) = \frac{A_0}{2} + \sum_{n=1}^{\infty} A_n \rho^n \cos n\phi,$$

where the constants A_n $(n = 0, 1, 2, \dots)$ are yet to be determined.

In view of the nonhomogeneous condition (3),

$$f(\phi) = \frac{A_0}{2} + \sum_{n=1}^{\infty} A_n a^n \cos n\phi \qquad (0 < \phi < \pi).$$

Consequently, $A_0 = a_0$ and $A_n a^n = a_n$ $(n = 0, 1, 2, \dots)$, where

$$(7) \qquad a_n = \frac{2}{\pi} \int_0^\pi f(\phi) \cos n\phi \, d\phi \qquad (n = 0, 1, 2, \dots).$$

The complete solution of our boundary value problem is, then,

$$(8) \qquad u(\rho, \phi) = \frac{a_0}{2} + \sum_{n=1}^{\infty} a_n \left(\frac{\rho}{a}\right)^n \cos n\phi,$$

with coefficients (7). This solution can, of course, be alternatively written as

$$(9) \quad u(\rho, \phi) = \frac{1}{\pi} \int_0^\pi f(\psi) \, d\psi + \frac{2}{\pi} \sum_{n=1}^{\infty} \left(\frac{\rho}{a}\right)^n \cos n\phi \int_0^\pi f(\psi) \cos n\psi \, d\psi,$$

where the variable of integration ψ is to be distinguished from the free variable ϕ.

PROBLEMS

1. A square plate has its faces and its edges $x = 0$ and $x = \pi$ $(0 < y < \pi)$ insulated. Its edges $y = 0$ and $y = \pi$ $(0 < x < \pi)$ are kept at temperatures zero and $f(x)$, respectively. Let $u(x, y)$ denote its steady temperatures. Derive the expression

$$u(x, y) = \frac{a_0}{2\pi}y + \sum_{n=1}^{\infty} a_n \frac{\sinh ny}{\sinh n\pi} \cos nx,$$

where

$$a_n = \frac{2}{\pi} \int_0^{\pi} f(x) \cos nx \, dx \qquad (n = 0, 1, 2, \dots).$$

Find $u(x, y)$ when $f(x) = u_0$, where u_0 is a constant.

2. One edge of a square plate with insulated faces is kept at a uniform temperature u_0, and the other three edges are kept at temperature zero. Without solving a boundary value problem, but by superposition of solutions of like problems to obtain the trivial case in which all four edges are at temperature u_0, show why the steady temperature at the center of the given plate must be $u_0/4$.

3. If A, B, and C are constants, the differential equation

$$Ax^2y'' + Bxy' + Cy = 0$$

is called a *Cauchy-Euler equation*. Show that, with the substitution $x = \exp s$, it can be transformed into the constant-coefficient differential equation

$$A\frac{d^2y}{ds^2} + (B - A)\frac{dy}{ds} + Cy = 0.$$

4. Let ρ, ϕ, z be cylindrical coordinates. Find the harmonic function $u(\rho, \phi)$ in the domain $1 < \rho < b, 0 < \phi < \pi/2$ of the plane $z = 0$ when $u = 0$ and $u = f(\phi)$ on the arcs $\rho = 1$ and $\rho = b$ $(0 < \phi < \pi/2)$, respectively, and $u_\phi(\rho, 0) = u_\phi(\rho, \pi/2) = 0$ $(1 < \rho < b)$. Give a physical interpretation of this problem.

$$\text{Answer: } u(\rho, \phi) = \frac{a_0}{2} \cdot \frac{\ln \rho}{\ln b} + \sum_{n=1}^{\infty} a_n \frac{\rho^{2n} - \rho^{-2n}}{b^{2n} - b^{-2n}} \cos 2n\phi,$$

where

$$a_n = \frac{4}{\pi} \int_0^{\pi/2} f(\phi) \cos 2n\phi \, d\phi \qquad (n = 0, 1, 2, \dots).$$

5. Let the faces of a plate in the shape of a wedge $0 \leqslant \rho \leqslant a, 0 \leqslant \phi \leqslant \alpha$ be insulated. Find the steady temperatures $u(\rho, \phi)$ in the plate when $u = 0$ on the two rays $\phi = 0$, $\phi = \alpha$ $(0 < \rho < a)$ and $u = f(\phi)$ on the arc $\rho = a$ $(0 < \phi < \alpha)$. Assume that f is piecewise smooth and that u is bounded.

$$\text{Answer: } u(\rho, \phi) = \frac{2}{\alpha} \sum_{n=1}^{\infty} \left(\frac{\rho}{a}\right)^{n\pi/\alpha} \sin \frac{n\pi\phi}{\alpha} \int_0^{\alpha} f(\psi) \sin \frac{n\pi\psi}{\alpha} \, d\psi.$$

6. The faces and edge $y = 0$ $(0 < x < \pi)$ of a rectangular plate $0 \leqslant x \leqslant \pi, 0 \leqslant y \leqslant y_0$ are insulated. The other three edges are maintained at the temperatures indicated in

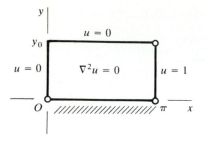

FIGURE 40

Fig. 40. By making the substitution $u(x, y) = U(x, y) + \Phi(x)$ in the boundary value problem for the steady temperatures $u(x, y)$ in the plate and using the method described in Example 2, Sec. 32, derive the temperature formula

$$ u(x, y) = \frac{1}{\pi}\left[x + 2 \sum_{n=1}^{\infty} \frac{(-1)^n}{n} \cdot \frac{\cosh ny}{\cosh ny_0} \sin nx \right]. $$

Suggestion: The series representation (Example 1, Sec. 14)

$$ x = 2 \sum_{n=1}^{\infty} \frac{(-1)^{n+1}}{n} \sin nx \qquad (0 < x < \pi) $$

is useful in finding $U(x, y)$.

7. Let $u(x, y)$ denote the bounded steady temperatures in the semi-infinite plate $x \geq 0$, $0 \leq y \leq \pi$, whose faces are insulated, when the edges are kept at the temperatures shown in Fig. 41. (The boundedness condition serves as a condition at the missing right-hand end of the plate.) Assuming that the function f is piecewise smooth, derive the temperature formula

$$ u(x, y) = \sum_{n=1}^{\infty} b_n e^{-nx} \sin ny, $$

where

$$ b_n = \frac{2}{\pi} \int_0^{\pi} f(y) \sin ny \, dy \qquad (n = 1, 2, \ldots). $$

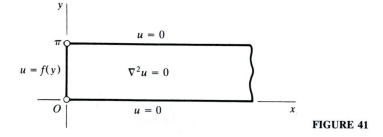

FIGURE 41

8. Suppose that in the plate described in Problem 7 there is a heat source depending on the variable y and that the entire boundary is kept at temperature zero. According to Sec. 3, the steady temperatures $u(x, y)$ in the plate must now satisfy *Poisson's equation*

$$u_{xx}(x, y) + u_{yy}(x, y) + q(y) = 0 \qquad (x > 0, 0 < y < \pi).$$

(*a*) By assuming a (bounded) solution of the form

$$u(x, y) = \sum_{n=1}^{\infty} B_n(x) \sin ny$$

of this temperature problem and using the method of variation of parameters (Sec. 33), show formally that

$$B_n(x) = \frac{q_n}{n^2}(1 - e^{-nx}) \qquad (n = 1, 2, \dots),$$

where q_n are the coefficients in the Fourier sine series for $q(y)$ on the interval $0 < y < \pi$.

(*b*) Show that when $q(y)$ is the constant function $q(y) = q_0$, the solution in part (*a*) becomes

$$u(x, y) = \frac{4q_0}{\pi} \sum_{n=1}^{\infty} \frac{1 - \exp[-(2n-1)x]}{(2n-1)^3} \sin(2n-1)y.$$

Suggestion: In part (*a*), recall that the general solution of a linear second-order equation $y'' + p(x)y = g(x)$ is of the form $y = y_c + y_p$, where y_p is any particular solution and y_c is the general solution of the complementary equation

$$y'' + p(x)y = 0.^{†}$$

9. Derive an expression for the bounded steady temperatures $u(x, y)$ in a semi-infinite slab $0 \leqslant x \leqslant c$, $y \geqslant 0$ whose faces in the planes $x = 0$ and $x = c$ are insulated and where $u(x, 0) = f(x)$. Assume that f is piecewise smooth on the interval $0 < x < c$.

36.   A STRING WITH PRESCRIBED INITIAL VELOCITY

When, initially, the string in Sec. 29 has some prescribed distribution of velocities $g(x)$ parallel to the y axis in its position of equilibrium $y = 0$, the boundary value problem for the displacements $y(x, t)$ becomes

(1) $$y_{tt}(x, t) = a^2 y_{xx}(x, t) \qquad (0 < x < c, t > 0),$$

(2) $$y(0, t) = 0, \qquad y(c, t) = 0,$$

(3) $$y(x, 0) = 0, \qquad y_t(x, 0) = g(x).$$

---

† See, for instance, the book by Boyce and DiPrima (1992, sec. 3.6), listed in the Bibliography.

If the xy plane, with the string lying on the x axis, is moving parallel to the y axis and is brought to rest at the instant $t = 0$, the function $g(x)$ is a constant. The hammer action in a piano may produce approximately a uniform initial velocity over a short span of a piano wire, in which case $g(x)$ may be considered to be a step function.

As in Sec. 29, we seek functions of the type $y = X(x)T(t)$ that satisfy all the homogeneous conditions in the boundary value problem. The Sturm-Liouville problem that arises is the same as the one in Sec. 29:

$$X''(x) + \lambda X(x) = 0, \qquad X(0) = 0, \qquad X(c) = 0.$$

We recall that the eigenvalues are $\lambda_n = (n\pi/c)^2$ $(n = 1, 2, \dots)$, with eigenfunctions $X_n(x) = \sin(n\pi x/c)$. Since the conditions on $T(t)$ are

$$T''(t) + \lambda a^2 T(t) = 0, \qquad T(0) = 0,$$

the corresponding functions of t are $T_n(t) = \sin(n\pi at/c)$.

The homogeneous conditions in the boundary value problem are, then, formally satisfied by the function

$$y(x,t) = \sum_{n=1}^{\infty} B_n \sin \frac{n\pi x}{c} \sin \frac{n\pi at}{c},$$

where the constants B_n are to be determined from the second of conditions (3):

(4) $$\sum_{n=1}^{\infty} B_n \frac{n\pi a}{c} \sin \frac{n\pi x}{c} = g(x) \qquad (0 < x < c).$$

Under the assumption that g is piecewise smooth, the series in equation (4) is the Fourier sine series representing $g(x)$ on the interval $0 < x < c$ if $B_n(n\pi a/c) = b_n$, where

(5) $$b_n = \frac{2}{c} \int_0^c g(x) \sin \frac{n\pi x}{c} \, dx.$$

Thus $B_n = (c/n\pi a)b_n$, and

(6) $$y(x,t) = \frac{c}{\pi a} \sum_{n=1}^{\infty} \frac{b_n}{n} \sin \frac{n\pi x}{c} \sin \frac{n\pi at}{c}.$$

We can sum the series here by first writing

$$y_t(x,t) = \sum_{n=1}^{\infty} b_n \sin \frac{n\pi x}{c} \cos \frac{n\pi at}{c} = \frac{1}{2}[G(x + at) + G(x - at)],$$

where G is the odd periodic extension, with period $2c$, of the given function g

(compare Sec. 30). Then, since $y(x,0) = 0$,[†]

$$y(x,t) = \frac{1}{2}\left[\int_0^t G(x+a\tau)\,d\tau + \int_0^t G(x-a\tau)\,d\tau\right]$$

$$= \frac{1}{2a}\left[\int_x^{x+at} G(s)\,ds - \int_x^{x-at} G(s)\,ds\right];$$

and, in terms of the periodic function

(7) $$I(x) = \int_0^x G(s)\,ds \qquad\qquad (-\infty < x < \infty),$$

(8) $$y(x,t) = \frac{1}{2a}[I(x+at) - I(x-at)].$$

If points on the string are given both nonzero initial displacements and nonzero initial velocities, so that

(9) $$y(x,0) = f(x) \qquad \text{and} \qquad y_t(x,0) = g(x),$$

the displacements $y(x,t)$ can be written as a superposition of solution (10), Sec. 30, and solution (8) above:

(10) $$y(x,t) = \frac{1}{2}[F(x+at) + F(x-at)] + \frac{1}{2a}[I(x+at) - I(x-at)].$$

Note that both terms satisfy the homogeneous conditions (1) and (2), while their sum clearly satisfies the nonhomogeneous conditions (9). (Compare Problem 5, Sec. 9.)

In general, the solution of a linear problem containing more than one nonhomogeneous condition can be written as a sum of solutions of problems each of which contains only one nonhomogeneous condition. The resolution of the original problem in this way, though not an essential step, often simplifies the process of solving it.

37. AN ELASTIC BAR

A cylindrical bar of natural length c is initially stretched by an amount bc (Fig. 42) and is at rest. The initial longitudinal displacements of its sections are then proportional to the distance from the fixed end $x = 0$. At the instant $t = 0$, both ends are released and left free. The longitudinal displacements

---

[†] See also the footnote with Problem 7, Sec. 33, regarding antiderivatives.

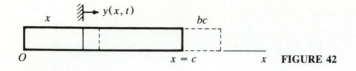

FIGURE 42

$y(x, t)$ satisfy the following boundary value problem, where $a^2 = E/\delta$ (Sec. 6):

$$(1) \qquad\qquad y_{tt}(x, t) = a^2 y_{xx}(x, t) \qquad (0 < x < c, t > 0),$$

$$(2) \qquad\qquad y_x(0, t) = 0, \qquad y_x(c, t) = 0,$$

$$(3) \qquad\qquad y(x, 0) = bx, \qquad y_t(x, 0) = 0.$$

The homogeneous two-point boundary conditions (2) state that the force per unit area on the end sections is zero.

The function $y(x, t)$ can also be interpreted as representing transverse displacements in a stretched string, released at rest from the position $y(x, 0) = bx$, when the ends are looped around perfectly smooth rods lying along the lines $x = 0$ and $x = c$. In that case, $a^2 = H/\delta$; and the boundary conditions (2) state that no forces act in the y direction at the ends of the string (see Sec. 5).

Functions $y = X(x)T(t)$ satisfy all the homogeneous conditions above when $X(x)$ is an eigenfunction of the problem

$$(4) \qquad X''(x) + \lambda X(x) = 0, \qquad X'(0) = 0, \qquad X'(c) = 0$$

and when, for the same eigenvalue λ,

$$(5) \qquad\qquad T''(t) + \lambda a^2 T(t) = 0, \qquad T'(0) = 0.$$

The eigenvalues are (Sec. 27) $\lambda_0 = 0$ and $\lambda_n = (n\pi/c)^2$ $(n = 1, 2, \ldots)$, with eigenfunctions $X_0(x) = \frac{1}{2}$ and $X_n(x) = \cos(n\pi x/c)$. The corresponding functions of t are $T_0(t) = 1$ and $T_n(t) = \cos(n\pi at/c)$.

Formally, then, the generalized linear combination

$$(6) \qquad y(x, t) = \frac{a_0}{2} + \sum_{n=1}^{\infty} a_n \cos\frac{n\pi x}{c} \cos\frac{n\pi at}{c}$$

satisfies conditions (1) through (3), provided that

$$(7) \qquad\qquad bx = \frac{a_0}{2} + \sum_{n=1}^{\infty} a_n \cos\frac{n\pi x}{c} \qquad (0 < x < c).$$

The function bx is such that it is represented by the Fourier cosine series (6) on the interval $0 \leqslant x \leqslant c$, where

$$a_n = \frac{2b}{c} \int_0^c x \cos\frac{n\pi x}{c} \, dx \qquad (n = 0, 2, \ldots).$$

Consequently,

$$(8) \qquad\qquad a_0 = bc, \qquad a_n = -\frac{2bc}{\pi^2} \cdot \frac{1 - (-1)^n}{n^2} \qquad (n = 1, 2, \ldots);$$

and we arrive at the solution

$$(9) \quad y(x,t) = \frac{bc}{2} - \frac{4bc}{\pi^2} \sum_{n=1}^{\infty} \frac{1}{(2n-1)^2} \cos \frac{(2n-1)\pi x}{c} \cos \frac{(2n-1)\pi at}{c}.$$

By a method already used in Secs. 30 and 36, we can put this series solution in closed form, involving the *even* periodic extension $P(x)$, with period $2c$, of the function bx $(0 \leqslant x \leqslant c)$. To be specific, we know from the trigonometric identity

$$2 \cos A \cos B = \cos (A + B) + \cos (A - B)$$

that

$$2 \cos \frac{n\pi x}{c} \cos \frac{n\pi at}{c} = \cos \frac{n\pi(x + at)}{c} + \cos \frac{n\pi(x - at)}{c}.$$

Hence expression (6) can be written as

$$(10) \qquad y(x,t) = \frac{1}{2} \left[\frac{a_0}{2} + \sum_{n=1}^{\infty} a_n \cos \frac{n\pi(x + at)}{c} \right.$$

$$\left. + \frac{a_0}{2} + \sum_{n=1}^{\infty} a_n \cos \frac{n\pi(x - at)}{c} \right].$$

But series (7) represents $P(x)$ for all values of x when the values (8) of the coefficients a_n $(n = 0, 1, 2, \ldots)$ are used. Hence expression (10), with those values of a_n, reduces to

$$(11) \qquad y(x,t) = \frac{1}{2} [P(x + at) + P(x - at)].$$

This is the desired closed form of solution (9).

38. RESONANCE

A stretched string, of length unity and with fixed ends, is initially at rest in its position of equilibrium. A simple periodic transverse force acts uniformly on all elements of the string, so that the transverse displacements $y(x,t)$ satisfy this modified form (see Sec. 5) of the wave equation:

$$(1) \qquad\qquad y_{tt}(x,t) = y_{xx}(x,t) + A \sin \omega t \qquad (0 < x < 1, t > 0),$$

where A is a constant. Equation (1), together with the boundary conditions

$$(2) \qquad\qquad y(0,t) = 0, \qquad y(1,t) = 0,$$

$$(3) \qquad\qquad y(x,0) = 0, \qquad y_t(x,0) = 0,$$

just described, make up a boundary value problem to which the method of variation of parameters (Sec. 33) can be applied.

We note that if the constant A were actually zero, the Sturm-Liouville problem arising would have eigenfunctions $\sin n\pi x$ $(n = 1, 2, \ldots)$. Hence we seek a solution of our boundary value problem having the form

$$(4) \qquad y(x,t) = \sum_{n=1}^{\infty} B_n(t) \sin n\pi x,$$

where the coefficients $B_n(t)$ are to be determined. Substituting series (4) into equation (1) and using the Fourier sine series representation

$$A = \sum_{n=1}^{\infty} \frac{2A\left[1 - (-1)^n\right]}{n\pi} \sin n\pi x \qquad (0 < x < 1),$$

we write

$$\sum_{n=1}^{\infty} B_n''(t) \sin n\pi x = \sum_{n=1}^{\infty} \left[-(n\pi)^2 B_n(t)\right] \sin n\pi x$$

$$+ \sum_{n=1}^{\infty} \frac{2A\left[1 - (-1)^n\right]}{n\pi} \sin \omega t \sin n\pi x.$$

Thus

$$(5) \qquad B_n''(t) + (n\pi)^2 B_n(t) = \frac{2A\left[1 - (-1)^n\right]}{n\pi} \sin \omega t \quad (n = 1, 2, \dots),$$

and conditions (3) yield the initial conditions

$$(6) \qquad B_n(0) = 0 \quad \text{and} \quad B_n'(0) = 0 \qquad (n = 1, 2, \dots)$$

on $B_n(t)$.

It is easy to see that $B_n(t) \equiv 0$ when n is even; that is, $B_{2n}(t) \equiv 0$ $(n = 1, 2, \dots)$. It remains to find $B_n(t)$ when n is odd. If we write

$$\omega_n = (2n - 1)\pi \qquad (n = 1, 2, \dots),$$

the initial value problem in ordinary differential equations for $B_{2n-1}(t)$ $(n = 1, 2, \dots)$ is

$$(7) \qquad B_{2n-1}''(t) + \omega_n^2 B_{2n-1}(t) = \frac{4A}{\omega_n} \sin \omega t,$$

$$(8) \qquad B_{2n-1}(0) = 0, \qquad B_{2n-1}'(0) = 0.$$

We may now refer to Problem 14 below, where methods learned in an introductory course in ordinary differential equations are used to solve the initial value problem consisting of the second-order equation

$$(9) \qquad y''(t) + a^2 y(t) = b \sin \omega t,$$

where a and b are constants, and the conditions

$$(10) \qquad y(0) = 0, \qquad y'(0) = 0.$$

To be specific, if $\omega \neq a$,

$$(11) \qquad y(t) = \frac{b}{\omega^2 - a^2}\left(\frac{\omega}{a}\sin at - \sin \omega t\right).$$

Thus we see that if $\omega \neq \omega_n$ for any value of n ($n = 1, 2, \ldots$), the solution of problem (7)–(8) is

$$B_{2n-1}(t) = \frac{4A}{\omega_n(\omega^2 - \omega_n^2)}\left(\frac{\omega}{\omega_n}\sin \omega_n t - \sin \omega t\right),$$

and it follows from equation (4) that

$$(12) \qquad y(x, t) = 4A \sum_{n=1}^{\infty} \frac{\sin \omega_n x}{\omega_n(\omega^2 - \omega_n^2)}\left(\frac{\omega}{\omega_n}\sin \omega_n t - \sin \omega t\right).$$

It is also shown in Problem 14 that if $\omega = a$, the solution of differential equation (9), with conditions (10), is

$$(13) \qquad y(t) = \frac{b}{2a}\left(\frac{1}{a}\sin at - t\cos at\right).$$

Hence, when there is a value N of n ($n = 1, 2, \ldots$) such that $\omega = \omega_N$,

$$(14) \qquad B_{2N-1}(t) = \frac{2A}{\omega_N^2}\left(\frac{1}{\omega_N}\sin \omega_N t - t\cos \omega_N t\right).$$

Because of the factor t with the cosine function here, this means that series (4) contains an *unstable* component. Such an unstable oscillation of sections of the string is called *resonance*. The periodic external force is evidently in resonance with the string when the frequency ω of that force coincides with any one of the resonant frequencies $\omega_n = (2n - 1)\pi$ ($n = 1, 2, \ldots$). Those frequencies depend, in general, on the physical properties of the string and the manner in which it is supported.

PROBLEMS

1. Show that, for each fixed x, the displacements y given by equation (10), Sec. 36, are periodic functions of t, with period $2c/a$.

2. Show that the motion of each cross section of the elastic bar treated in Sec. 37 is periodic in t, with period $2c/a$.

3. A string, stretched between the points 0 and π on the x axis, is initially straight with velocity $y_t(x, 0) = b\sin x$, where b is a constant. Write the boundary value problem in $y(x, t)$, solve it, and verify the solution.

$$Answer: y(x, t) = \frac{b}{a}\sin x \sin at.$$

4. Display graphically the periodic functions $G(x)$ and $I(x)$ in Sec. 36 when all points of the string there have the same initial velocity $g(x) = v_0$. Then, using expression

(8), Sec. 36, indicate some instantaneous positions of the string by means of line segments similar to those in Fig. 32 (Problem 8, Sec. 30).

5. In Sec. 36, we used the fact that

$$G(s) = \sum_{n=1}^{\infty} b_n \sin \frac{n\pi s}{c} \qquad (-\infty < s < \infty).$$

Integrate that series (Sec. 23) to show that

$$I(x) = \frac{c}{\pi} \sum_{n=1}^{\infty} \frac{b_n}{n}\left(1 - \cos \frac{n\pi x}{c}\right) \qquad (-\infty < x < \infty)$$

and hence that the function (8) in Sec. 36 is represented by the series (6) there.

6. From expression (11), Sec. 37, show that $y(0,t) = P(at)$ and hence that the end $x = 0$ of the bar moves with the constant velocity ab during the half period $0 < t < c/a$ and with velocity $-ab$ during the next half period.

7. A string, stretched between the points 0 and π on the x axis and initially at rest, is released from the position $y = f(x)$. Its motion is opposed by air resistance, which is proportional to the velocity at each point (Sec. 5). Let the unit of time be chosen so that the equation of motion becomes

$$y_{tt}(x,t) = y_{xx}(x,t) - 2\beta y_t(x,t) \qquad (0 < x < \pi, t > 0),$$

where β is a positive constant. Assuming that $0 < \beta < 1$, derive the expression

$$y(x,t) = e^{-\beta t} \sum_{n=1}^{\infty} b_n\left(\cos \alpha_n t + \frac{\beta}{\alpha_n} \sin \alpha_n t\right) \sin nx,$$

where

$$\alpha_n = \sqrt{n^2 - \beta^2}, \qquad b_n = \frac{2}{\pi}\int_0^{\pi} f(x) \sin nx\, dx \qquad (n = 1, 2, \ldots),$$

for the transverse displacements.

8. Suppose that the string in Problem 7 is initially straight with a uniform velocity in the direction of the y axis, as if a moving frame supporting the end points is brought to rest at the instant $t = 0$. The transverse displacements $y(x,t)$ thus satisfy the same differential equation, where $0 < \beta < 1$, and the boundary conditions

$$y(0,t) = y(\pi,t) = 0, \qquad y(x,0) = 0, \qquad y_t(x,0) = v_0.$$

Derive this expression for those displacements:

$$y(x,t) = \frac{4v_0}{\pi} e^{-\beta t} \sum_{n=1}^{\infty} \frac{\sin(2n-1)x}{(2n-1)\alpha_n} \sin \alpha_n t,$$

where $\alpha_n = \sqrt{(2n-1)^2 - \beta^2}$.

9. The ends of a stretched string are fixed at the origin and at the point $x = \pi$ on the horizontal x axis. The string is initially at rest along the x axis and then drops under its own weight. The vertical displacements $y(x,t)$ thus satisfy the differential equation (Sec. 5)

$$y_{tt}(x,t) = a^2 y_{xx}(x,t) - g \qquad (0 < x < \pi, t > 0),$$

where g is the acceleration due to gravity.

(a) Use the method of variation of parameters to derive the expression

$$y(x,t) = \frac{4g}{\pi a^2}\left[\sum_{n=1}^{\infty}\frac{\sin(2n-1)x}{(2n-1)^3}\cos(2n-1)at - \frac{\pi}{8}x(\pi-x)\right]$$

for these displacements.

(b) With the aid of the trigonometric identity

$$2\sin A \cos B = \sin(A+B) + \sin(A-B),$$

show that the expression found in part (a) can be put in the closed form

$$y(x,t) = \frac{g}{2a^2}\left[\frac{P(x+at)+P(x-at)}{2} - x(\pi-x)\right],$$

where $P(x)$ is the odd periodic extension, with period 2π, of the function $x(\pi-x)$ $(0 \leqslant x \leqslant \pi)$.

Suggestion: In both parts (a) and (b), the Fourier sine series representation (Problem 5, Sec. 14)

$$x(\pi-x) = \frac{8}{\pi}\sum_{n=1}^{\infty}\frac{\sin(2n-1)x}{(2n-1)^3} \qquad (0 \leqslant x \leqslant \pi)$$

is needed. Also, for part (a), see the suggestion with Problem 8, Sec. 35.

10. The end $x = 0$ of an elastic bar is free, and a constant longitudinal force F_0 per unit area is applied at the end $x = c$ (Fig. 43). The bar is initially unstrained and at rest. Set up the boundary value problem for the longitudinal displacements $y(x,t)$, the conditions at the ends of the bar being $y_x(0,t) = 0$ and $y_x(c,t) = F_0/E$ (Sec. 6). After noting that the method of separation of variables cannot be applied directly, follow the steps below to find $y(x,t)$.

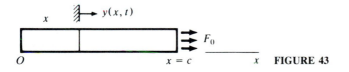

$x = c$ x **FIGURE 43**

(a) By writing $y(x,t) = Y(x,t) + Ax^2$, so that $y_x(x,t)$ and $Y_x(x,t)$ differ by a linear function of x (compare Problem 3, Sec. 33), determine a value of A that leads to the new boundary value problem

$$Y_{tt}(x,t) = a^2 Y_{xx}(x,t) + \frac{F_0 a^2}{cE} \qquad (0 < x < c, t > 0),$$

$$Y_x(0,t) = 0, \qquad Y_x(c,t) = 0,$$

$$Y(x,0) = -\frac{F_0}{2cE}x^2, \qquad Y_t(x,0) = 0.$$

(b) Point out why it is reasonable to expect that the boundary value problem in part (a) has a solution of the form

$$Y(x, t) = \frac{A_0(t)}{2} + \sum_{n=1}^{\infty} A_n(t) \cos \frac{n\pi x}{c}.$$

Then use the method of variation of parameters to find $Y(x, t)$ and thereby derive the solution

$$y(x, t) = \frac{F_0}{6cE} \left[3(x^2 + a^2 t^2) - c^2 - \frac{12c^2}{\pi^2} \sum_{n=1}^{\infty} \frac{(-1)^n}{n^2} \cos \frac{n\pi a t}{c} \cos \frac{n\pi x}{c} \right]$$

of the original problem.

(c) Use the trigonometric identity

$$2 \cos A \cos B = \cos (A + B) + \cos (A - B)$$

and the series representation [Problem 5(b), Sec. 21]

$$x^2 = \frac{c^2}{3} + \frac{4c^2}{\pi^2} \sum_{n=1}^{\infty} \frac{(-1)^n}{n^2} \cos \frac{n\pi x}{c} \qquad (-c \leqslant x \leqslant c)$$

to write the expression for $y(x, t)$ in part (b) as

$$y(x, t) = \frac{F_0}{2cE} \left[x^2 + a^2 t^2 - \frac{P(x + at) + P(x - at)}{2} \right],$$

where $P(x)$ is the periodic extension, with period $2c$, of the function x^2 $(-c \leqslant x \leqslant c)$.

11. Show how it follows from the expression for $y(x, t)$ in Problem 10(c) that the end $x = 0$ of the bar remains at rest until time $t = c/a$ and then moves with velocity $v_0 = 2aF_0/E$ when $c/a < t < 3c/a$, with velocity $2v_0$ when $3c/a < t < 5c/a$, etc.

12. The boundary value problem

$$y_{tt}(x, t) = a^2 y_{xx}(x, t) + Ax \sin \omega t \qquad (0 < x < c, t > 0),$$
$$y(0, t) = y(c, t) = 0, \qquad y(x, 0) = y_t(x, 0) = 0$$

describes transverse displacements in a vibrating string (compare Sec. 38). Show that resonance occurs when ω has one of the values $\omega_n = (n\pi a)/c$ $(n = 1, 2, \ldots)$.

13. Let a, b, and ω denote nonzero constants. The general solution of the ordinary differential equation

$$y''(t) + a^2 y(t) = b \sin \omega t$$

is of the form $y = y_c + y_p$, where y_c is the general solution of the complementary equation $y''(t) + a^2 y(t) = 0$ and y_p is any particular solution of the original nonhomogeneous equation.[†]

[†] For the method of solution to be used here, which is known as the *method of undetermined coefficients*, see, for instance, the book by Boyce and DiPrima (1992) or the one by Rainville and Bedient (1989). Both books are listed in the Bibliography.

(a) Suppose that $\omega \neq a$. After substituting $y_p = A \cos \omega t + B \sin \omega t$, where A and B are constants, into the given differential equation, determine values of A and B such that y_p is a solution. Thus derive the general solution

$$y(t) = C_1 \cos at + C_2 \sin at + \frac{b}{a^2 - \omega^2} \sin \omega t$$

of that equation.

(b) Suppose that $\omega = a$, and find constants A and B such that $y_p = At \cos \omega t + Bt \sin \omega t$ is a particular solution of the given differential equation. Thus obtain the general solution

$$y(t) = C_1 \cos at + C_2 \sin at - \frac{b}{2a} t \cos at.$$

14. Use the general solutions derived in Problem 13 to obtain the following solutions of the initial value problem

$$y''(t) + a^2 y(t) = b \sin \omega t, \qquad y(0) = 0, \qquad y'(0) = 0:$$

$$y(t) = \begin{cases} \dfrac{b}{\omega^2 - a^2}\left(\dfrac{\omega}{a} \sin at - \sin \omega t\right) & \text{when } \omega \neq a; \\[4mm] \dfrac{b}{2a}\left(\dfrac{1}{a} \sin at - t \cos at\right) & \text{when } \omega = a. \end{cases}$$

39. FOURIER SERIES IN TWO VARIABLES

Let $z(x, y, t)$ denote the transverse displacement at each point (x, y) at time t in a membrane that is stretched across a rigid square frame in the xy plane. To simplify the notation, we select the origin and the point (π, π) as ends of a diagonal of the frame. If the membrane is released at rest with a given initial displacement $f(x, y)$ that is continuous and vanishes on the boundary of the square, then (Sec. 6)

(1) $$z_{tt} = a^2(z_{xx} + z_{yy})$$

in the three-dimensional domain $0 < x < \pi, 0 < y < \pi, t > 0$; and

(2) $$z(0, y, t) = z(\pi, y, t) = z(x, 0, t) = z(x, \pi, t) = 0,$$

(3) $$z(x, y, 0) = f(x, y), \qquad z_t(x, y, 0) = 0,$$

where $0 \leqslant x \leqslant \pi, 0 \leqslant y \leqslant \pi$. We assume that the partial derivatives $f_x(x, y)$ and $f_y(x, y)$ are also continuous.

Functions of the type $z = X(x)Y(y)T(t)$ satisfy equation (1) if

(4) $$\frac{T''(t)}{a^2 T(t)} = \frac{X''(x)}{X(x)} + \frac{Y''(y)}{Y(y)} = -\lambda.$$

Separating variables again, in the second of equations (4), we find that

$$\frac{Y''(y)}{Y(y)} = -\lambda - \frac{X''(x)}{X(x)} = -\mu,$$

where μ is another separation constant. So we are led to two Sturm-Liouville problems,

$$X''(x) + (\lambda - \mu)X(x) = 0, \qquad X(0) = 0, \qquad X(\pi) = 0$$

and

$$Y''(y) + \mu Y(y) = 0, \qquad Y(0) = 0, \qquad Y(\pi) = 0,$$

and to the conditions

$$T''(t) + \lambda a^2 T(t) = 0, \qquad T'(0) = 0$$

on T.

We turn to the Sturm-Liouville problem in Y first since it involves only one of the separation constants. According to Sec. 29, the values $\mu = m^2$ $(m = 1, 2, \ldots)$ of μ give rise to the eigenfunctions $Y_m(y) = \sin my$; and, when $\lambda - \mu = n^2$ $(n = 1, 2, \ldots)$, the eigenfunctions $X_n(x) = \sin nx$ of the problem in X are obtained. The conditions on T thus become

$$T''(t) + a^2(m^2 + n^2)T(t) = 0, \qquad T'(0) = 0,$$

where $m = 1, 2, \ldots$ and $n = 1, 2, \ldots$. For any fixed positive integers m and n, the solution of this problem in T is, except for a constant factor, $T_{mn}(t) = \cos(at\sqrt{m^2 + n^2})$.

The formal solution of our boundary value problem is, therefore,

$$(5) \qquad z(x, y, t) = \sum_{n=1}^{\infty} \sum_{m=1}^{\infty} B_{mn} \sin nx \sin my \cos\left(at\sqrt{m^2 + n^2}\right),$$

where the coefficients B_{mn} need to be determined so that

$$(6) \qquad f(x, y) = \sum_{n=1}^{\infty} \sum_{m=1}^{\infty} B_{mn} \sin nx \sin my$$

when $0 \leqslant x \leqslant \pi$ and $0 \leqslant y \leqslant \pi$. By grouping terms in this double sine series so as to display the total coefficient of $\sin nx$ for each n, one can write, formally,

$$(7) \qquad f(x, y) = \sum_{n=1}^{\infty} \left(\sum_{m=1}^{\infty} B_{mn} \sin my \right) \sin nx.$$

For each fixed y $(0 \leqslant y \leqslant \pi)$, equation (7) is a Fourier sine series representation of the function $f(x, y)$, with variable x $(0 \leqslant x \leqslant \pi)$, provided that

$$(8) \qquad \sum_{m=1}^{\infty} B_{mn} \sin my = \frac{2}{\pi} \int_0^\pi f(x, y) \sin nx \, dx \qquad (n = 1, 2, \ldots).$$

The right-hand side here is a sequence of functions $F_n(y)$ $(n = 1, 2, \ldots)$, each represented by its Fourier sine series (8) on the interval $0 \leqslant y \leqslant \pi$ when

$$B_{mn} = \frac{2}{\pi} \int_0^\pi F_n(y) \sin my \, dy \qquad (m = 1, 2, \dots).$$

Hence the coefficients B_{mn} have the values

(9) $$B_{mn} = \frac{4}{\pi^2} \int_0^\pi \sin my \int_0^\pi f(x, y) \sin nx \, dx \, dy.$$

A formal solution of our membrane problem is now given by equation (5) with the coefficients defined by equation (9).

Since the numbers $\sqrt{m^2 + n^2}$ do not change by integral multiples of some fixed number as m and n vary through integral values, the cosine functions in equation (5) have no common period in the variable t; so the displacement z is not generally a periodic function of t. Consequently, the vibrating membrane, in contrast to the vibrating string, generally does not produce a musical note. It can be made to do so, however, by giving it the proper initial displacement. If, for example,

$$z(x, y, 0) = A \sin x \sin y,$$

where A is a constant, the displacements (5) are given by a single term:

$$z(x, y, t) = A \sin x \sin y \cos \left(a\sqrt{2}\, t \right).$$

Then z is periodic in t, with period $\pi\sqrt{2}/a$.

40. PERIODIC BOUNDARY CONDITIONS

The solutions of the boundary value problems in this chapter have been based on the solutions of just two Sturm-Liouville problems, which lead to Fourier cosine and sine series representations of prescribed functions. Although Chap. 5 is devoted to the theory and application of many other Sturm-Liouville problems, as well as to the precise definition of such a problem, we conclude this chapter by considering a third problem that arises in certain boundary value problems for regions with circular boundaries:

(1) $\quad X''(x) + \lambda X(x) = 0, \qquad X(-\pi) = X(\pi), \qquad X'(-\pi) = X'(\pi).$

We include it here since its solutions also lead to Fourier series representations, but now involving *both* cosines and sines on the interval $-\pi < x < \pi$, and since the general theory of Sturm-Liouville problems is not actually required. We need accept only the fact, to be verified in Chap. 5 (Sec. 43), that each eigenvalue, or value of λ for which problem (1) has a nontrivial solution, is a real number. In anticipation of Chap. 5, we continue to refer to such values of λ as eigenvalues and to the nontrivial solutions as eigenfunctions.

To solve problem (1), we consider first the case in which $\lambda = 0$. Then $X(x) = Ax + B$, where A and B are constants; and the boundary conditions are satisfied if $A = 0$. Since the conditions in problem (1) are all homogeneous, we thus find that, except for a constant factor, $X(x) = \frac{1}{2}$.

When $\lambda > 0$, we write $\lambda = \alpha^2$ ($\alpha > 0$) and note that the general solution of the differential equation in problem (1) is

$$X(x) = C_1 \cos \alpha x + C_2 \sin \alpha x.$$

It is straightforward to show that, in order for the boundary conditions to be satisfied,

$$C_2 \sin \alpha\pi = 0 \quad \text{and} \quad C_1 \sin \alpha\pi = 0.$$

Since C_1 and C_2 cannot both vanish if $X(x)$ is to be nontrivial, it follows that the positive number α must, in fact, be a positive integer n. Thus $\lambda = n^2$ ($n = 1, 2, \ldots$), and the corresponding general solution of problem (1) is an arbitrary linear combination of two linearly independent eigenfunctions, $\cos nx$ and $\sin nx$.

It is left to the reader (Problem 4) to show that there are no negative eigenvalues.

We now illustrate the use of this Sturm-Liouville problem, involving periodic boundary conditions.

EXAMPLE. Let $u(\rho, \phi)$ denote the steady temperatures in a thin disk $\rho \leqslant 1$, with insulated surfaces, when its edge $\rho = 1$ is kept at temperatures $f(\phi)$. The variables ρ and ϕ are, of course, polar coordinates, and u satisfies Laplace's equation $\nabla^2 u = 0$. That is,

$$(2) \quad \rho^2 u_{\rho\rho}(\rho, \phi) + \rho u_{\rho}(\rho, \phi) + u_{\phi\phi}(\rho, \phi) = 0 \quad (0 < \rho < 1, \ -\pi < \phi < \pi),$$

where

$$(3) \qquad\qquad\qquad u(1, \phi) = f(\phi) \qquad\qquad (-\pi < \phi < \pi).$$

Also, u and its partial derivatives of the first and second order are continuous and bounded in the interior of the disk. In particular, u and its first-order partial derivatives are continuous on the ray $\phi = \pi$ (Fig. 44).

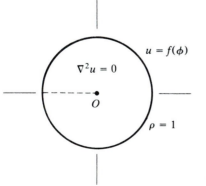

$u = f(\phi)$

$\nabla^2 u = 0$

O

$\rho = 1$

FIGURE 44

If functions of the type $u = R(\rho)\Phi(\phi)$ are to satisfy condition (2) and the continuity requirements, then

(4) $$\rho^2 R''(\rho) + \rho R'(\rho) - \lambda R(\rho) = 0 \qquad (0 < \rho < 1)$$

and

(5) $$\Phi''(\phi) + \lambda\Phi(\phi) = 0, \qquad \Phi(-\pi) = \Phi(\pi), \qquad \Phi'(-\pi) = \Phi'(\pi),$$

where λ is a separation constant. We now recognize that conditions (5) constitute a Sturm-Liouville problem in Φ, with eigenvalues $\lambda_0 = 0$ and $\lambda_n = n^2$ $(n = 1, 2, \dots)$. The corresponding eigenfunctions are $\frac{1}{2}$ and linear combinations of $\cos n\phi$ and $\sin n\phi$. Equation (4) is a Cauchy-Euler equation, and we know from Sec. 35 that its bounded solutions are $R_0(\rho) = 1$ when $\lambda = 0$ and $R_n(\rho) = \rho^n$ when $\lambda = n^2$ $(n = 1, 2, \dots)$. Hence the generalized linear combination of our continuous functions $R\Phi$ can be written

(6) $$u(\rho, \phi) = \frac{a_0}{2} + \sum_{n=1}^{\infty} \rho^n (a_n \cos n\phi + b_n \sin n\phi).$$

This satisfies the boundary condition (3) if a_n and b_n, including a_0, are the coefficients in the Fourier series for f on the interval $-\pi < x < \pi$:

(7) $$a_n = \frac{1}{\pi}\int_{-\pi}^{\pi} f(\phi) \cos n\phi \, d\phi, \qquad b_n = \frac{1}{\pi}\int_{-\pi}^{\pi} f(\phi) \sin n\phi \, d\phi.$$

We assume that f is piecewise smooth.

PROBLEMS

1. All four faces of an infinitely long rectangular prism, formed by the planes $x = 0$, $x = a$, $y = 0$, and $y = b$, are kept at temperature zero. Let the initial temperature distribution be $f(x, y)$, and derive this expression for the temperatures $u(x, y, t)$ in the prism:

$$u(x, y, t) = \sum_{n=1}^{\infty}\sum_{m=1}^{\infty} B_{mn} \exp\left[-\pi^2 kt\left(\frac{m^2}{a^2} + \frac{n^2}{b^2}\right)\right] \sin\frac{n\pi x}{a} \sin\frac{m\pi y}{b},$$

where

$$B_{mn} = \frac{4}{ab}\int_0^b \sin\frac{m\pi y}{b}\int_0^a f(x, y)\sin\frac{n\pi x}{a}\, dx\, dy.$$

2. Write $f(x, y) = g(x)h(y)$ in Problem 1 and show that the double series obtained there for u reduces to the product

$$u(x, y, t) = v(x, t)w(y, t)$$

of two series, where v and w represent temperatures in the slabs $0 \leqslant x \leqslant a$ and $0 \leqslant y \leqslant b$ with faces at temperature zero and with initial temperatures $g(x)$ and $h(y)$, respectively.

3. Let the functions $v(x, t)$ and $w(y, t)$ satisfy the heat equation for one-dimensional flow:

$$v_t = k v_{xx}, \qquad w_t = k w_{yy}.$$

Show by differentiation that their product $u = vw$ satisfies the two-dimensional heat equation

$$u_t = k(u_{xx} + u_{yy}).$$

Use this result to arrive at the expression for $u(x, y, t)$ in Problem 2.

4. Write $\lambda = -\alpha^2$ $(\alpha > 0)$, and show that the Sturm-Liouville problem

$$X''(x) + \lambda X(x) = 0, \qquad X(-\pi) = X(\pi), \qquad X'(-\pi) = X'(\pi)$$

in Sec. 40 has no negative eigenvalues.

5. Using the cylindrical coordinates ρ, ϕ, and z, let $u(\rho, \phi)$ denote steady temperatures in a long hollow cylinder $a \le \rho \le b$, $-\infty < z < \infty$ when the temperatures on the inner surface $\rho = a$ are $f(\phi)$ and the temperature of the outer surface $\rho = b$ is zero. (a) Derive the temperature formula

$$u(\rho, \phi) = \frac{\ln (b/\rho)}{\ln (b/a)} \cdot \frac{a_0}{2} + \sum_{n=1}^{\infty} \frac{(b/\rho)^n - (\rho/b)^n}{(b/a)^n - (a/b)^n} (a_n \cos n\phi + b_n \sin n\phi),$$

where the coefficients a_n and b_n, including a_0, are given by equations (7), Sec. 40. (b) Use the result in part (a) to show that if $f(\phi) = A + B \sin \phi$, where A and B are constants, then

$$u(\rho, \phi) = A \frac{\ln (b/\rho)}{\ln (b/a)} + \frac{Bab}{b^2 - a^2} \left(\frac{b}{\rho} - \frac{\rho}{b} \right) \sin \phi.$$

6. Solve the boundary value problem

$$u_t(x, t) = k u_{xx}(x, t) \qquad\qquad (-\pi < x < \pi, t > 0),$$

$$u(-\pi, t) = u(\pi, t), \qquad u_x(-\pi, t) = u_x(\pi, t), \qquad u(x, 0) = f(x).$$

The solution $u(x, t)$ represents, for example, temperatures in an insulated wire of length 2π that is bent into a unit circle and has a given temperature distribution along it. For convenience, the wire is thought of as being cut at one point and laid on the x axis between $x = -\pi$ and $x = \pi$. The variable x then measures the distance along the wire, starting at the point $x = -\pi$, and the points $x = -\pi$ and $x = \pi$ denote the same point on the circle. The first two boundary conditions in the problem state that the temperatures and the flux must be the same for each of those values of x. This problem was of considerable interest to Fourier himself, and the wire has come to be known as *Fourier's ring*.

$$\text{Answer: } u(x, t) = \frac{a_0}{2} + \sum_{n=1}^{\infty} (a_n \cos nx + b_n \sin nx) e^{-n^2 kt},$$

where

$$a_n = \frac{1}{\pi} \int_{-\pi}^{\pi} f(x) \cos nx \, dx, \qquad b_n = \frac{1}{\pi} \int_{-\pi}^{\pi} f(x) \sin nx \, dx.$$

7. (*a*) By writing $A = n\theta$ and $B = \theta$ in the trigonometric identity

$$2 \cos A \cos B = \cos (A + B) + \cos (A - B),$$

multiplying through the resulting equation by a^n ($-1 < a < 1$), and then summing each side from $n = 1$ to $n = \infty$, derive the summation formula

$$\sum_{n=1}^{\infty} a^n \cos n\theta = \frac{a \cos \theta - a^2}{1 - 2a \cos \theta + a^2} \qquad (-1 < a < 1).$$

[One can readily see that this series is absolutely convergent by comparing it with the geometric series whose terms are a^n ($n = 1, 2, \ldots$).]

(*b*) Write expression (6), with coefficients (7), in Sec. 40 for steady temperatures in a disk as

$$u(\rho, \phi) = \frac{1}{2\pi} \int_{-\pi}^{\pi} f(\psi) \left[1 + 2 \sum_{n=1}^{\infty} \rho^n \cos n(\phi - \psi)\right] d\psi.$$

Then, with the aid of the summation formula in part (*a*), derive *Poisson's integral formula* for those temperatures:

$$u(\rho, \phi) = \frac{1}{2\pi} \int_{-\pi}^{\pi} f(\psi) \frac{1 - \rho^2}{1 - 2\rho \cos (\phi - \psi) + \rho^2} d\psi \qquad (\rho < 1).$$

CHAPTER
5

STURM-LIOUVILLE PROBLEMS AND APPLICATIONS

We turn now to a careful presentation of the rudiments of the theory of Sturm-Liouville problems and their solutions. Once that is done, we shall illustrate the Fourier method in solving physical problems involving eigenfunctions not encountered in earlier chapters.

41. REGULAR STURM-LIOUVILLE PROBLEMS

In Chap. 4, we found solutions of various boundary value problems by the Fourier method. Except in Sec. 40, the method always led to the need for a Fourier cosine or sine series representation of a given function. The cosine and sine functions in the series were the eigenfunctions of one of the following two Sturm-Liouville problems on an interval $0 \leqslant x \leqslant c$:

$$(1) \qquad X''(x) + \lambda X(x) = 0, \qquad X'(0) = 0, \qquad X'(c) = 0,$$

$$(2) \qquad X''(x) + \lambda X(x) = 0, \qquad X(0) = 0, \qquad X(c) = 0.$$

When applied to many other boundary value problems in partial differential equations, the Fourier method continues to involve a Sturm-Liouville problem consisting of a homogeneous ordinary differential equation of the type

$$(3) \qquad X''(x) + R(x)X'(x) + [Q(x) + \lambda P(x)]X(x) = 0$$

168

on a finite interval $a < x < b$, together with a pair of homogeneous boundary conditions at the end points of the interval. The functions P, Q, and R and the boundary conditions are prescribed by the original boundary value problem involving a partial differential equation. Values of the parameter λ, which appears in equation (3) only as indicated, and corresponding nontrivial solutions $X(x)$ are to be determined. We now give a precise definition of such a Sturm-Liouville problem, which includes problems (1) and (2) as special cases.

Note that a function

$$r(x) = \exp \int R(x)\, dx$$

is an integrating factor for the sum of the first two terms in equation (3); that is,

$$r(x)[X''(x) + R(x)X'(x)] = [r(x)X'(x)]'.$$

Consequently, when each of its terms is multiplied by $r(x)$, equation (3) takes the standard form

(4) $$[r(x)X'(x)]' + [q(x) + \lambda p(x)]X(x) = 0 \qquad (a < x < b),$$

where the functions p, q, and r are independent of λ. *We assume here that p, q, r, and r' are real-valued functions of the real variable x which are continuous on the closed bounded interval $a \leqslant x \leqslant b$ and that $p(x) > 0$ and $r(x) > 0$ when $a \leqslant x \leqslant b$.* Also, $X(x)$ is required to satisfy the homogeneous *separated* boundary conditions

(5) $$a_1 X(a) + a_2 X'(a) = 0, \qquad b_1 X(b) + b_2 X'(b) = 0.$$

The constants a_1, a_2, b_1, b_2 are all real numbers, independent of λ. Moreover, a_1 and a_2 are not both zero; and the same is true of the constants b_1 and b_2. The differential equation (4) and boundary conditions (5) make up a *regular Sturm-Liouville problem.*[†] Sturm-Liouville problems other than regular ones will be noted in Sec. 42.

EXAMPLES. Problems (1) and (2) are both regular Sturm-Liouville problems. Two other examples, to be solved later in this chapter, are

$$X''(x) + \lambda X(x) = 0 \qquad\qquad (0 < x < c),$$
$$X'(0) = 0, \qquad hX(c) + X'(c) = 0,$$

where h denotes a positive constant, and

$$[x^2 X'(x)]' + \lambda X(x) = 0 \qquad\qquad (1 < x < b),$$
$$X(1) = 0, \qquad X(b) = 0.$$

[†] Papers by J. C. F. Sturm and J. Liouville giving the first extensive development of the theory of this problem appeared in vols. 1–3 of the *Journal de mathématique* (1836–1838).

As was the case with problems (1) and (2), a value of λ for which problem (4)–(5) has a nontrivial solution is called an *eigenvalue*; and the nontrivial solution is called an *eigenfunction*. Note that if $X(x)$ is an eigenfunction, then so is $CX(x)$, where C is any nonzero constant. It is understood that in order for $X(x)$ to be an eigenfunction, $X(x)$ and $X'(x)$ *must be continuous on the closed interval* $a \leqslant x \leqslant b$. Such continuity conditions are usually required of solutions of boundary value problems in ordinary differential equations.

The set of eigenvalues of problem (4)–(5) is called the *spectrum* of the problem. It can be shown that the spectrum of a regular Sturm-Liouville problem consists of an infinite number of eigenvalues $\lambda_1, \lambda_2, \ldots$. We state this fact without proof, which is quite involved.[†] In special cases, the eigenvalues will be found; and so their existence will not be in doubt. When eigenvalues are sought, however, it is useful to know that they are all real and hence that there is no possibility of discovering others in the complex plane. The proof that the eigenvalues must be real is given in Sec. 43; and we agree that they are to be arranged in ascending order of magnitude, so that $\lambda_n < \lambda_{n+1}$ $(n = 1, 2, \ldots)$. It can be shown that $\lambda_n \to \infty$ as $n \to \infty$.

If we define the differential operator $\mathscr{L}$ by means of the equation

(6) $$\mathscr{L}[X(x)] = [r(x)X'(x)]' + q(x)X(x),$$

then the Sturm-Liouville equation (4) takes the form

(7) $$\mathscr{L}[X(x)] + \lambda p(x)X(x) = 0.$$

This operator $\mathscr{L}$ is a special case of the general second-order linear differential operator L defined by the equation

(8) $$L[X(x)] = A(x)X''(x) + B(x)X'(x) + C(x)X(x).$$

In discussing such operators, we tacitly agree that they are to be defined on function spaces (Sec. 10) in which the functions X are suitably differentiable.

The *adjoint* of L is the operator L^* such that

(9) $$L^*[X(x)] = [A(x)X(x)]'' - [B(x)X(x)]' + C(x)X(x),$$

and L is called *self-adjoint* when $L^* = L$. It can be shown (Problem 2, Sec. 43) that a necessary and sufficient condition for L to be self-adjoint is that $B(x) = A'(x)$, in which case equation (8) becomes

$$L[X(x)] = [A(x)X'(x)]' + C(x)X(x).$$

Except for notation, this is equation (6); and so we use $\mathscr{L}$ to denote the general self-adjoint linear differential operator of the second order.

The Sturm-Liouville differential equation (4) is said to be in *self-adjoint form* since it is the same as equation (7), which involves the self-adjoint operator

---

[†] For verification of statements in this section that we do not prove, see the book by Churchill (1972, chap. 9), which contains proofs when $a_2 = b_2 = 0$ in conditions (5), and the one by Birkhoff and Rota (1989). Also, extensive treatments of Sturm-Liouville theory appear in the books by Ince (1956) and Titchmarsh (1962). These references are all listed in the Bibliography.

$\mathscr{L}$. This form turns out to be especially useful because of properties of the operator $\mathscr{L}$. Some of those properties are noted explicitly in the problems.

42. MODIFICATIONS

Although we are mainly concerned in this chapter with the theory and application of *regular* Sturm-Liouville problems, described in Sec. 41, certain important modifications are also of interest in practice. We mention them here since some of their theory is conveniently included in the discussion of regular Sturm-Liouville problems in Sec. 43.

A Sturm-Liouville problem

$$(1) \qquad [r(x)X'(x)]' + [q(x) + \lambda p(x)]X(x) = 0 \qquad (a < x < b),$$

$$(2) \qquad a_1 X(a) + a_2 X'(a) = 0, \qquad b_1 X(b) + b_2 X'(b) = 0$$

is *singular* when at least one of the regularity conditions stated in Sec. 41 fails to be satisfied. The function q, for example, may have an infinite discontinuity at an end point of the interval $a \leqslant x \leqslant b$. The problem is also singular if $p(x)$ or $r(x)$ vanishes at an end point. When $r(x)$ does this, we drop the boundary condition at the end point in question. Note that the dropping of the boundary condition at $x = a$ is the same as letting both of the coefficients a_1 and a_2 in that condition be zero; a similar remark can be made when the condition at $x = b$ is to be dropped.

EXAMPLE 1. One singular Sturm-Liouville problem to be studied in Chap. 7 consists of the differential equation

$$[xX'(x)]' + \left(-\frac{n^2}{x} + \lambda x\right)X(x) = 0 \qquad (0 < x < c),$$

where $n = 0, 1, 2, \ldots$, and the single boundary condition $X(c) = 0$. Observe that the functions $p(x) = x$ and $r(x) = x$ both vanish at $x = 0$ and that the function $q(x) = -n^2/x$ has an infinite discontinuity there when n is positive.

EXAMPLE 2. The differential equation

$$[(1 - x^2)X'(x)]' + \lambda X(x) = 0 \qquad (-1 < x < 1),$$

with no boundary conditions, constitutes a singular Sturm-Liouville problem. Here the function $r(x) = 1 - x^2$ vanishes at both ends $x = \pm 1$ of the interval $-1 \leqslant x \leqslant 1$. This problem is the main one that is solved and used in Chap. 8.

Although it will turn out that the problems in Examples 1 and 2 have *discrete* spectra, where the eigenvalues may be indexed with the positive or nonnegative integers, this is not always the case with singular problems. Such problems may, in fact, have no eigenvalues at all. Moreover, other types of singular problems, defined on infinite or semi-infinite intervals and to be encountered in Chap. 6, have *continuous* spectra containing all nonnegative values of λ. As indicated in Sec. 41, the nature of the spectrum of any particular problem will be determined by actually finding the eigenvalues.

Finally, in addition to singular problems, another modification of problem (1)–(2) occurs when $r(a) = r(b)$ and conditions (2) are replaced by the *periodic* boundary conditions

$$(3) \qquad\qquad X(a) = X(b), \qquad X'(a) = X'(b).$$

EXAMPLE 3. The problem

$$X''(x) + \lambda X(x) = 0, \qquad X(-\pi) = X(\pi), \qquad X'(-\pi) = X'(\pi),$$

already solved in Sec. 40, has periodic boundary conditions.

43. ORTHOGONALITY OF EIGENFUNCTIONS

As pointed out in Sec. 41, a regular Sturm-Liouville problem

$$(1) \qquad\qquad [r(x)X'(x)]' + [q(x) + \lambda p(x)]X(x) = 0 \qquad (a < x < b),$$

$$(2) \qquad a_1 X(a) + a_2 X'(a) = 0, \qquad b_1 X(b) + b_2 X'(b) = 0$$

always has an infinite number of eigenvalues $\lambda_1, \lambda_2, \ldots$. In this section, we shall establish the orthogonality of eigenfunctions corresponding to *distinct* eigenvalues. The concept of orthogonality to be used here is, however, a slight generalization of the one originally introduced in Sec. 11. To be specific, a set $\{\psi_n(x)\}$ $(n = 1, 2, \ldots)$ is orthogonal on an interval $a < x < b$ with respect to a *weight function $p(x)$*, which is piecewise continuous and positive on that interval, if

$$\int_a^b p(x)\psi_m(x)\psi_n(x)\, dx = 0 \qquad\qquad \text{when } m \neq n.$$

The integral here represents an inner product (ψ_m, ψ_n) with respect to the weight function. The set is normalized by dividing each $\psi_n(x)$ by $\|\psi_n\|$, where

$$\|\psi_n\|^2 = (\psi_n, \psi_n) = \int_a^b p(x)[\psi_n(x)]^2\, dx$$

and where it is assumed that $\|\psi_n\| \neq 0$. This type of orthogonality can, of course, be reduced to that in Sec. 11 by using the products $\sqrt{p(x)}\,\psi_n(x)$ as functions of the set. Orthogonal sets with respect to weight functions that are not piecewise continuous, or ones where the fundamental interval is unbounded, also occur in applied mathematics.

The theorem below states that eigenfunctions associated with distinct eigenvalues are orthogonal on the interval $a < x < b$ with respect to the weight function $p(x)$, where $p(x)$ is the same function as in equation (1). If such eigenfunctions are denoted by $X_n(x)$ $(n = 1, 2, \dots)$, the normalized eigenfunctions are

$$\phi_n(x) = \frac{X_n(x)}{\|X_n\|}, \qquad \text{where} \qquad \|X_n\|^2 = \int_a^b p(x)[X_n(x)]^2 \, dx;$$

and a generalized Fourier series corresponding to a given function $f(x)$ in $C_p(a,b)$ is (compare Sec. 12)

$$f(x) \sim \sum_{n=1}^{\infty} c_n \phi_n(x) \qquad\qquad (a < x < b),$$

where

$$c_n = (f, \phi_n) = \int_a^b p(x) f(x) \phi_n(x) \, dx \qquad (n = 1, 2, \dots).$$

Examples of such series will be given in Sec. 46.

In presenting the theorem, we relax the conditions of regularity on the coefficients in the differential equation (1) so that the result can also be applied to eigenfunctions that are found for the modifications of regular Sturm-Liouville problems mentioned in Sec. 42. We retain all the conditions for a regular problem, stated in Sec. 41, except that now q may be discontinuous at an end point of the interval $a \leqslant x \leqslant b$, and $p(x)$ and $r(x)$ may vanish at an end point. That is, p, r, and r' *are continuous on the closed interval $a \leqslant x \leqslant b$, q is continuous on the open interval $a < x < b$, and $p(x) > 0$ and $r(x) > 0$ when* $a < x < b$.

Theorem. *If λ_m and λ_n are distinct eigenvalues of the Sturm-Liouville problem (1)–(2), then corresponding eigenfunctions $X_m(x)$ and $X_n(x)$ are orthogonal with respect to the weight function $p(x)$ on the interval $a < x < b$. The orthogonality also holds in each of the following cases*:

(a) *When $r(a) = 0$ and the first of boundary conditions (2) is dropped from the problem*;
(b) *When $r(b) = 0$ and the second of conditions (2) is dropped*;
(c) *When $r(a) = r(b)$ and conditions (2) are replaced by the conditions*

$$X(a) = X(b), \qquad X'(a) = X'(b).$$

Note that both cases (a) and (b) here may apply to a given Sturm-Liouville problem (see Example 2, Sec. 42).

To prove the theorem, we first observe that

$$(rX_m')' + qX_m = -\lambda_m p X_m, \qquad (rX_n')' + qX_n = -\lambda_n p X_n$$

since each eigenfunction satisfies equation (1) when λ is the eigenvalue to which it corresponds. We then multiply each side of these two equations by X_n and

X_m, respectively, and subtract:

$$(3) \qquad (\lambda_m - \lambda_n) p X_m X_n = X_m (r X_n')' - X_n (r X_m')'$$

$$= \frac{d}{dx} \left[r (X_m X_n' - X_m' X_n) \right].$$

While the final reduction here to an exact derivative is elementary, it is made possible by the special nature of the self-adjoint operator $\mathscr{L}$ defined in equation (6), Sec. 41. Details regarding this point are left to the problems.

The function q has been eliminated, and the continuity conditions on the remaining functions allow us to write

$$(4) \qquad (\lambda_m - \lambda_n) \int_a^b p X_m X_n \, dx = [r(x) \Delta(x)]_a^b,$$

where $\Delta(x)$ is the determinant

$$(5) \qquad \Delta(x) = \begin{vmatrix} X_m(x) & X_m'(x) \\ X_n(x) & X_n'(x) \end{vmatrix}.$$

That is,

$$(6) \qquad (\lambda_m - \lambda_n) \int_a^b p X_m X_n \, dx = r(b) \Delta(b) - r(a) \Delta(a).$$

The first of boundary conditions (2) requires that

$$a_1 X_m(a) + a_2 X_m'(a) = 0,$$

$$a_1 X_n(a) + a_2 X_n'(a) = 0;$$

and for this pair of linear homogeneous equations in a_1 and a_2 to be satisfied by numbers a_1 and a_2, not both zero, it is necessary that the determinant $\Delta(a)$ be zero. Similarly, from the second boundary condition, where b_1 and b_2 are not both zero, we see that $\Delta(b) = 0$. Then, according to equation (6),

$$(7) \qquad (\lambda_m - \lambda_n) \int_a^b p X_m X_n \, dx = 0;$$

and, since $\lambda_m \neq \lambda_n$, the desired orthogonality property follows:

$$(8) \qquad \int_a^b p(x) X_m(x) X_n(x) \, dx = 0.$$

If $r(a) = 0$, property (8) follows from equation (6) even when $\Delta(a) \neq 0$, or when $a_1 = a_2 = 0$, in which case the first of boundary conditions (2) disappears. Similarly, if $r(b) = 0$, the second of those conditions is not used.

When $r(a) = r(b)$ and the periodic boundary conditions

$$X(a) = X(b), \qquad X'(a) = X'(b)$$

are used in place of conditions (2), then

$$r(b)\Delta(b) = r(a)\Delta(a);$$

and, again, property (8) follows. This completes the proof of the theorem.

EXAMPLE 1. The eigenfunctions of the regular Sturm-Liouville problem

$$X''(x) + \lambda X(x) = 0 \qquad\qquad (0 < x < \pi),$$

$$X(0) = 0, \qquad X(\pi) = 0$$

are $X_n(x) = \sin nx$ ($n = 1, 2, \ldots$), and they correspond to the distinct eigenvalues $\lambda_n = n^2$ (Sec. 29). The theorem tells us that the functions $X_n(x)$ are orthogonal on the interval $0 < x < \pi$ with weight function $p(x) = 1$:

(9) $$\int_0^\pi \sin mx \sin nx\, dx = 0 \qquad\qquad (m \neq n).$$

We recall that this orthogonality was established earlier in Example 1, Sec. 11, where integral (9) was evaluated directly.

EXAMPLE 2. Eigenfunctions corresponding to distinct eigenvalues of the regular Sturm-Liouville problem

$$[xX'(x)]' + \frac{\lambda}{x}X(x) = 0 \qquad\qquad (1 < x < b),$$

$$X(1) = 0, \qquad X(b) = 0$$

are, according to the theorem, orthogonal on the interval $1 < x < b$ with weight function $p(x) = 1/x$. In Problem 1 the eigenfunctions are actually found, and the orthogonality is verified.

The following corollary is an immediate consequence of the theorem.

Corollary. *If λ is an eigenvalue of the Sturm-Liouville problem (1)–(2), then it must be a real number; and the same is true in cases (a), (b), and (c), treated in the theorem.*

We begin the proof by writing the eigenvalue as $\lambda = \alpha + i\beta$, where α and β are real numbers. If X denotes a corresponding eigenfunction, which is nontrivial and may be complex-valued, conditions (1) and (2) are satisfied. Now the complex conjugate of λ is the number $\bar{\lambda} = \alpha - i\beta$; and $\bar{X} = u - iv$ and $\bar{X}' = u' + iv'$ if $X = u + iv$. Also, the conjugate of a sum or product of two complex numbers is the sum or product, respectively, of the conjugates of those numbers. Hence, by taking the conjugates of both sides of the equations in conditions (1) and (2) and keeping in mind that the functions p, q, and r are real-valued and that the coefficients in conditions (2) are real numbers, we see that

$$(r\bar{X}')' + (q + \bar{\lambda}p)\bar{X} = 0,$$

$$a_1\bar{X}(a) + a_2\bar{X}'(a) = 0, \qquad b_1\bar{X}(b) + b_2\bar{X}'(b) = 0.$$

Thus the nontrivial function $\bar{X}$ is an eigenfunction corresponding to $\bar{\lambda}$.

If we assume that $\beta \neq 0$, then $\bar{\lambda} \neq \lambda$; and the theorem tells us that X and $\bar{X}$ are orthogonal on the interval $a < x < b$ with respect to the weight function $p(x)$, even in cases (a), (b), and (c):

(10)
$$\int_a^b p(x)X(x)\bar{X}(x)\,dx = 0.$$

But $p(x) > 0$ when $a < x < b$. Moreover,

$$X\bar{X} = u^2 + v^2 = |X|^2 \geq 0$$

when $a \leq x \leq b$; and $|X|^2$ is not identically zero since X is an eigenfunction. So integral (10) has positive value, and our assumption that $\beta \neq 0$ has led us to a contradiction. Hence we must conclude that $\beta = 0$, or that λ is real.

PROBLEMS

1. (a) After writing the differential equation in the regular Sturm-Liouville problem

$$[xX'(x)]' + \frac{\lambda}{x}X(x) = 0 \qquad (1 < x < b),$$

$$X(1) = 0, \qquad X(b) = 0$$

in Cauchy-Euler form (see Problem 3, Sec. 35), use the substitution $x = \exp s$ to transform the problem into one consisting of the differential equation

$$\frac{d^2X}{ds^2} + \lambda X = 0 \qquad (0 < s < \ln b)$$

and the boundary conditions

$$X = 0 \qquad \text{when} \qquad s = 0 \text{ and } s = \ln b.$$

Then, by simply referring to the solutions of the Sturm-Liouville problem in Sec. 29, show that the eigenvalues and eigenfunctions of the original problem here are

$$\lambda_n = \alpha_n^2, \qquad X_n(x) = \sin(\alpha_n \ln x) \qquad (n = 1, 2, \dots),$$

where $\alpha_n = n\pi/\ln b$.

(b) By making the substitution $s = (\pi/\ln b)\ln x$ in the integral involved and then referring to the integration formula (9) in Example 1, Sec. 43, give a direct verification that the eigenfunctions $X_n(x)$ obtained in part (a) are orthogonal on the interval $1 < x < b$ with weight function $p(x) = 1/x$, as ensured by the theorem in Sec. 43.

2. Let L be the general second-order linear differential operator defined by the equation

$$L[X] = AX'' + BX' + CX,$$

where A, B, and C are functions of x. In Sec. 41, the adjoint of L was defined as the operator L^* such that

$$L^*[X] = (AX)'' - (BX)' + CX.$$

Show that a necessary and sufficient condition for L to be self-adjoint ($L^* = L$) is that $B = A'$.

   Suggestion: Note that if $L^*[X] = L[X]$, then $L^*[1] = L[1]$ and $L^*[x] = L[x]$, in particular. The condition $B = A'$ follows from these last two equations.

3. (*a*) Verify the identity

$$X(rY')' - Y(rX')' = \frac{d}{dx}[r(XY' - X'Y)],$$

   where r, X, and Y denote functions of x.

   (*b*) Show that if $\mathscr{L}$ is the self-adjoint differential operator defined by the equation (Sec. 41)

$$\mathscr{L}[X] = (rX')' + qX,$$

   then the identity in part (*a*) can be written

$$X\mathscr{L}[Y] - Y\mathscr{L}[X] = \frac{d}{dx}[r(XY' - X'Y)].$$

   This is called *Lagrange's identity* for the operator $\mathscr{L}$.

4. (*a*) Suppose that the self-adjoint operator $\mathscr{L}$ in Problem 3(*b*) is defined on a space of functions satisfying the conditions

$$a_1 X(a) + a_2 X'(a) = 0, \qquad b_1 X(b) + b_2 X'(b) = 0,$$

   where a_1 and a_2 are not both zero and where the same is true of b_1 and b_2. Use Lagrange's identity, obtained in that problem, to show that

$$(X, \mathscr{L}[Y]) = (\mathscr{L}[X], Y),$$

   where these inner products are on the interval $a < x < b$ and with weight function unity.

   (*b*) Let λ_m and λ_n denote distinct eigenvalues of a regular Sturm-Liouville problem, whose differential equation is (Sec. 41)

$$\mathscr{L}[X] + \lambda pX = 0.$$

   Use the result in part (*a*) to prove that if X_m and X_n are eigenfunctions corresponding to λ_m and λ_n, then

$$(pX_m, X_n) = 0.$$

   Thus show that X_m and X_n are orthogonal on the interval $a < x < b$ with weight function p, as already demonstrated in Sec. 43.

5. Show that if L is the general second-order linear differential operator, where

$$L[X] = AX'' + BX' + CX,$$

   and if L^* is its adjoint, defined by the equation (Sec. 41)

$$L^*[X] = (AX)'' - (BX)' + CX,$$

   then the adjoint of L^* is L. That is, show that $L^{**} = L$.

6. Show that if $\mathscr{D}$ is the operator d^4/dx^4, then

$$X\mathscr{D}[Y] - Y\mathscr{D}[X] = \frac{d}{dx}(XY''' - YX''' - X'Y'' + Y'X'').$$

Thus show that if X_1 and X_2 are eigenfunctions of the fourth-order eigenvalue problem

$$\mathscr{D}[X] + \lambda X = 0, \qquad X(0) = X''(0) = 0, \qquad X(c) = X''(c) = 0,$$

corresponding to distinct eigenvalues λ_1 and λ_2, then X_1 is orthogonal to X_2 on the interval $0 < x < c$ with weight function unity.

44.   UNIQUENESS OF EIGENFUNCTIONS

The theory of ordinary differential equations ensures the existence and uniqueness of solutions of certain types of *initial value problems*, problems in which all boundary data are given at one point. We state here as a lemma, without proof, a fundamental result from the theory of ordinary second-order linear equations that we shall use in discussing the uniqueness of eigenfunctions.[†]

Lemma. Let A, B, and C denote continuous functions of x on an interval $a \leqslant x \leqslant b$. If x_0 is a point in that interval and y_0 and y_0' are prescribed constants, then there is one and only one function y, continuous together with its derivative y' when $a \leqslant x \leqslant b$, that satisfies the differential equation

$$y''(x) + A(x)y'(x) + B(x)y(x) = C(x) \qquad (a < x < b)$$

and the two initial conditions

$$y(x_0) = y_0, \qquad y'(x_0) = y_0'.$$

Note that since $y'' = C - Ay' - By$, y'' is continuous when $a < x < b$. Also, since any values can be assigned to the constants y_0 and y_0', the general solution of the differential equation has two arbitrary constants.

Suppose now that X and Y are two eigenfunctions corresponding to the same eigenvalue λ of the regular Sturm-Liouville problem

(1) $$(rX')' + (q + \lambda p)X = 0 \qquad (a < x < b),$$

(2) $$a_1 X(a) + a_2 X'(a) = 0, \qquad b_1 X(b) + b_2 X'(b) = 0.$$

As stated in Sec. 41, *the functions p, q, r, and r' are continuous on the interval $a \leqslant x \leqslant b$; also, $p(x) > 0$ and $r(x) > 0$ when $a \leqslant x \leqslant b$.* The above lemma

---

[†] A proof of the lemma can be found in, for instance, the book by Coddington (1989, chap. 6) that is listed in the Bibliography.

enables us to prove that X and Y can differ by at most a constant factor; that is,

(3) $$Y(x) = CX(x),$$

where C is a nonzero constant.

We start the proof by observing that, in view of the principle of superposition, the linear combination

(4) $$Z(x) = Y'(a)X(x) - X'(a)Y(x)$$

satisfies the linear homogeneous differential equation

(5) $$(rZ')' + (q + \lambda p)Z = 0 \qquad (a < x < b);$$

in addition, $Z'(a) = 0$. Since X and Y satisfy the conditions

$$a_1 X(a) + a_2 X'(a) = 0,$$

$$a_1 Y(a) + a_2 Y'(a) = 0,$$

where a_1 and a_2 are not both zero, and since $Z(a)$ is the determinant of that pair of linear homogeneous equations in a_1 and a_2, we also know that $Z(a) = 0$. According to the lemma, then, $Z(z) = 0$ when $a \leqslant x \leqslant b$. That is,

(6) $$Y'(a)X(x) - X'(a)Y(x) = 0 \qquad (a \leqslant x \leqslant b).$$

Since eigenfunctions cannot be identically zero, it is clear from relation (6) that if either of the values $X'(a)$ or $Y'(a)$ is zero, then so is the other.

Relation (3) now follows from equation (6), provided that $X'(a)$ and $Y'(a)$ are nonzero. Suppose, on the other hand, that $X'(a) = Y'(a) = 0$. Then $X(a)$ and $Y(a)$ are nonzero since, otherwise, X and Y would be identically zero, according to the lemma; and zero is not an eigenfunction. The procedure applied to $Z(x)$ may now be used to show that the linear combination

(7) $$W(x) = Y(a)X(x) - X(a)Y(x)$$

is zero when $a \leqslant x \leqslant b$ and hence that relation (3) still holds.

It follows immediately from relation (3) that, except possibly for a nonzero constant factor, any eigenfunction X of problem (1)–(2) is real-valued. To show this, we first recall from the corollary in Sec. 43 that the eigenvalue λ to which X corresponds must be real. So if we make the substitution $X = U + iV$, where U and V are real-valued functions, in problem (1)–(2) and separate real and imaginary parts, we find that U and V are themselves eigenfunctions corresponding to λ. Hence there is a nonzero constant β such that $V = \beta U$. Here β is real since U and V are real-valued, and we may conclude that

$$X = U + i\beta U = (1 + i\beta)U.$$

That is, X can be expressed as a nonzero constant times a real-valued function. Note, too, how one can write

$$U = \left(\frac{1}{1 + i\beta} \right) X.$$

We collect our results as follows.

Theorem. *If X and Y are eigenfunctions corresponding to the same eigenvalue of a regular Sturm-Liouville problem, then $Y = CX$, where C is a nonzero constant. Also, each eigenfunction can be made real-valued by multiplying it by an appropriate nonzero constant.*

According to the theorem, a regular Sturm-Liouville problem cannot have two linearly independent eigenfunctions corresponding to the same eigenvalue. For certain modifications of regular Sturm-Liouville problems, however, it is possible to have an eigenvalue with linearly independent eigenfunctions (see Sec. 40).

The following corollary, which uses the fact that there is always a real-valued eigenfunction corresponding to a given eigenvalue of problem (1)–(2), is an additional aid in determining eigenvalues since it often eliminates the possibility that there are negative ones. We already know from the corollary in Sec. 43 that each eigenvalue must be real.

Corollary. *If λ is an eigenvalue of the regular Sturm-Liouville problem (1)–(2) and if the conditions $q(x) \leqslant 0$ $(a \leqslant x \leqslant b)$ and $a_1 a_2 \leqslant 0$, $b_1 b_2 \geqslant 0$ are satisfied, then $\lambda \geqslant 0$.*

To prove this, we let X denote a real-valued eigenfunction corresponding to the eigenvalue λ. Equation (1) is thus satisfied, and we multiply each term of that equation by X and integrate each of the resulting terms from $x = a$ to $x = b$:

$$(8) \qquad \int_a^b X(rX')' \, dx + \int_a^b qX^2 \, dx + \lambda \int_a^b pX^2 \, dx = 0.$$

After applying integration by parts to the first of these integrals, we can write equation (8) in the form

$$(9) \qquad \lambda \int_a^b pX^2 \, dx = \int_a^b (-qX^2) \, dx + \int_a^b r(X')^2 \, dx$$

$$+ \, r(a)X(a)X'(a) - r(b)X(b)X'(b).$$

Let us now assume that the conditions stated in the corollary are satisfied. Since $-q(x) \geqslant 0$ and $r(x) > 0$ when $a \leqslant x \leqslant b$, the values of the two integrals on the right in equation (9) are clearly nonnegative. As for the third term on the right, we note that if $a_1 = 0$ or $a_2 = 0$ in the first of conditions (2), then $X'(a) = 0$ or $X(a) = 0$, respectively. In either case, the third term is zero. If, on the other hand, neither a_1 nor a_2 is zero, then

$$r(a)X(a)X'(a) = \frac{r(a)[a_1 X(a)]^2}{-a_1 a_2} \geqslant 0.$$

Similarly, $-r(b)X(b)X'(b) \geqslant 0$; and it follows that all the terms on the right-hand side of equation (9) are nonnegative. Consequently,

$$\lambda \int_a^b p(x)[X(x)]^2 \, dx \geq 0.$$

But this integral has positive value, and so $\lambda \geq 0$.

45. METHODS OF SOLUTION

We turn now to two examples that illustrate methods to be used in finding the eigenvalues and eigenfunctions of the problems that follow this section. The basic method has already been touched on in Secs. 27 and 40, where simpler Sturm-Liouville problems were solved. It consists of first finding the general solution of a differential equation and then applying the boundary conditions in order to determine the eigenvalues.

EXAMPLE 1. Let us solve the regular Sturm-Liouville problem

(1) $$X'' + \lambda X = 0 \qquad\qquad (0 < x < c),$$

(2) $$X'(0) = 0, \qquad hX(c) + X'(c) = 0,$$

where h is a positive constant.

From the corollary in Sec. 44, we know that there are no negative eigenvalues. If $\lambda = 0$, the general solution of equation (1) is $X(x) = Ax + B$, where A and B are constants; and it follows from boundary conditions (2) that $A = 0$ and $B = 0$. But eigenfunctions cannot be identically zero. Consequently, the number $\lambda = 0$ is not an eigenvalue. This leaves only the possibility that $\lambda > 0$.

If $\lambda > 0$, we write $\lambda = \alpha^2$ ($\alpha > 0$). The general solution of equation (1) this time is

$$X(x) = C_1 \cos \alpha x + C_2 \sin \alpha x.$$

It reduces to

(3) $$X(x) = C_1 \cos \alpha x$$

when the first of boundary conditions (2) is applied. The second boundary condition then requires that

(4) $$C_1(h \cos \alpha c - \alpha \sin \alpha c) = 0.$$

If the function (3) is to be nontrivial, the constant C_1 must be nonzero. Hence the factor in parentheses in equation (4) must be equal to zero. That is, if we are to have an eigenvalue $\lambda = \alpha^2$ ($\alpha > 0$), the number α must be a positive root of the equation

(5) $$\tan \alpha c = \frac{h}{\alpha}.$$

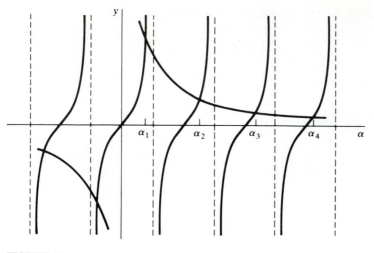

FIGURE 45

Figure 45, where the graphs of $y = \tan \alpha c$ and $y = h/\alpha$ are plotted, shows that equation (5) has an infinite number of positive roots $\alpha_1, \alpha_2, \ldots$, where

$$\alpha_n < \alpha_{n+1} \qquad\qquad (n = 1, 2, \ldots);$$

they are the positive values of α for which those graphs intersect. The eigenvalues are, then, the numbers $\lambda_n = \alpha_n^2$ ($n = 1, 2, \ldots$). We identify them by simply writing

$$(6) \qquad\qquad \lambda_n = \alpha_n^2, \qquad \text{where} \qquad \tan \alpha_n c = \frac{h}{\alpha_n} \qquad\qquad (\alpha_n > 0).$$

Note that the dashed vertical lines in Fig. 45 are equally spaced π/c units apart. Also, as n tends to infinity, the numbers α_n tend to be the positive roots of the equation $\tan \alpha c = 0$. More precisely, we see from Fig. 45 that when n is large, α_n is approximately $(n-1)\pi/c$. For various values of the constant $a = hc$, the first few positive roots $x_1, x_2, \ldots$ of the equation $\tan x = a/x$ have been tabulated, and it follows from equation (5) that $\alpha_1 = x_1/c, \alpha_2 = x_2/c, \ldots$.[†] In view of the above remarks, the eigenvalues $\lambda_n = \alpha_n^2$ are approximately $[(n-1)\pi/c]^2$ when n is large. This is in agreement with the statement, made earlier in Sec. 41, that if λ_n ($n = 1, 2, \ldots$) are the eigenvalues of a regular Sturm-Liouville problem, arranged in ascending order of magnitude, then it is always true that $\lambda_n \to \infty$ as $n \to \infty$.

[†] Roots of this and the related equation $\tan x = ax$, arising in some of the problems of this section, are tabulated in, for example, the handbook edited by Abramowitz and Stegun (1972, pp. 224–225) that is listed in the Bibliography.

Expression (3) now tells us that, except for constant factors, the corresponding eigenfunctions are $X_n(x) = \cos \alpha_n x$ $(n = 1, 2, \ldots)$. Let us put these eigenfunctions in *normalized* form, the form that we shall need in the applications. To accomplish this, we note that the functions $X_n(x)$ are orthogonal on the interval $0 < x < c$ with weight function unity, according to the theorem in Sec. 43. Thus

$$\|X_n\|^2 = \int_0^c \cos^2 \alpha_n x\, dx = \frac{1}{2} \int_0^c (1 + \cos 2\alpha_n x)\, dx = \frac{1}{2}\left(c + \frac{\sin 2\alpha_n c}{2\alpha_n}\right);$$

and the relations $\sin 2\alpha_n c = 2 \sin \alpha_n c \cos \alpha_n c$ and $\alpha_n = h/\tan \alpha_n c$ enable us to write this expression for $\|X_n\|^2$ in the form

$$(7) \qquad\qquad \|X_n\|^2 = \frac{hc + \sin^2 \alpha_n c}{2h},$$

which is obviously positive since h and c are positive. Dividing each $X_n(x)$ by $\|X_n\|$, we then arrive at the normalized eigenfunctions

$$(8) \qquad\qquad \phi_n(x) = \sqrt{\frac{2h}{hc + \sin^2 \alpha_n c}} \cos \alpha_n x \qquad (n = 1, 2, \ldots).$$

Sometimes the solutions of a given Sturm-Liouville problem are most easily obtained by transforming the problem into one whose solutions are known. This has already been indicated in Problem 1(a), Sec. 43, and the next example illustrates the method more fully.

EXAMPLE 2. We consider here the problem

$$(9) \qquad\qquad (xX')' + \frac{\lambda}{x} X = 0 \qquad\qquad (1 < x < b),$$

$$(10) \qquad\qquad X'(1) = 0, \qquad hX(b) + X'(b) = 0,$$

where h is a positive constant.

Since equation (9) can be put in Cauchy-Euler form (see Problem 3, Sec. 35),

$$x^2 X'' + x X' + \lambda X = 0,$$

the substitution $x = \exp s$ transforms it into the equation

$$(11) \qquad\qquad \frac{d^2 X}{ds^2} + \lambda X = 0 \qquad\qquad (0 < s < \ln b).$$

Also, since

$$\frac{dX}{dx} = \frac{dX}{ds} e^{-s},$$

the boundary conditions (10) become

$$(12) \quad \frac{dX}{ds} = 0 \quad \text{when} \quad s = 0, \quad (hb)X + \frac{dX}{ds} = 0 \quad \text{when} \quad s = \ln b.$$

Hence, by referring to Example 1, we see immediately that the eigenvalues of problem (11)–(12), and therefore of problem (9)–(10), are the numbers

$$(13) \quad \lambda_n = \alpha_n^2, \quad \text{where} \quad \tan(\alpha_n \ln b) = \frac{hb}{\alpha_n} \quad (\alpha_n > 0).$$

The corresponding eigenfunctions are evidently

$$X_n = \cos \alpha_n s = \cos(\alpha_n \ln x) \quad (n = 1, 2, \dots).$$

From equations (9) and (11), we know that the weight functions for the eigenfunctions $X_n = \cos(\alpha_n \ln x)$ and $X_n = \cos \alpha_n s$ are $1/x$ and 1, respectively. The value of the norm $\|X_n\|$ is, however, the same regardless of whether we think of X_n as a function of x or s. For the substitution $x = \exp s$ ($s = \ln x$) shows that

$$\int_1^b \frac{1}{x} \cos^2(\alpha_n \ln x) \, dx = \int_0^{\ln b} \cos^2 \alpha_n s \, ds.$$

So, in view of expression (7), the normalized eigenfunctions of problem (9)–(10) are

$$(14) \quad \phi_n(x) = \sqrt{\frac{2hb}{hb \ln b + \sin^2(\alpha_n \ln b)}} \cos(\alpha_n \ln x) \quad (n = 1, 2, \dots).$$

PROBLEMS

In Problems 1 through 5, solve directly (without referring to any other problems) for the eigenvalues and normalized eigenfunctions.

1. $X'' + \lambda X = 0,$ $X(0) = 0,$ $X'(1) = 0.$

   Answer: $\lambda_n = \alpha_n^2,$ $\phi_n(x) = \sqrt{2} \sin \alpha_n x$ $(n = 1, 2, \dots)$; $\alpha_n = \dfrac{(2n-1)\pi}{2}.$

2. $X'' + \lambda X = 0,$ $X(0) = 0,$ $hX(1) + X'(1) = 0$ $(h > 0).$

   Answer: $\lambda_n = \alpha_n^2,$ $\phi_n(x) = \sqrt{\dfrac{2h}{h + \cos^2 \alpha_n}} \sin \alpha_n x$ $(n = 1, 2, \dots);$

   $\tan \alpha_n = \dfrac{-\alpha_n}{h}$ $(\alpha_n > 0).$

3. $X'' + \lambda X = 0,$ $X'(0) = 0,$ $X(c) = 0.$

   Answer: $\lambda_n = \alpha_n^2,$ $\phi_n(x) = \sqrt{\dfrac{2}{c}} \cos \alpha_n x$ $(n = 1, 2, \dots);$ $\alpha_n = \dfrac{(2n-1)\pi}{2c}.$

4. $X'' + \lambda X = 0,$ $X(0) = 0,$ $X(1) - X'(1) = 0.$

 Suggestion: The trigonometric identity $\cos^2 A = 1/(1 + \tan^2 A)$ is useful in putting $\|X_n\|^2$ in a form that leads to the expression for $\phi_n(x)$ in the answer below.

$$\text{Answer: } \lambda_0 = 0, \lambda_n = \alpha_n^2, \phi_0(x) = \sqrt{3}\,x, \phi_n(x) = \frac{\sqrt{2(\alpha_n^2 + 1)}}{\alpha_n}\sin \alpha_n x$$

$(n = 1, 2, \dots)$; $\tan \alpha_n = \alpha_n$ $(\alpha_n > 0).$

5. $X'' + \lambda X = 0,$ $hX(0) - X'(0) = 0,$ $X(1) = 0$ $(h > 0).$

$$\text{Answer: } \lambda_n = \alpha_n^2, \phi_n(x) = \sqrt{\frac{2h}{h + \cos^2 \alpha_n}}\sin \alpha_n(1 - x) \quad (n = 1, 2, \dots);$$

$\tan \alpha_n = -\dfrac{\alpha_n}{h}$ $(\alpha_n > 0).$

6. In Problem 1(a), Sec. 43, the eigenvalues and eigenfunctions of the Sturm-Liouville problem

$$(xX')' + \frac{\lambda}{x}X = 0,\qquad X(1) = 0,\qquad X(b) = 0$$

were found to be

$$\lambda_n = \alpha_n^2,\qquad X_n(x) = \sin(\alpha_n \ln x) \qquad (n = 1, 2, \dots),$$

where $\alpha_n = n\pi/\ln b$. Show that the *normalized* eigenfunctions are

$$\phi_n(x) = \sqrt{\frac{2}{\ln b}}\sin(\alpha_n \ln x) \qquad (n = 1, 2, \dots).$$

 Suggestion: The integral that arises can be evaluated by making the substitution $s = (\pi/\ln b)\ln x$ and referring to the integration formula (10) in Example 1, Sec. 11.

7. Find the eigenvalues and normalized eigenfunctions of the Sturm-Liouville problem

$$X'' + \lambda X = 0,\qquad X(0) = 0,\qquad X'(c) = 0$$

by making the substitution $s = x/c$ and referring to the solutions of Problem 1.

$$\text{Answer: } \lambda_n = \alpha_n^2, \phi_n(x) = \sqrt{\frac{2}{c}}\sin \alpha_n x \ (n = 1, 2, \dots); \alpha_n = \frac{(2n - 1)\pi}{2c}.$$

8. (a) Show that the solutions obtained in Problem 2 can be written

$$\lambda_n = \alpha_n^2,\qquad \phi_n(x) = \sqrt{\frac{2(\alpha_n^2 + h^2)}{\alpha_n^2 + h^2 + h}}\sin \alpha_n x \quad (n = 1, 2, \dots),$$

where $\alpha_n \cos \alpha_n = -h \sin \alpha_n$ $(\alpha_n > 0).$

(b) By referring to the solutions of Problem 1, point out why the solutions in part (a) here are actually valid solutions of Problem 2 when $h \geqslant 0$, not just when $h > 0$. *Suggestion:* In part (a), use the trigonometric identity

$$\cos^2 A = \frac{1}{1 + \tan^2 A}.$$

9. Use the solutions obtained in Problem 3 to find the eigenvalues and normalized eigenfunctions of the Sturm-Liouville problem

$$(xX')' + \frac{\lambda}{x}X = 0, \qquad X'(1) = 0, \qquad X(b) = 0.$$

Answer:

$$\lambda_n = \alpha_n^2, \; \phi_n(x) = \sqrt{\frac{2}{\ln b}} \; \cos(\alpha_n \ln x) \; (n = 1, 2, \ldots); \; \alpha_n = \frac{(2n-1)\pi}{2 \ln b}.$$

10. By making an appropriate substitution and referring to the known solutions of the same problem on a different interval in the section indicated, find the eigenfunctions of the Sturm-Liouville problem
   (a) $X'' + \lambda X = 0$, $X'(-\pi) = 0$, $X'(\pi) = 0$ (Sec. 27);
   (b) $X'' + \lambda X = 0$, $X(-c) = X(c)$, $X'(-c) = X'(c)$ (Sec. 40).

   Answers: (a) $\dfrac{1}{2}$, $\cos \dfrac{n(x+\pi)}{2}$ $(n = 1, 2, \ldots)$;

   (b) $\dfrac{1}{2}$, $\cos \dfrac{n\pi x}{c}$, $\sin \dfrac{n\pi x}{c}$ $(n = 1, 2, \ldots)$.

11. (a) By making the substitutions

$$X = \frac{Y}{\sqrt{x}} \qquad \text{and} \qquad \lambda = \frac{1}{4} + \mu,$$

   transform the regular Sturm-Liouville problem

$$(x^2 X')' + \lambda X = 0, \qquad X(1) = 0, \qquad X(b) = 0,$$

   where $b > 1$, into the problem

$$(xY')' + \frac{\mu}{x}Y = 0, \qquad Y(1) = 0, \qquad Y(b) = 0.$$

(b) Write the eigenvalues and normalized eigenfunctions of the new problem in part (a) by referring to Problem 6. Then substitute back to show that, for the original problem in part (a), the eigenvalues and normalized eigenfunctions are

$$\lambda_n = \frac{1}{4} + \alpha_n^2, \qquad \phi_n(x) = \sqrt{\frac{2}{x \ln b}} \; \sin(\alpha_n \ln x) \qquad (n = 1, 2, \ldots),$$

   where $\alpha_n = n\pi/\ln b$.

12. Find the eigenfunctions of each of the following Sturm-Liouville problems:
(a) $X'' + \lambda X = 0$, $X(0) = 0$, $hX(1) + X'(1) = 0$ $(h < -1)$;
(b) $(x^3 X')' + \lambda x X = 0$, $X(1) = 0$, $X(e) = 0$.

Answers: (a) $X_0(x) = \sinh \alpha_0 x$, where $\tanh \alpha_0 = -\dfrac{\alpha_0}{h}$ $(\alpha_0 > 0)$,

$X_n(x) = \sin \alpha_n x$ $(n = 1, 2, \ldots)$, where $\tan \alpha_n = -\dfrac{\alpha_n}{h}$ $(\alpha_n > 0)$;

(b) $X_n(x) = \dfrac{1}{x} \sin (n\pi \ln x)$ $(n = 1, 2, \ldots)$.

13. Give details showing that the function $W(x)$ defined by equation (7), Sec. 44, is identically zero on the interval $a \leqslant x \leqslant b$.

46. EXAMPLES OF EIGENFUNCTION EXPANSIONS

We now illustrate how generalized Fourier series representations

$$(1) \qquad\qquad f(x) = \sum_{n=1}^{\infty} c_n \phi_n(x) \qquad\qquad (a < x < b)$$

are obtained when the functions $\phi_n(x)$ $(n = 1, 2, \ldots)$ are the normalized eigenfunctions of specific Sturm-Liouville problems. We have, of course, already illustrated the method when the eigenfunctions are ones leading to Fourier cosine and sine series on the interval $0 < x < \pi$, as well as Fourier series on $-\pi < x < \pi$ (see Secs. 12, 14, and 15). Except for a few cases in which it is easy to establish the validity of an expansion by transforming it into a known Fourier series representation, in this book we do not treat the convergence of series (1) for other eigenfunctions. We merely accept the fact that results analogous to the Fourier theorem and its corollary in Sec. 19 exist when specific eigenfunctions are used. Such results are often obtained with the aid of the theory of residues of functions of a complex variable.[†] Proofs are complicated by the fact that explicit solutions of the Sturm-Liouville differential equation with arbitrary coefficients cannot be written.

EXAMPLE 1. According to Problem 6, Sec. 45, the Sturm-Liouville problem

$$(xX')' + \frac{\lambda}{x}X = 0, \qquad X(1) = 0, \qquad X(b) = 0$$

has eigenvalues and normalized eigenfunctions

$$\lambda_n = \alpha_n^2, \qquad \phi_n(x) = \sqrt{\frac{2}{\ln b}} \sin (\alpha_n \ln x) \qquad (n = 1, 2, \ldots),$$

[†] The theory of eigenfunction expansions is extensively developed in the volumes by Titchmarsh (1962, 1958) that are listed in the Bibliography.

where $\alpha_n = n\pi/\ln b$. Since the orthogonality of the set $\{\phi_n(x)\}$ $(n = 1, 2, \ldots)$ is with respect to the weight function $p(x) = 1/x$, the coefficients in the expansion

$$(2) \qquad\qquad 1 = \sum_{n=1}^{\infty} c_n \phi_n(x) \qquad\qquad (1 < x < b)$$

are

$$c_n = (f, \phi_n) = \sqrt{\frac{2}{\ln b}} \int_1^b \frac{1}{x} \sin(\alpha_n \ln x) \, dx.$$

Making the substitution $s = \ln x$ here and noting that

$$\cos(\alpha_n \ln b) = \cos n\pi = (-1)^n,$$

we readily see that

$$\int_1^b \frac{1}{x} \sin(\alpha_n \ln x) \, dx = \int_0^{\ln b} \sin \alpha_n s \, ds = \frac{1 - (-1)^n}{\alpha_n}.$$

Thus

$$c_n = \sqrt{\frac{2}{\ln b}} \cdot \frac{1 - (-1)^n}{\alpha_n} \qquad\qquad (n = 1, 2, \ldots),$$

and expansion (2) becomes

$$(3) \qquad\qquad 1 = \frac{4}{\ln b} \sum_{n=1}^{\infty} \frac{\sin(\alpha_{2n-1} \ln x)}{\alpha_{2n-1}} \qquad\qquad (1 < x < b).$$

The validity of this representation is evident if we make the substitution

$$(2n - 1)s = \alpha_{2n-1} \ln x$$

and note that $\alpha_{2n-1} = (2n - 1)\pi/\ln b$. For the result,

$$1 = \frac{4}{\pi} \sum_{n=1}^{\infty} \frac{\sin(2n - 1)s}{2n - 1} \qquad\qquad (0 < s < \pi),$$

is a known [Problem 1(b), Sec. 14] Fourier sine series representation on the indicated interval.

EXAMPLE 2. The eigenvalues and normalized eigenfunctions of the Sturm-Liouville problem

$$X'' + \lambda X = 0, \qquad X(0) = 0, \qquad X'(c) = 0$$

are (Problem 7, Sec. 45)

$$\lambda_n = \alpha_n^2, \qquad \phi_n(x) = \sqrt{\frac{2}{c}} \sin \alpha_n x \qquad (n = 1, 2, \ldots),$$

where

$$\alpha_n = \frac{(2n-1)\pi}{2c}.$$

The weight function is $p(x) = 1$, and we may find the coefficients in the expansion

$$x = \sum_{n=1}^{\infty} c_n \phi_n(x) \qquad (0 < x < c)$$

by writing

$$c_n = (f, \phi_n) = \sqrt{\frac{2}{c}} \int_0^c x \sin \alpha_n x \, dx = \sqrt{\frac{2}{c}} \left[-\frac{x \cos \alpha_n x}{\alpha_n} + \frac{\sin \alpha_n x}{\alpha_n^2} \right]_0^c.$$

Since $\cos \alpha_n c = 0$ and $\sin \alpha_n c = (-1)^{n+1}$, this expression for c_n reduces to

$$c_n = \sqrt{\frac{2}{c}} \cdot \frac{(-1)^{n+1}}{\alpha_n^2} \qquad (n = 1, 2, \dots).$$

Hence

(4)
$$x = \frac{2}{c} \sum_{n=1}^{\infty} \frac{(-1)^{n+1}}{\alpha_n^2} \sin \alpha_n x \qquad (0 < x < c).$$

After putting expansion (4) in the form

$$x = \frac{8c}{\pi^2} \sum_{n=1}^{\infty} \frac{(-1)^{n+1}}{(2n-1)^2} \sin \frac{(2n-1)\pi x}{2c} \qquad (0 < x < c),$$

we see from Problem 12, Sec. 21, that it is actually valid on the closed interval $-c \leq x \leq c$. Furthermore, since $\sin \alpha_n(x + 2c) = -\sin \alpha_n x$ $(n = 1, 2, \dots)$, series (4) converges for all x; and if $H(x)$ denotes the sum of that series for each value of x, it is clear that $H(x)$ represents the triangular wave function defined by means of the equations (see Fig. 46)

(5)
$$H(x) = x \qquad (-c \leq x \leq c),$$
$$H(x + 2c) = -H(x) \qquad (-\infty < x < \infty).$$

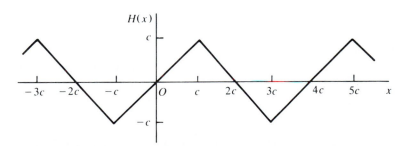

FIGURE 46

Thus $H(x)$ is an *antiperiodic* function, with period $2c$. It is also periodic, with period $4c$, as is seen by writing

$$H(x + 4c) = H(x + 2c + 2c) = -H(x + 2c) = H(x).$$

Note, too, that

$$H(2c - x) = -H(x - 2c) = -H(x + 2c) = H(x).$$

We conclude with an example in which the series obtained is a sine series that cannot be transformed into an ordinary Fourier sine series. We must, therefore, accept the representation without verification.

EXAMPLE 3. We consider here the eigenvalues and normalized eigenfunctions of the Sturm-Liouville problem

$$X'' + \lambda X = 0, \qquad X(0) = 0, \qquad X(1) - X'(1) = 0.$$

According to Problem 4, Sec. 45, they are

$$\lambda_0 = 0, \qquad \lambda_n = \alpha_n^2 \qquad\qquad (n = 1, 2, \ldots)$$

and

$$\phi_0(x) = \sqrt{3}\, x, \qquad \phi_n(x) = \frac{\sqrt{2(\alpha_n^2 + 1)}}{\alpha_n} \sin \alpha_n x \quad (n = 1, 2, \ldots),$$

where $\tan \alpha_n = \alpha_n$ ($\alpha_n > 0$), the weight function being unity. The coefficients in the representation

$$f(x) = c_0 \phi_0(x) + \sum_{n=1}^{\infty} c_n \phi_n(x) \qquad (0 < x < 1)$$

of a piecewise smooth function $f(x)$ are

$$c_0 = (f, \phi_0) = \sqrt{3} \int_0^1 x f(x)\, dx$$

and

$$c_n = (f, \phi_n) = \frac{\sqrt{2(\alpha_n^2 + 1)}}{\alpha_n} \int_0^1 f(x) \sin \alpha_n x\, dx \quad (n = 1, 2, \ldots).$$

Consequently,

$$f(x) = B_0 x + \sum_{n=1}^{\infty} B_n \sin \alpha_n x,$$

where

$$B_0 = 3 \int_0^1 x f(x)\, dx \quad \text{and} \quad B_n = \frac{2(\alpha_n^2 + 1)}{\alpha_n^2} \int_0^1 f(x) \sin \alpha_n x\, dx$$

$$(n = 1, 2, \dots).$$

PROBLEMS

1. Use the normalized eigenfunctions in Problem 3, Sec. 45, to derive the representation

$$1 = \frac{2}{c} \sum_{n=1}^{\infty} \frac{(-1)^{n+1}}{\alpha_n} \cos \alpha_n x \qquad (0 < x < c),$$

where

$$\alpha_n = \frac{(2n-1)\pi}{2c}.$$

2. Derive the expansion

$$1 = \frac{2}{c} \sum_{n=1}^{\infty} \frac{\sin \alpha_n x}{\alpha_n} \qquad (0 < x < c),$$

where

$$\alpha_n = \frac{(2n-1)\pi}{2c},$$

using the normalized eigenfunctions in Problem 7, Sec. 45.

3. Use the normalized eigenfunctions in Problem 2, Sec. 45, to derive the expansion

$$1 = 2h \sum_{n=1}^{\infty} \frac{1 - \cos \alpha_n}{\alpha_n (h + \cos^2 \alpha_n)} \sin \alpha_n x \qquad (0 < x < 1),$$

where $\tan \alpha_n = -\alpha_n / h$ $(\alpha_n > 0)$.

4. Using the normalized eigenfunctions in Problem 3, Sec. 45, when $c = \pi$, show that

$$\pi^2 - x^2 = \frac{4}{\pi} \sum_{n=1}^{\infty} \frac{(-1)^{n+1}}{\alpha_n^3} \cos \alpha_n x \qquad (0 < x < \pi),$$

where $\alpha_n = (2n-1)/2$.

5. (a) Use the normalized eigenfunctions in Problem 7, Sec. 45, to obtain the expansion

$$x(2c - x) = \frac{4}{c} \sum_{n=1}^{\infty} \frac{\sin \alpha_n x}{\alpha_n^3} \qquad (0 < x < c),$$

where

$$\alpha_n = \frac{(2n-1)\pi}{2c}.$$

(b) Show how it follows from the result in Problem 8, Sec. 21, that the series in part (a) converges for all x and that its sum is the antiperiodic function $Q(x)$ (see Example 2, Sec. 46), with period $2c$, that can be described by means of the equations

$$Q(x) = x(2c - x) \ (0 \leqslant x \leqslant 2c), \qquad Q(x + 2c) = -Q(x) \ (-\infty < x < \infty).$$

6. Using the normalized eigenfunctions in Problem 2, Sec. 45, derive the representation

$$x\left(\frac{2 + h}{1 + h} - x\right) = 4h \sum_{n=1}^{\infty} \frac{1 - \cos \alpha_n}{\alpha_n^3 (h + \cos^2 \alpha_n)} \sin \alpha_n x \qquad (0 < x < 1),$$

where $\tan \alpha_n = -\alpha_n / h \ (\alpha_n > 0)$.

Suggestion: In the simplifications, it is useful to note that

$$-h \sin \alpha_n = \alpha_n \cos \alpha_n.$$

7. Use the normalized eigenfunctions in Problem 1, Sec. 45, to show that

$$\sin \omega x = 2\omega \cos \omega \sum_{n=1}^{\infty} \frac{(-1)^n}{\omega^2 - \omega_n^2} \sin \omega_n x \qquad (0 < x < 1),$$

where $\omega_n = (2n - 1)\pi/2$ and $\omega \neq \omega_n$ for any value of n.

Suggestion: The trigonometric identity

$$2 \sin A \sin B = \cos (A - B) - \cos (A + B)$$

is useful in evaluating the integrals that arise.

8. Find the Fourier constants c_n for the function $f(x) = x \ (1 < x < b)$ with respect to the normalized eigenfunctions in Problem 6, Sec. 45, and reduce them to the form

$$c_n = \sqrt{2 \ln b} \ \frac{n\pi\left[1 + (-1)^{n+1}b\right]}{(\ln b)^2 + (n\pi)^2} \qquad (n = 1, 2, \ldots).$$

Suggestion: The integration formula

$$\int e^x \sin ax \, dx = \frac{e^x(\sin ax - a \cos ax)}{1 + a^2},$$

derived in calculus, is useful here.

9. Let f be a piecewise smooth function defined on the interval $1 < x < b$.
   (a) Use the normalized eigenfunctions in Problem 6, Sec. 45, to show formally that if $\alpha_n = n\pi/\ln b$, then

$$f(x) = \sum_{n=1}^{\infty} B_n \sin (\alpha_n \ln x) \qquad (1 < x < b),$$

where

$$B_n = \frac{2}{\ln b} \int_1^b \frac{1}{x} f(x) \sin (\alpha_n \ln x) \, dx \qquad (n = 1, 2, \ldots).$$

   (b) By making the substitution $x = \exp s$ in the series and integral in part (a) and then referring to the corollary in Sec. 21, verify that the series representation in part (a) is valid at all points in the interval $1 < x < b$ at which f is continuous. (Compare Example 1, Sec. 46.)

10. Suppose that a function f, defined on the interval $0 < x < c$, is piecewise smooth there.

(a) Use the normalized eigenfunctions (Problem 7, Sec. 45)

$$\phi_n(x) = \sqrt{\frac{2}{c}} \sin \alpha_n x \qquad\qquad (n = 1, 2, \dots),$$

where

$$\alpha_n = \frac{(2n - 1)\pi}{2c},$$

to show formally that

$$f(x) = \sum_{n=1}^{\infty} B_n \sin \alpha_n x \qquad\qquad (0 < x < c),$$

where

$$B_n = \frac{2}{c} \int_0^c f(x) \sin \alpha_n x \, dx \qquad\qquad (n = 1, 2, \dots).$$

(b) Note that, according to Problem 11, Sec. 21, the series in part (a) is actually a Fourier sine series for an extension of f on the interval $0 < x < 2c$. Then, with the aid of the corollary in Sec. 21, state why the representation in part (a) is valid for each point x ($0 < x < c$) at which f is continuous.

11. (a) Use the normalized eigenfunctions

$$\phi_n(x) = \sqrt{2} \sin \alpha_n x \qquad\qquad (n = 1, 2, \dots),$$

where

$$\alpha_n = \frac{(2n - 1)\pi}{2},$$

in Problem 1, Sec. 45, to show formally that

$$x\left(1 - \frac{1}{3}x^2\right) = 4 \sum_{n=1}^{\infty} \frac{(-1)^{n+1}}{\alpha_n^4} \sin \alpha_n x \qquad\qquad (0 < x < 1).$$

(b) Note that, according to Problem 10(a) above and Problem 11, Sec. 21, the series in part (a) here is a Fourier sine series on the interval $0 < x < 2$. Then, with the aid of the corollary in Sec. 21, show that the series in part (a) converges for all x and that its sum is the antiperiodic function $Q(x)$, with period 2, that is described by the equations

$$Q(x) = x\left(1 - \frac{1}{3}x^2\right) \ (-1 \leqslant x \leqslant 1), \qquad Q(x + 2) = -Q(x) \ (-\infty < x < \infty).$$

47. SURFACE HEAT TRANSFER

The following two examples illustrate the Fourier method in solving temperature problems in rectangular coordinates when Sturm-Liouville problems other than those used in Chap. 4 arise. Here, and in the rest of the chapter, we seek only *formal* solutions of the boundary value problems.

EXAMPLE 1. Let $u(x,t)$ denote temperatures in a slab $0 \leqslant x \leqslant 1$, initially at temperatures $f(x)$, when the face $x = 0$ is insulated and surface heat transfer takes place at the face $x = 1$ into a medium at temperature zero (Fig. 47). According to Newton's law of cooling (Sec. 3), the condition on u at the face $x = 1$ is $u_x(1, t) = -hu(1, t)$, where h is a positive constant. The boundary value problem to be solved is, then,

$$(1) \qquad\qquad u_t(x,t) = ku_{xx}(x,t) \qquad\qquad (0 < x < 1, t > 0),$$

$$(2) \qquad u_x(0, t) = 0, \qquad u_x(1, t) = -hu(1, t), \qquad u(x,0) = f(x).$$

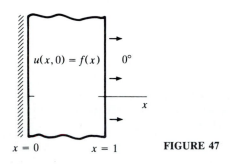

$u(x,0) = f(x)$ $0°$

$x = 0$ $x = 1$ **FIGURE 47**

Writing $u = X(x)T(t)$ and separating variables, we arrive at the Sturm-Liouville problem

$$(3) \qquad X''(x) + \lambda X(x) = 0, \qquad X'(0) = 0, \qquad hX(1) + X'(1) = 0,$$

along with the condition $T'(t) + \lambda kT(t) = 0$. The eigenvalues and normalized eigenfunctions of problem (3) are, according to Example 1, Sec. 45, $\lambda_n = \alpha_n^2$ and

$$(4) \qquad\qquad X_n = \phi_n(x) = \sqrt{\frac{2h}{h + \sin^2 \alpha_n}} \cos \alpha_n x \qquad (n = 1, 2, \dots),$$

where $\tan \alpha_n = h/\alpha_n$ ($\alpha_n > 0$). The corresponding functions of t are evidently constant multiples of

$$T_n(t) = \exp\left(-\alpha_n^2 kt\right) \qquad\qquad (n = 1, 2, \dots).$$

Hence the formal solution of our temperature problem is

$$(5) \qquad\qquad u(x,t) = \sum_{n=1}^{\infty} c_n \exp\left(-\alpha_n^2 kt\right) \phi_n(x),$$

where, in order that $u(x,0) = f(x)$ $(0 < x < 1)$,

(6) $$c_n = (f, \phi_n) = \int_0^1 f(x)\phi_n(x)\, dx$$

$$= \sqrt{\frac{2h}{h + \sin^2 \alpha_n}} \int_0^1 f(x) \cos \alpha_n x\, dx \qquad (n = 1, 2, \dots).$$

Observe that series (5), when expression (4) for $\phi_n(x)$ is substituted, is

$$u(x,t) = \sum_{n=1}^{\infty} \left(\sqrt{\frac{2h}{h + \sin^2 \alpha_n}}\, c_n \right) \exp\left(-\alpha_n^2 kt\right) \cos \alpha_n x.$$

Hence the solution just obtained can be written

(7) $$u(x,t) = \sum_{n=1}^{\infty} A_n \exp\left(-\alpha_n^2 kt\right) \cos \alpha_n x,$$

where

(8) $$A_n = \frac{2h}{h + \sin^2 \alpha_n} \int_0^1 f(x) \cos \alpha_n x\, dx \qquad (n = 1, 2, \dots).$$

It is easy to show that solution (7), with coefficients (8), also satisfies the boundary value problem

(9) $$u_t(x,t) = ku_{xx}(x,t) \qquad (-1 < x < 1, t > 0),$$

(10) $$u_x(-1,t) = hu(-1,t), \qquad u_x(1,t) = -hu(1,t) \qquad (t > 0),$$

(11) $$u(x,0) = f(x) \qquad (-1 < x < 1)$$

when f is an *even* function, or when $f(-x) = f(x)$ $(-1 < x < 1)$. For we already know that u satisfies the heat equation and the second of boundary conditions (10). Since the cosine function is even, it is clear from expression (7) that u is even in x; and its partial derivative u_x is odd in x. Hence the first of boundary conditions (10) is also satisfied:

$$u_x(-1,t) = -u_x(1,t) = hu(1,t) = hu(-1,t).$$

Finally, we already know that $u(x,0) = f(x)$ when $0 < x < 1$; furthermore, when $-1 < x < 0$, the fact that u and f are even in x enables us to write

$$u(x,0) = u(-x,0) = f(-x) = f(x).$$

The boundary value problem (9)–(11) is, of course, a temperature problem for a slab $-1 \leqslant x \leqslant 1$ with initial temperatures (11) and with surface heat transfer at both faces into a medium at temperature zero (Fig. 48). The solution when f is not necessarily even is obtained in the problems.

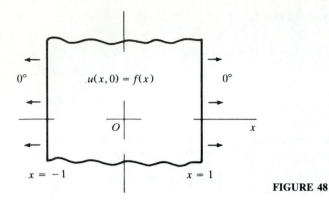

FIGURE 48

EXAMPLE 2. Let $u(x, y)$ denote the bounded steady temperatures in a semi-infinite slab bounded by the planes $x = 0$, $x = \pi$, and $y = 0$ (Fig. 49), whose faces are subject to the following conditions. The face in the plane $x = 0$ is insulated, the face in the plane $x = \pi$ is kept at temperature zero, and the flux inward through the face in the plane $y = 0$ (see Sec. 3) is a prescribed function $f(x)$. The boundary value problem for this slab is

$$(12) \qquad u_{xx}(x, y) + u_{yy}(x, y) = 0 \qquad (0 < x < \pi, \, y > 0),$$

$$(13) \qquad u_x(0, y) = 0, \qquad u(\pi, y) = 0 \qquad (y > 0),$$

$$(14) \qquad -Ku_y(x, 0) = f(x) \qquad (0 < x < \pi),$$

where K is a positive constant.

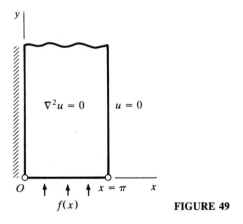

FIGURE 49

By assuming a product solution $u = X(x)Y(y)$ of conditions (12) and (13) and separating variables, we find that

$$(15) \qquad X''(x) + \lambda X(x) = 0, \qquad X'(0) = 0, \qquad X(\pi) = 0$$

and that $Y(y)$ is to be a bounded solution of the differential equation

(16) $$Y''(y) - \lambda Y(y) = 0.$$

According to Problem 3, Sec. 45, the eigenvalues and normalized eigenfunctions of the Sturm-Liouville problem (15) are

$$\lambda_n = \alpha_n^2, \qquad X_n = \phi_n(x) = \sqrt{\frac{2}{\pi}} \cos \alpha_n x \quad (n = 1, 2, \ldots),$$

where $\alpha_n = (2n - 1)/2$. The corresponding *bounded* solutions of equation (16) are constant multiples of the functions

$$Y_n(y) = \exp(-\alpha_n y) \qquad (n = 1, 2, \ldots).$$

Consequently,

(17) $$u(x, y) = \sum_{n=1}^{\infty} c_n \exp(-\alpha_n y) \phi_n(x).$$

Applying the nonhomogeneous condition (14) to this expression, we see that the constants c_n must be such that

$$f(x) = \sum_{n=1}^{\infty} (Kc_n \alpha_n) \phi_n(x) \qquad (0 < x < \pi).$$

That is,

(18) $$Kc_n \alpha_n = (f, \phi_n) = \sqrt{\frac{2}{\pi}} \int_0^\pi f(x) \cos \alpha_n x \, dx \quad (n = 1, 2, \ldots).$$

Finally, it follows from expressions (17) and (18) that

(19) $$u(x, y) = \frac{1}{K} \sum_{n=1}^{\infty} A_n \frac{\exp(-\alpha_n y)}{\alpha_n} \cos \alpha_n x,$$

where

(20) $$A_n = \frac{2}{\pi} \int_0^\pi f(x) \cos \alpha_n x \, dx \qquad (n = 1, 2, \ldots).$$

Since $\alpha_n = (2n - 1)/2$, equations (19) and (20) can, of course, be written in the form

$$u(x, y) = \frac{2}{K} \sum_{n=1}^{\infty} A_n \frac{\exp[-(2n-1)y/2]}{2n - 1} \cos \frac{(2n-1)x}{2},$$

where

$$A_n = \frac{2}{\pi} \int_0^\pi f(x) \cos \frac{(2n-1)x}{2} \, dx \qquad (n = 1, 2, \ldots).$$

48. POLAR COORDINATES

We consider here a Dirichlet problem (Sec. 7) for a function $u(\rho, \phi)$, involving polar coordinates, that satisfies Laplace's equation

$$(1) \qquad \rho^2 u_{\rho\rho}(\rho, \phi) + \rho u_\rho(\rho, \phi) + u_{\phi\phi}(\rho, \phi) = 0 \qquad (1 < \rho < b, 0 < \phi < \pi)$$

and the boundary conditions (Fig. 50)

$$(2) \qquad\qquad u(\rho, 0) = 0, \qquad u(\rho, \pi) = u_0 \qquad\qquad (1 < \rho < b),$$

$$(3) \qquad\qquad u(1, \phi) = 0, \qquad u(b, \phi) = 0 \qquad\qquad (0 < \phi < \pi),$$

where u_0 is a constant.

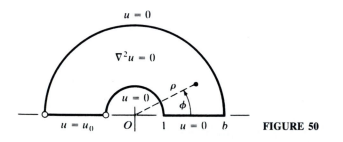

FIGURE 50

    Substitution of the product $u = R(\rho)\Phi(\phi)$ into the homogeneous conditions here yields these conditions on R and Φ:

$$(4) \qquad [\rho R'(\rho)]' + \frac{\lambda}{\rho} R(\rho) = 0, \qquad R(1) = 0, \qquad R(b) = 0,$$

$$(5) \qquad\qquad \Phi''(\phi) - \lambda\Phi(\phi) = 0, \qquad \Phi(0) = 0.$$

Conditions (4) on R make up a Sturm-Liouville problem whose eigenvalues are $\lambda_n = \alpha_n^2$ $(n = 1, 2, \ldots)$, where $\alpha_n = n\pi/\ln b$, and whose normalized eigenfunctions are

$$R_n = \phi_n(\rho) = \sqrt{\frac{2}{\ln b}} \, \sin(\alpha_n \ln \rho) \qquad (n = 1, 2, \ldots).$$

(See Problem 6, Sec. 45.) Note that the weight function for these eigenfunctions is $1/\rho$. Except for constant factors, the corresponding functions of ϕ, arising from conditions (5), are

$$\Phi_n(\phi) = \sinh \alpha_n \phi \qquad (n = 1, 2, \ldots).$$

Hence

$$(6) \qquad\qquad u(\rho, \phi) = \sum_{n=1}^{\infty} c_n \sinh \alpha_n \phi \, \phi_n(\rho).$$

Turning to the nonhomogeneous condition $u(\rho, \pi) = u_0$, we set $\phi = \pi$ in expression (6) and write

$$u_0 = \sum_{n=1}^{\infty} (c_n \sinh \alpha_n \pi) \phi_n(\rho) \qquad (1 < \rho < b).$$

Evidently, then,

$$c_n \sinh \alpha_n \pi = (u_0, \phi_n) = u_0 \sqrt{\frac{2}{\ln b}} \int_1^b \frac{1}{\rho} \sin (\alpha_n \ln \rho) \, d\rho.$$

This integral is readily evaluated by making the substitution $\rho = \exp s$; and, by recalling that $\alpha_n = n\pi/\ln b$, one can simplify the result to show that

$$(7) \qquad\qquad c_n \sinh \alpha_n \pi = \frac{u_0 \sqrt{2 \ln b}}{\pi} \cdot \frac{1 - (-1)^n}{n}.$$

So, in view of expressions (6) and (7),

$$u(\rho, \phi) = \frac{2u_0}{\pi} \sum_{n=1}^{\infty} \frac{1 - (-1)^n}{n} \cdot \frac{\sinh \alpha_n \phi}{\sinh \alpha_n \pi} \sin (\alpha_n \ln \rho).$$

That is,

$$(8) \qquad\qquad u(\rho, \phi) = \frac{4u_0}{\pi} \sum_{n=1}^{\infty} \frac{\sinh \alpha_{2n-1} \phi}{\sinh \alpha_{2n-1} \pi} \cdot \frac{\sin (\alpha_{2n-1} \ln \rho)}{2n - 1}.$$

It is interesting to contrast this solution with the one obtained in Example 2, Sec. 34, for a Dirichlet problem involving the same region but with the nonhomogeneous condition $u = u_0$ occurring when $\rho = b$ instead of when $\phi = \pi$.

PROBLEMS[†]

1. Show that when $f(x) = 1 \ (0 < x < 1)$ in the boundary value problem in Example 1, Sec. 47, the solution (7)–(8) there reduces to

$$u(x, t) = 2h \sum_{n=1}^{\infty} \frac{\sin \alpha_n}{\alpha_n (h + \sin^2 \alpha_n)} \exp (-\alpha_n^2 kt) \cos \alpha_n x,$$

where $\tan \alpha_n = h/\alpha_n \ (\alpha_n > 0)$.

2. Show that if the condition $u(\rho, \pi) = u_0 \ (1 < \rho < b)$ in Sec. 48 is replaced by the condition $u(\rho, \pi) = \rho \ (1 < \rho < b)$, then

$$u(\rho, \phi) = 2\pi \sum_{n=1}^{\infty} \frac{n[1 + (-1)^{n+1} b]}{(\ln b)^2 + (n\pi)^2} \cdot \frac{\sinh \alpha_n \phi}{\sinh \alpha_n \pi} \sin (\alpha_n \ln \rho),$$

where $\alpha_n = n\pi/\ln b$.

[†] The eigenvalues and (normalized) eigenfunctions of any Sturm-Liouville problem that arises have already been found in Sec. 45 or in one of the problems of that section.

Suggestion: The Fourier constants found in Problem 8, Sec. 46, can be used here.

3. Use the normalized eigenfunctions of the Sturm-Liouville problem

$$X'' + \lambda X = 0, \qquad X(0) = 0, \qquad X'(\pi) = 0$$

to solve the boundary value problem

$$u_t(x, t) = k u_{xx}(x, t) \qquad (0 < x < \pi, t > 0),$$

$$u(0, t) = 0, \qquad u_x(\pi, t) = 0, \qquad u(x, 0) = f(x).$$

Show that the solution can be written

$$u(x, t) = \sum_{n=1}^{\infty} b_{2n-1} \exp\left[-\frac{(2n-1)^2 k}{4} t \right] \sin \frac{(2n-1)x}{2},$$

where

$$b_{2n-1} = \frac{2}{\pi} \int_0^{\pi} f(x) \sin \frac{(2n-1)x}{2} \, dx \qquad (n = 1, 2, \dots).$$

(The solution in this form was obtained in another way in Example 3, Sec. 32.)

4. Solve the boundary value problem

$$u_{xx}(x, y) + u_{yy}(x, y) = 0 \qquad (0 < x < a, 0 < y < b),$$

$$u_x(0, y) = 0, \qquad u_x(a, y) = -hu(a, y) \qquad (0 < y < b),$$

$$u(x, 0) = 0, \qquad u(x, b) = f(x) \qquad (0 < x < a),$$

where h is a positive constant, and interpret $u(x, y)$ physically.

Answer: $u(x, y) = 2h \sum_{n=1}^{\infty} \dfrac{\cos \alpha_n x}{ha + \sin^2 \alpha_n a} \cdot \dfrac{\sinh \alpha_n y}{\sinh \alpha_n b} \displaystyle\int_0^a f(s) \cos \alpha_n s \, ds,$

where $\tan \alpha_n a = h/\alpha_n$ $(\alpha_n > 0)$.

5. A bounded harmonic function $u(x, y)$ in the semi-infinite strip $x > 0, 0 < y < 1$ is to satisfy the boundary conditions

$$u(x, 0) = 0, \qquad u_y(x, 1) = -hu(x, 1), \qquad u(0, y) = u_0,$$

where h $(h > 0)$ and u_0 are constants. Derive the expression

$$u(x, y) = 2hu_0 \sum_{n=1}^{\infty} \frac{1 - \cos \alpha_n}{\alpha_n (h + \cos^2 \alpha_n)} \exp(-\alpha_n x) \sin \alpha_n y,$$

where $\tan \alpha_n = -\alpha_n/h$ $(\alpha_n > 0)$. Interpret $u(x, y)$ physically.

6. Find the bounded harmonic function $u(x, y)$ in the semi-infinite strip $0 < x < 1$, $y > 0$ that satisfies the boundary conditions

$$u_x(0, y) = 0, \qquad u_x(1, y) = -hu(1, y), \qquad u(x, 0) = f(x),$$

where h is a positive constant, and interpret $u(x, y)$ physically.

Answer: $u(x, y) = \sum_{n=1}^{\infty} A_n \exp(-\alpha_n y) \cos \alpha_n x,$

where $\tan \alpha_n = h/\alpha_n$ $(\alpha_n > 0)$ and

$$A_n = \frac{2h}{h + \sin^2 \alpha_n} \int_0^1 f(x) \cos \alpha_n x \, dx \qquad (n = 1, 2, \dots).$$

7. Find the bounded solution of this boundary value problem, where b and h are positive constants:

$$u_{xx}(x, y) + u_{yy}(x, y) - bu(x, y) = 0 \quad (0 < x < 1, \, y > 0),$$

$$u(0, y) = 0, \qquad u_x(1, y) = -hu(1, y), \qquad u(x, 0) = f(x).$$

Answer: $u(x, y) = \displaystyle\sum_{n=1}^{\infty} B_n \frac{\sin \alpha_n x}{\exp\left(y\sqrt{b + \alpha_n^2}\right)},$

where $\tan \alpha_n = -\alpha_n/h$ $(\alpha_n > 0)$ and

$$B_n = \frac{2h}{h + \cos^2 \alpha_n} \int_0^1 f(x) \sin \alpha_n x \, dx \qquad (n = 1, 2, \dots).$$

8. Let ρ, ϕ, z denote cylindrical coordinates, and solve the following boundary value problem in the region $1 \leqslant \rho \leqslant b, 0 \leqslant \phi \leqslant \pi$ of the plane $z = 0$:

$$\rho^2 u_{\rho\rho}(\rho, \phi) + \rho u_\rho(\rho, \phi) + u_{\phi\phi}(\rho, \phi) = 0 \quad (1 < \rho < b, 0 < \phi < \pi),$$

$$u_\rho(1, \phi) = 0, \qquad u_\rho(b, \phi) = -hu(b, \phi) \qquad (0 < \phi < \pi),$$

$$u_\phi(\rho, 0) = 0, \qquad u(\rho, \pi) = u_0 \qquad (1 < \rho < b),$$

where h $(h > 0)$ and u_0 are constants. Interpret the function $u(\rho, \phi)$ physically.

Answer: $u(\rho, \phi) = 2hbu_0 \displaystyle\sum_{n=1}^{\infty} \frac{\cosh \alpha_n \phi}{\cosh \alpha_n \pi} \cdot \frac{\sin(\alpha_n \ln b) \cos(\alpha_n \ln \rho)}{\alpha_n \left[hb \ln b + \sin^2(\alpha_n \ln b) \right]},$

where $\tan(\alpha_n \ln b) = hb/\alpha_n$ $(\alpha_n > 0)$.

9. Give a full physical interpretation of the following temperature problem, involving a time-dependent diffusivity, and derive its solution:

$$(t + 1)u_t(x, t) = u_{xx}(x, t) \qquad (0 < x < 1, \, t > 0),$$

$$u(0, t) = 0, \qquad u_x(1, t) = 0, \qquad u(x, 0) = 1.$$

Answer: $u(x, t) = 2 \displaystyle\sum_{n=1}^{\infty} (t + 1)^{-\alpha_n^2} \frac{\sin \alpha_n x}{\alpha_n},$ where $\alpha_n = \dfrac{(2n - 1)\pi}{2}.$

10. (a) Give a physical interpretation of the boundary value problem

$$u_t(x, t) = ku_{xx}(x, t) \qquad (0 < x < 1, \, t > 0),$$

$$u(0, t) = 0, \qquad u_x(1, t) = -hu(1, t), \qquad u(x, 0) = f(x),$$

where h is a positive constant. Then derive the solution

$$u(x, t) = \sum_{n=1}^{\infty} B_n \exp\left(-\alpha_n^2 kt\right) \sin \alpha_n x,$$

where $\tan \alpha_n = -\alpha_n/h$ $(\alpha_n > 0)$ and

$$B_n = \frac{2h}{h + \cos^2 \alpha_n} \int_0^1 f(x) \sin \alpha_n x \, dx \qquad (n = 1, 2, \dots).$$

(*b*) Use an argument similar to the one at the end of Example 1 in Sec. 47 to show that the solution found in part (*a*) formally satisfies the boundary value problem (9)–(11) in that example when the function f there is *odd*, or when $f(-x) = -f(x)$ $(-1 < x < 1)$.

11. Use the following method to solve the temperature problem (see Fig. 48 in Sec. 47)

$$u_t(x,t) = ku_{xx}(x,t) \qquad (-1 < x < 1, t > 0),$$
$$u_x(-1,t) = hu(-1,t), \qquad u_x(1,t) = -hu(1,t) \qquad (t > 0),$$
$$u(x,0) = f(x) \qquad (-1 < x < 1)$$

when the function f is not necessarily even or odd, as it was in Example 1, Sec. 47, and Problem 10(*b*).

(*a*) Show that if $v(x,t)$ is the solution of the problem when $f(x)$ is replaced by the function

$$G(x) = \frac{f(x) + f(-x)}{2}$$

and if $w(x,t)$ is the solution when $f(x)$ is replaced by

$$H(x) = \frac{f(x) - f(-x)}{2},$$

then the sum $u(x,t) = v(x,t) + w(x,t)$ satisfies the above boundary value problem.

(*b*) After noting that the functions G and H in part (*a*) are even and odd, respectively, apply the result there, together with results in Example 1, Sec. 47, and Problem 10, to show that

$$u(x,t) = \sum_{n=1}^{\infty} A_n \exp\left(-\alpha_n^2 kt\right) \cos \alpha_n x + \sum_{n=1}^{\infty} B_n \exp\left(-\beta_n^2 kt\right) \sin \beta_n x,$$

where $\tan \alpha_n = h/\alpha_n$ $(\alpha_n > 0)$, $\tan \beta_n = -\beta_n/h$ $(\beta_n > 0)$ and

$$A_n = \frac{h}{h + \sin^2 \alpha_n} \int_{-1}^1 f(x) \cos \alpha_n x \, dx,$$

$$B_n = \frac{h}{h + \cos^2 \beta_n} \int_{-1}^1 f(x) \sin \beta_n x \, dx.$$

Suggestion: A procedure similar to the one in the suggestion with Problem 12, Sec. 14, can be used in obtaining the expressions for A_n and B_n in part (*b*).

49. MODIFICATIONS OF THE METHOD

In this section, we illustrate two modifications of the Fourier method involving normalized eigenfunctions. Both such modifications were used in Chap. 4 when only ordinary Fourier cosine and sine series arose.

EXAMPLE 1. If heat is introduced through the face $x = 1$ of a slab $0 \leqslant x \leqslant 1$ at a uniform rate A $(A > 0)$ per unit area (Sec. 3) while the face $x = 0$ is kept at the initial temperature zero of the slab, then the temperature function $u(x, t)$ satisfies the conditions

(1) $u_t(x, t) = k u_{xx}(x, t)$ $(0 < x < 1, t > 0)$,

(2) $u(0, t) = 0, \qquad K u_x(1, t) = A$ $(t > 0)$,

(3) $u(x, 0) = 0$ $(0 < x < 1)$.

Because the second of conditions (2) is nonhomogeneous, we do not have two-point boundary conditions leading to a Sturm-Liouville problem. But, by writing

(4) $u(x, t) = U(x, t) + \Phi(x)$

(compare Example 2, Sec. 32), we find that conditions (1)–(3) become

$$U_t(x, t) = k[U_{xx}(x, t) + \Phi''(x)],$$

$$U(0, t) + \Phi(0) = 0, \qquad K[U_x(1, t) + \Phi'(1)] = A,$$

and

$$U(x, 0) + \Phi(x) = 0.$$

Hence, if we require that

(5) $\Phi'' = 0$ and $\Phi(0) = 0, \qquad K\Phi'(1) = A,$

we have a boundary value problem for $U(x, t)$ that does have two-point boundary conditions leading to a Sturm-Liouville problem:

(6) $U_t = k U_{xx}, \qquad U(0, t) = 0, \qquad U_x(1, t) = 0, \qquad U(x, 0) = -\Phi(x).$

It follows readily from conditions (5) that

(7) $\Phi(x) = \dfrac{A}{K} x.$

Also, by assuming a product solution $U = X(x)T(t)$ of the homogeneous conditions in problem (6), we see that

(8) $X''(x) + \lambda X(x) = 0, \qquad X(0) = 0, \qquad X'(1) = 0$

and $T'(t) + \lambda k T(t) = 0$. According to Problem 1, Sec. 45, the Sturm-Liouville problem (8) has the eigenvalues and normalized eigenfunctions

$$\lambda_n = \alpha_n^2, \qquad X_n = \phi_n(x) = \sqrt{2} \sin \alpha_n x \qquad (n = 1, 2, \ldots),$$

where $\alpha_n = (2n - 1)\pi/2$; and the corresponding functions of t are $T_n(t) = \exp(-\alpha_n^2 k t)$ $(n = 1, 2, \ldots)$. Hence

(9) $U(x, t) = \displaystyle\sum_{n=1}^{\infty} c_n \exp\left(-\alpha_n^2 k t\right) \phi_n(x),$

where, in view of the last of conditions (6),

$$-\frac{A}{K}x = \sum_{n=1}^{\infty} c_n \phi_n(x) \qquad (0 < x < 1).$$

Now the Fourier constants for x $(0 < x < 1)$ with respect to the normalized eigenfunctions here are already known to us (see Example 2, Sec. 46, when $c = 1$), and that earlier result tells us that

$$c_n = \sqrt{2}\,\frac{A}{K} \cdot \frac{(-1)^n}{\alpha_n^2}.$$

After substituting these values of c_n into expression (9) and simplifying and combining the result with expression (7), as indicated in equation (4), we arrive at the desired temperature function:

$$(10) \qquad u(x,t) = \frac{A}{K}\left[x + 2\sum_{n=1}^{\infty} \frac{(-1)^n}{\alpha_n^2} \exp\left(-\alpha_n^2 kt\right)\sin \alpha_n x \right],$$

where $\alpha_n = (2n - 1)\pi/2$.

EXAMPLE 2. Let $u(x,t)$ denote temperatures in a slab $0 \leqslant x \leqslant \pi$ (Fig. 51) that is initially at temperature zero and whose face $x = 0$ is insulated, while the face $x = \pi$ has temperatures $u(\pi,t) = t$ $(t \geqslant 0)$. If the unit of time is chosen so that the thermal diffusivity k in the heat equation is unity, the boundary value problem for $u(x,t)$ is

$$(11) \qquad\qquad u_t(x,t) = u_{xx}(x,t) \qquad\qquad (0 < x < \pi, t > 0),$$

$$(12) \qquad u_x(0,t) = 0, \qquad u(\pi,t) = t, \qquad u(x,0) = 0.$$

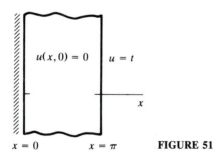

$x = 0$ $\qquad\qquad$ $x = \pi$ $\qquad\qquad$ **FIGURE 51**

Observe that if $u(x,t)$ satisfies the first two of conditions (12), then the related function $U(x,t) = u(x,t) - t$ satisfies the conditions

$$(13) \qquad\qquad U_x(0,t) = 0 \qquad \text{and} \qquad U(\pi,t) = 0,$$

both of which are homogeneous. In fact, by writing

$$u(x,t) = U(x,t) + t,$$

we have the related boundary value problem consisting of the differential equation

(14) $$U_t(x,t) = U_{xx}(x,t) - 1$$

and conditions (13), along with the condition

(15) $$U(x,0) = 0.$$

The nonhomogeneity in the second of conditions (12) is now transferred to the differential equation in the new boundary value problem, consisting of equations (13)–(15), and this suggests applying the method of variation of parameters, first used in Sec. 33.

We begin by noting that when the method of separation of variables is applied to the homogeneous differential equation $U_t(x,t) = U_{xx}(x,t)$, which is equation (14) with the -1 term deleted, and conditions (13), the Sturm-Liouville problem

$$X''(x) + \lambda X(x) = 0, \qquad X'(0) = 0, \qquad X(\pi) = 0$$

arises. Furthermore, from Problem 3, Sec. 45, we know that the eigenfunctions of this problem are the cosine functions $\cos \alpha_n x$ ($n = 1, 2, \ldots$), where $\alpha_n = (2n - 1)/2$. We thus seek a solution of the boundary value problem (13)–(15) having the form

(16) $$U(x,t) = \sum_{n=1}^{\infty} A_n(t) \cos \alpha_n x.$$

By substituting series (16) into equation (14) and referring to Problem 1, Sec. 46, for the expansion

$$1 = \frac{2}{\pi} \sum_{n=1}^{\infty} \frac{(-1)^{n+1}}{\alpha_n} \cos \alpha_n x \qquad (0 < x < \pi),$$

we find that

$$\sum_{n=1}^{\infty} \left[A_n'(t) + \alpha_n^2 A_n(t) \right] \cos \alpha_n x = \sum_{n=1}^{\infty} \frac{2(-1)^n}{\pi \alpha_n} \cos \alpha_n x.$$

Then, by identifying the coefficients in the eigenfunction expansions on each side here, we have the differential equation

(17) $$A_n'(t) + \alpha_n^2 A_n(t) = \frac{2(-1)^n}{\pi \alpha_n} \qquad (n = 1, 2, \ldots).$$

Also, condition (15) tells us that

$$\sum_{n=1}^{\infty} A_n(0) \cos \alpha_n x = 0,$$

or $A_n(0) = 0$ ($n = 1, 2, \ldots$).

Now an integrating factor for the linear first-order differential equation (17) is

$$\exp \int \alpha_n^2 \, dt = \exp \alpha_n^2 t.$$

Hence, if we multiply through the differential equation by this integrating factor, we have

$$\frac{d}{dt}\left[(\exp \alpha_n^2 t) \, A_n(t)\right] = \frac{2(-1)^n}{\pi \alpha_n} \exp \alpha_n^2 t.$$

By replacing the variable t here by τ, integrating the result from $\tau = 0$ to $\tau = t$, and keeping in mind the requirement that $A_n(0) = 0$, we see that

$$(\exp \alpha_n^2 t) \, A_n(t) = \frac{2(-1)^n}{\pi \alpha_n^3} (\exp \alpha_n^2 t - 1),$$

or

$$A_n(t) = \frac{2(-1)^n}{\pi \alpha_n^3}\left[1 - \exp\left(-\alpha_n^2 t\right)\right] \qquad (n = 1, 2, \dots).$$

Finally, by substituting this expression for $A_n(t)$ into equation (16) and then recalling that $u(x, t) = U(x, t) + t$, we obtain the solution of the original boundary value problem:

$$(18) \qquad u(x, t) = t + \frac{2}{\pi} \sum_{n=1}^{\infty} \frac{(-1)^n}{\alpha_n^3}\left[1 - \exp\left(-\alpha_n^2 t\right)\right] \cos \alpha_n x,$$

where $\alpha_n = (2n - 1)/2$.

Note that, in view of the representation found in Problem 4, Sec. 46, this solution can also be written as

$$(19) \qquad u(x, t) = t - \frac{\pi^2 - x^2}{2} - \frac{2}{\pi} \sum_{n=1}^{\infty} \frac{(-1)^n}{\alpha_n^3} \exp\left(-\alpha_n^2 t\right) \cos \alpha_n x.$$

PROURLEMS[†]

1. With the aid of representation (4) in Example 2, Sec. 46, show that the temperature function (10) in Example 1, Sec. 49, can be written in the form

$$u(x, t) = \frac{2A}{K} \sum_{n=1}^{\infty} \frac{(-1)^{n+1}}{\alpha_n^2}\left[1 - \exp\left(-\alpha_n^2 kt\right)\right] \sin \alpha_n x,$$

where $\alpha_n = (2n - 1)\pi/2$.

[†] The footnote with the problem set for Sec. 48 also applies here.

2. Heat transfer takes place at the surface $x = 0$ of a slab $0 \leqslant x \leqslant 1$ into a medium at temperature zero according to the linear law of surface heat transfer, so that (Sec. 3)

$$u_x(0, t) = hu(0, t) \qquad\qquad (h > 0).$$

The other boundary conditions are as indicated in Fig. 52, and the unit of time is chosen so that $k = 1$ in the heat equation. By proceeding as in Example 1, Sec. 49, derive the temperature formula

$$u(x, t) = \frac{hx + 1}{h + 1} - 2h \sum_{n=1}^{\infty} \frac{\sin \alpha_n (1 - x)}{\alpha_n (h + \cos^2 \alpha_n)} \exp\left(-\alpha_n^2 t\right),$$

where $\tan \alpha_n = -\alpha_n / h \; (\alpha_n > 0)$.

Suggestion: In simplifying the expression for the Fourier constants that arise, it is useful to note that

$$-\frac{h \sin \alpha_n}{\alpha_n^2} = \frac{\cos \alpha_n}{\alpha_n}.$$

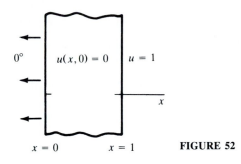

$0°$ | $u(x, 0) = 0$ | $u = 1$

x

$x = 0 \qquad\qquad x = 1 \qquad\qquad$ **FIGURE 52**

3. Use the method of variation of parameters to solve the boundary value problem

$$u_t(x, t) = ku_{xx}(x, t) + q(t) \qquad\qquad (0 < x < 1, t > 0),$$

$$u_x(0, t) = 0, \qquad u(1, t) = 0, \qquad u(x, 0) = 0$$

for temperatures in an internally heated slab.

Suggestion: The representation, with $c = 1$, that was found in Problem 1, Sec. 46, is needed here.

$$\text{Answer:} \; u(x, t) = 2 \sum_{n=1}^{\infty} \frac{(-1)^{n+1}}{\alpha_n} \cos \alpha_n x \int_0^t \exp\left[-\alpha_n^2 k (t - \tau)\right] q(\tau) \, d\tau,$$

where $\alpha_n = (2n - 1)\pi / 2$.

4. Solve the temperature problem

$$u_t(x, t) = u_{xx}(x, t) \qquad\qquad (0 < x < 1, t > 0),$$

$$u_x(0, t) = 0, \qquad u(1, t) = F(t), \qquad u(x, 0) = 0,$$

where F is continuous and $F(0) = 0$. (Compare Example 2, Sec. 49.) Express the answer in the form

$$u(x,t) = F(t) + 2\sum_{n=1}^{\infty} \frac{(-1)^n}{\alpha_n} \cos \alpha_n x \int_0^t \exp\left[-\alpha_n^2(t-\tau)\right] F'(\tau)\, d\tau,$$

where $\alpha_n = (2n-1)\pi/2$.

5. Use the method in Example 1, Sec. 49, to solve the boundary value problem

$$(t+1)u_t(x,t) = u_{xx}(x,t) \qquad (0 < x < 1, t > 0),$$

$$u_x(0,t) = -1, \qquad u(1,t) = 0, \qquad u(x,0) = 0.$$

Interpret this problem physically (compare Problem 9, Sec. 48).

Answer: $u(x,t) = 1 - x - 2\displaystyle\sum_{n=1}^{\infty} (t+1)^{-\alpha_n^2} \frac{\cos \alpha_n x}{\alpha_n^2}$,

where $\alpha_n = (2n-1)\pi/2$.

6. By using the method of variation of parameters, derive the bounded solution of this problem:

$$u_{xx}(x,y) + u_{yy}(x,y) + q_0 = 0 \qquad (0 < x < 1, y > 0),$$

$$u(0,y) = 0, \qquad u_x(1,y) = -hu(1,y) \quad (h > 0), \qquad u(x,0) = 0,$$

where q_0 and h are constants. Interpret the problem physically.

Suggestion: The representation found in Problem 3, Sec. 46, is needed here. Also, for general comments on solving a nonhomogeneous linear second-order differential equation that arises, see the suggestion with Problem 8, Sec. 35.

Answer: $u(x,y) = 2q_0 h \displaystyle\sum_{n=1}^{\infty} \frac{1 - \cos \alpha_n}{\alpha_n^3(h + \cos^2 \alpha_n)}\left[1 - \exp(-\alpha_n y)\right] \sin \alpha_n x$,

where $\tan \alpha_n = -\alpha_n/h$ $(\alpha_n > 0)$.

7. With the aid of the representation found in Problem 6, Sec. 46, write the solution in Problem 6 above as

$$u(x,t) = \frac{q_0}{2}\left[x\left(\frac{2+h}{1+h} - x\right) - 4h\sum_{n=1}^{\infty} \frac{1 - \cos \alpha_n}{\alpha_n^3(h + \cos^2 \alpha_n)} \exp(-\alpha_n y) \sin \alpha_n x\right],$$

where $\tan \alpha_n = -\alpha_n/h$ $(\alpha_n > 0)$. Then observe how it follows that

$$\lim_{y \to \infty} u(x,y) = \frac{q_0}{2}x\left(\frac{2+h}{1+h} - x\right).$$

8. The boundary $r = 1$ of a *solid sphere* is kept insulated, and that solid is initially at temperatures $f(r)$. If $u(r,t)$ denotes subsequent temperatures, then

$$\frac{\partial u}{\partial t} = \frac{k}{r}\frac{\partial^2}{\partial r^2}(ru), \qquad u_r(1,t) = 0, \qquad u(r,0) = f(r).$$

By writing $v(r,t) = ru(r,t)$ and noting that u is continuous when $r = 0$ (compare Problem 6, Sec. 32), set up a boundary value problem in v, involving the boundary conditions

$$v(0,t) = 0, \qquad v(1,t) = v_r(1,t), \qquad v(r,0) = rf(r).$$

Then derive the temperature formula

$$u(r,t) = B_0 + \sum_{n=1}^{\infty} B_n \exp\left(-\alpha_n^2 kt\right) \frac{\sin \alpha_n r}{\alpha_n r},$$

where $\tan \alpha_n = \alpha_n$ ($\alpha_n > 0$) and

$$B_0 = 3 \int_0^1 r^2 f(r)\, dr, \qquad B_n = 2\left(\alpha_n + \frac{1}{\alpha_n}\right) \int_0^1 rf(r) \sin \alpha_n r\, dr \quad (n = 1, 2, \dots).$$

(An eigenfunction expansion similar to the one required here was found in Example 3, Sec. 46.)

50. A VERTICALLY HUNG ELASTIC BAR

An unstrained elastic bar, or heavy coiled spring, is clamped along its length c so as to prevent longitudinal displacements and then hung from its end $x = 0$ (Fig. 53). At the instant $t = 0$, the clamp is released and the bar vibrates longitudinally because of its own weight. If $y(x,t)$ denotes longitudinal displacements in the bar once it is released, then $y(x,t)$ satisfies the modified form

$$(1) \qquad\qquad y_{tt}(x,t) = a^2 y_{xx}(x,t) + g \qquad\qquad (0 < x < c, t > 0)$$

of the wave equation, where g is the acceleration due to gravity. The stated conditions at the ends of the bar tell us that

$$(2) \qquad\qquad y(0,t) = 0, \qquad y_x(c,t) = 0,$$

the initial conditions being

$$(3) \qquad\qquad y(x,0) = 0, \qquad y_t(x,0) = 0.$$

The fact that equation (1) is nonhomogeneous suggests that we use the method of variation of parameters. More precisely, we seek a solution of our

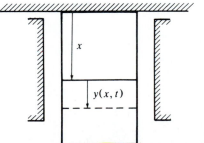

FIGURE 53

boundary value problem having the form

$$(4) \qquad y(x,t) = \sum_{n=1}^{\infty} B_n(t) \sin \alpha_n x,$$

where

$$\alpha_n = \frac{(2n-1)\pi}{2c}.$$

We have chosen the sine functions $\sin \alpha_n x$ $(n = 1, 2, \ldots)$ here since they are the eigenfunctions (Problem 7, Sec. 45) of the Sturm-Liouville problem

$$X''(x) + \lambda X(x) = 0, \qquad X(0) = 0, \qquad X'(c) = 0,$$

which arises when the method of separation of variables is applied to the homogeneous wave equation $y_{xx}(x, t) = a^2 y_{tt}(x, t)$ and conditions (2). Substituting series (4) into equation (1) and recalling the representation (Problem 2, Sec. 46)

$$1 = \frac{2}{c} \sum_{n=1}^{\infty} \frac{\sin \alpha_n x}{\alpha_n} \qquad (0 < x < c),$$

we find that

$$\sum_{n=1}^{\infty} \left[B_n''(t) + (\alpha_n a)^2 B_n(t) \right] \sin \alpha_n x = \sum_{n=1}^{\infty} \frac{2g}{c\alpha_n} \sin \alpha_n x.$$

That is,

$$(5) \qquad B_n''(t) + (\alpha_n a)^2 B_n(t) = \frac{2g}{c\alpha_n} \qquad (n = 1, 2, \ldots).$$

It follows, moreover, from conditions (3) that

$$(6) \qquad B_n(0) = 0, \qquad B_n'(0) = 0 \qquad (n = 1, 2, \ldots).$$

Now the general solution of the complementary equation

$$B_n''(t) + (\alpha_n a)^2 B_n(t) = 0$$

is

$$B_n(t) = C_1 \cos \alpha_n at + C_2 \sin \alpha_n at,$$

where C_1 and C_2 are arbitrary constants, and it is easy to see that a particular solution of equation (5) is

$$B_n(t) = \frac{2g}{a^2 c \alpha_n^3}.$$

Hence the general solution of equation (5) is

$$(7) \qquad B_n(t) = C_1 \cos \alpha_n at + C_2 \sin \alpha_n at + \frac{2g}{a^2 c \alpha_n^3}.$$

The constants C_1 and C_2 are readily determined by imposing conditions (6) on expression (7). The result is

$$B_n(t) = \frac{2g}{a^2 c \alpha_n^3}(1 - \cos \alpha_n at);$$

and, in view of equation (4), this means that

(8) $$y(x,t) = \frac{2g}{a^2 c}\sum_{n=1}^{\infty}\frac{\sin \alpha_n x}{\alpha_n^3}(1 - \cos \alpha_n at).$$

This solution can actually be written in closed form in the following way. We first recall from Problem 5, Sec. 46, that

(9) $$\frac{4}{c}\sum_{n=1}^{\infty}\frac{\sin \alpha_n x}{\alpha_n^3} = Q(x) \qquad\qquad (-\infty < x < \infty),$$

where $Q(x)$ is the antiperiodic function, with period $2c$, described by means of the equations

(10)
$$Q(x) = x(2c - x) \qquad\qquad (0 \leqslant x \leqslant 2c),$$
$$Q(x + 2c) = -Q(x) \qquad\qquad (-\infty < x < \infty).$$

Thus we can put expression (8) in the form

(11) $$y(x,t) = \frac{g}{2a^2}\left[x(2c - x) - \frac{4}{c}\sum_{n=1}^{\infty}\frac{\sin \alpha_n x \cos \alpha_n at}{\alpha_n^3}\right].$$

As for the remaining series here, the trigonometric identity

$$2\sin A \cos B = \sin(A + B) + \sin(A - B)$$

enables us to write

$$2\sum_{n=1}^{\infty}\frac{\sin \alpha_n x \cos \alpha_n at}{\alpha_n^3} = \sum_{n=1}^{\infty}\frac{\sin \alpha_n(x + at)}{\alpha_n^3} + \sum_{n=1}^{\infty}\frac{\sin \alpha_n(x - at)}{\alpha_n^3},$$

or

$$\frac{4}{c}\sum_{n=1}^{\infty}\frac{\sin \alpha_n x \cos \alpha_n at}{\alpha_n^3} = \frac{Q(x + at) + Q(x - at)}{2}.$$

Finally, then,

(12) $$y(x,t) = \frac{g}{2a^2}\left[x(2c - x) - \frac{Q(x + at) + Q(x - at)}{2}\right].$$

PROBLEMS†

1. A horizontal elastic bar, with its end $x = 0$ kept fixed, is initially stretched so that its longitudinal displacements are $y(x, 0) = bx$ $(0 \leqslant x \leqslant c)$. It is released from rest in that position at the instant $t = 0$; and its end $x = c$ is kept free, so that $y_x(c, t) = 0$. Derive this expression for the displacements:

$$y(x, t) = \frac{b}{2}[H(x + at) + H(x - at)],$$

where $H(x)$ is the triangular wave function (5) in Example 2, Sec. 46. (Except for the condition at $x = 0$, the boundary value problem here is the same as the one solved in Sec. 37.)

2. Suppose that the end $x = 0$ of a horizontal elastic bar of length c is kept fixed and that a constant force F_0 per unit area acts parallel to the bar at the end $x = c$. Let all parts of the bar be initially unstrained and at rest. The displacements $y(x, t)$ then satisfy the boundary value problem

$$y_{tt}(x, t) = a^2 y_{xx}(x, t) \qquad (0 < x < c, t > 0),$$

$$y(0, t) = 0, \qquad E y_x(c, t) = F_0,$$

$$y(x, 0) = 0, \qquad y_t(x, 0) = 0,$$

where $a^2 = E/\delta$, E is Young's modulus of elasticity, and δ is the mass per unit volume of the material (see Sec. 6).

(a) Write $y(x, t) = Y(x, t) + \Phi(x)$ (compare Example 1, Sec. 49) and determine $\Phi(x)$ such that $Y(x, t)$ satisfies a boundary value problem whose solution is obtained by simply referring to the solution in Problem 1. Thus show that

$$y(x, t) = \frac{F_0}{E}\left[x - \frac{H(x + at) + H(x - at)}{2}\right],$$

where $H(x)$ is the same triangular wave function as in Problem 1.

(b) Use the expression for $y(x, t)$ in part (a) to show that those displacements are periodic in t, with period

$$T_0 = \frac{4c}{a} = 4c\sqrt{\frac{\delta}{E}}.$$

That is, show that $y(x, t + T_0) = y(x, t)$.

3. Show that the displacements at the end $x = c$ of the bar in Problem 2 are

$$y(c, t) = \frac{F_0}{E}[c + H(at - c)]$$

and that the graph of this function is as shown in Fig. 54.

---

† The footnote with the problem set for Sec. 48 applies here as well.

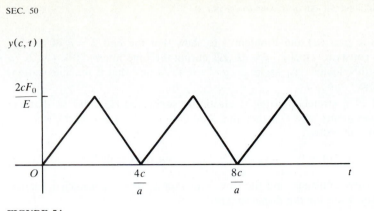

FIGURE 54

4. Show that the force per unit area exerted by the bar in Problem 2 on the support at the end $x = 0$ is the function (see Sec. 6)

$$Ey_x(0, t) = F_0[1 - H'(at)]$$

and that the graph of this function is as shown in Fig. 55. (Note that this force becomes twice the applied force during regularly spaced intervals of time.)

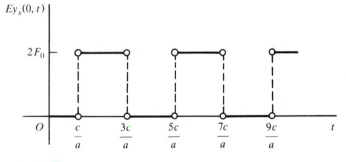

FIGURE 55

5. Let the constant F_0 in Problem 2 be replaced by a finite impulse of duration $4c/a$:

$$F(t) = \begin{cases} F_0 & \text{when } 0 < t < \dfrac{4c}{a}, \\ 0 & \text{when } \quad t > \dfrac{4c}{a}. \end{cases}$$

(a) State why the displacements $y(x, t)$ are the same as in Problem 2 during the time interval $0 < t < 4c/a$. Then, after showing that

$$y(x, 4c/a) = 0 \quad \text{and} \quad y_t(x, 4c/a) = 0$$

when $y(x, t)$ is the solution in Problem 2, state why there is no motion in the bar here after the time $t = 4c/a$.

(b) Use results in part (a) and Problem 3 to show that the end $x = c$ of the bar moves with constant velocity $v_0 = aF_0/E$ during the time interval $0 < t < 2c/a$ and then with velocity $-v_0$ when $2c/a < t < 4c/a$ and that it remains stationary after time $t = 4c/a$.

6. The end $x = 1$ of a stretched string is elastically supported (Fig. 56) so that the transverse displacements $y(x, t)$ satisfy the condition $y_x(1, t) = -hy(1, t)$, where h is a positive constant. Also,

$$y(0, t) = 0, \qquad y(x, 0) = bx, \qquad y_t(x, 0) = 0,$$

where b is a positive constant; and the wave equation $y_{tt} = y_{xx}$ is satisfied. Derive the following expression for the displacements:

$$y(x, t) = 2bh(h + 1) \sum_{n=1}^{\infty} \frac{\sin \alpha_n \sin \alpha_n x}{\alpha_n^2(h + \cos^2 \alpha_n)} \cos \alpha_n t,$$

where $\tan \alpha_n = -\alpha_n/h \ (\alpha_n > 0)$.

Suggestion: In simplifying the solution to the form given here, note that

$$-\frac{\cos \alpha_n}{\alpha_n} = \frac{h \sin \alpha_n}{\alpha_n^2}.$$

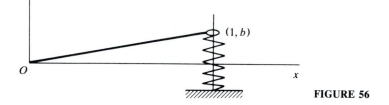

FIGURE 56

7. An unstrained elastic bar of length c, whose cross sections have area A and whose modulus of elasticity (Sec. 6) is E, is moving lengthwise with velocity v_0 when at the instant $t = 0$ its right-hand end $x = c$ meets and adheres to a rigid support (Fig. 57). The displacements $y(x, t)$ thus satisfy the wave equation $y_{tt} = a^2 y_{xx}$ and the end conditions $y_x(0, t) = y(c, t) = 0$, as well as the initial conditions

$$y(x, 0) = 0, \qquad y_t(x, 0) = v_0.$$

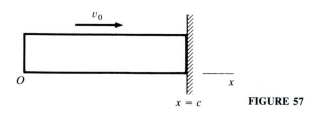

FIGURE 57

(a) Derive this expression for the displacements:

$$y(x,t) = \frac{2v_0}{ac} \sum_{n=1}^{\infty} \frac{(-1)^{n+1}}{\alpha_n^2} \cos \alpha_n x \sin \alpha_n a t,$$

where $\alpha_n = (2n - 1)\pi/2c$ $(n = 1, 2, \ldots)$.

(b) Use the expression for $y(x, t)$ in part (a) to show that

$$y\left(x, \frac{2c}{a}\right) = 0 \quad \text{and} \quad y_t\left(x, \frac{2c}{a}\right) = -v_0 \quad (0 < x < c).$$

According to these two equations, if the end $x = c$ of the bar is suddenly freed from the support at time $t = 2c/a$, the bar will move after that time as a rigid unstrained body with velocity $-v_0$.

(c) Show how it follows from the expression in part (a) that, as long as the end of the bar continues to adhere to the support, the force on the support can be written

$$-AEy_x(c, t) = \frac{AEv_0}{a} M\left(\frac{2c}{a}, t\right),$$

where $M(c, t)$ $(t > 0)$ is the square wave represented by the series (see Problem 9, Sec. 21)

$$M(c, t) = \frac{4}{\pi} \sum_{n=1}^{\infty} \frac{1}{2n - 1} \sin \frac{(2n - 1)\pi t}{c} \quad (t \neq c, 2c, 3c, \ldots).$$

8. Let $y(x, t)$ denote longitudinal displacements in an elastic bar of length unity whose end $x = 0$ is fixed and at whose end $x = 1$ a force proportional to t^2 acts longitudinally (Fig. 58), so that

$$y(0, t) = 0 \quad \text{and} \quad y_x(1, t) = At^2 \qquad (A \neq 0).$$

The bar is initially unstrained and at rest, and the unit of time is such that $a = 1$ in the wave equation.

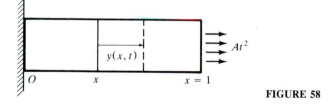

FIGURE 58

(a) Write out the complete boundary value problem for $y(x, t)$; and observe that if $Y(x, t) = y(x, t) - At^2x$, then

$$Y(0, t) = 0 \quad \text{and} \quad Y_x(1, t) = 0.$$

Set up the complete boundary value problem for $Y(x, t)$, the differential equation being

$$Y_{tt}(x, t) = Y_{xx}(x, t) - 2Ax \qquad (0 < x < 1, t > 0).$$

Then, with the aid of representation (4) in Example 2, Sec. 46, apply the method of variation of parameters to solve the boundary value problem for $Y(x, t)$ and thus derive this solution of the original problem:

$$y(x, t) = A\left[xt^2 - 4\sum_{n=1}^{\infty} \frac{(-1)^{n+1}}{\alpha_n^4} (1 - \cos \alpha_n t) \sin \alpha_n x \right],$$

where $\alpha_n = (2n - 1)\pi/2$.

(*b*) Use the result in Problem 11(*b*), Sec. 46, to write the solution in part (*a*) here in the form

$$y(x, t) = A\left[x(t^2 - 1) + \frac{1}{3}x^3 + \frac{Q(x + t) + Q(x - t)}{2} \right],$$

where $Q(x)$ is the antiperiodic function, with period 2, described by the equations

$$Q(x) = x\left(1 - \frac{1}{3}x^2\right) \qquad\qquad (-1 \leqslant x \leqslant 1),$$

$$Q(x + 2) = -Q(x) \qquad\qquad (-\infty < x < \infty).$$

9. Consider the same boundary value problem as in Problem 8 except that the condition at the end $x = 1$ of the bar is now replaced by the condition

$$y_x(1, t) = \sin \omega t.$$

(*a*) By proceeding in the same manner as in Problem 8, show that if

$$\omega_n = \frac{(2n - 1)\pi}{2} \qquad\qquad (n = 1, 2, \dots)$$

and $\omega \neq \omega_n$ for any value of n, then

$$y(x, t) = x \sin \omega t + 2\omega \sum_{n=1}^{\infty} \frac{(-1)^n}{\omega_n(\omega^2 - \omega_n^2)}\left(\frac{\omega}{\omega_n} \sin \omega t - \sin \omega_n t\right) \sin \omega_n x.$$

(*b*) Modify part (*a*) to show that resonance (Sec. 38) occurs when $\omega = \omega_N$ for any value of N.

 Suggestion: In each part of this problem, it is helpful to refer to the general solution of a certain ordinary differential equation in Problem 13, Sec. 38.

10. By referring to expansion (4) in Example 2, Sec. 46, and the expansion found in Problem 7, Sec. 46, write the solution in Problem 9(*a*) here in the form

$$y(x, t) = \frac{\sin \omega x \sin \omega t}{\omega \cos \omega} + 2\omega \sum_{n=1}^{\infty} \frac{(-1)^{n+1}}{\omega_n(\omega^2 - \omega_n^2)} \sin \omega_n x \sin \omega_n t,$$

where $\omega_n = (2n - 1)\pi/2$.

CHAPTER

6

FOURIER
INTEGRALS
AND
APPLICATIONS

In Chap. 2 (Sec. 21), we saw that a periodic function, with period $2c$, has a Fourier series representation that is valid for all x when it satisfies certain conditions on the fundamental interval $-c < x < c$. In this chapter, we develop the theory of trigonometric representations for functions, defined for all x, that are *not* periodic. Such representations, which are analogous to Fourier series representations, involve improper integrals, rather than infinite series.

51. THE FOURIER INTEGRAL FORMULA

From Problem 13, Sec. 21, we know that the Fourier series corresponding to a function $f(x)$ on an interval $-c < x < c$ can be written

(1) $$\frac{1}{2c} \int_{-c}^{c} f(s) \, ds + \frac{1}{c} \sum_{n=1}^{\infty} \int_{-c}^{c} f(s) \cos\left[\frac{n\pi}{c}(s - x)\right] ds;$$

and, from the corollary in Sec. 21, we know conditions under which this series converges to $f(x)$ everywhere in the interval $-c < x < c$. Namely, it is sufficient that f be piecewise smooth on the interval and that the value of f at each of its points of discontinuity be the mean value of the one-sided limits $f(x +)$ and $f(x -)$.

Suppose now that f satisfies such conditions on *every* bounded interval $-c < x < c$. Here c may be given any fixed positive value, arbitrarily large but

217

finite, and series (1) will represent $f(x)$ over the large segment $-c < x < c$ of the x axis. But that series representation cannot apply over the rest of the x axis unless f is periodic, with period $2c$, because the sum of the series has that periodicity.

In seeking a representation that is valid for all real x when f is not periodic, it is natural to try to modify series (1) by letting c tend to infinity. The first term in the series will then vanish, provided that the improper integral

$$\int_{-\infty}^{\infty} f(s)\, ds$$

exists. If we write $\Delta\alpha = \pi/c$, the remaining terms take the form

$$(2) \qquad \frac{1}{\pi} \sum_{n=1}^{\infty} \Delta\alpha \int_{-c}^{c} f(s) \cos[n\,\Delta\alpha(s-x)]\, ds,$$

which is the same as

$$(3) \qquad \frac{1}{\pi} \sum_{n=1}^{\infty} F_c(n\,\Delta\alpha, x)\,\Delta\alpha \qquad\qquad \left(\Delta\alpha = \frac{\pi}{c}\right),$$

where

$$(4) \qquad F_c(\alpha, x) = \int_{-c}^{c} f(s) \cos\alpha(s-x)\, ds.$$

Let the value of x be fixed and c be large, so that $\Delta\alpha$ is a small positive number. The points $n\,\Delta\alpha$ $(n = 1, 2, \ldots)$ are equally spaced along the entire positive α axis; and, because of the resemblance of the series in expression (3) to a sum of the type used in defining definite integrals, one might expect that the sum of that series tends to

$$(5) \qquad \int_{0}^{\infty} F_c(\alpha, x)\, d\alpha,$$

or possibly

$$(6) \qquad \int_{0}^{\infty} F_\infty(\alpha, x)\, d\alpha,$$

as $\Delta\alpha$ tends to zero. As integral (6) indicates, however, the function $F_c(\alpha, x)$ changes with $\Delta\alpha$ because $c = \pi/\Delta\alpha$. Also, the limit of the series in expression (3) as $\Delta\alpha$ tends to zero is not, in fact, the definition of the improper integral (5) even if c could be kept fixed.

The above manipulations merely *suggest* that, under appropriate conditions on f, the function may have the representation

$$(7) \qquad f(x) = \frac{1}{\pi} \int_{0}^{\infty} \int_{-\infty}^{\infty} f(s) \cos\alpha(s-x)\, ds\, d\alpha \qquad (-\infty < x < \infty).$$

This is the *Fourier integral formula* for the function f, to be established in Sec. 54.

The formula can be written in terms of separate cosine and sine functions as follows:

$$(8) \qquad f(x) = \int_0^\infty [A(\alpha) \cos \alpha x + B(\alpha) \sin \alpha x] \, d\alpha \quad (-\infty < x < \infty),$$

where

$$(9) \quad A(\alpha) = \frac{1}{\pi} \int_{-\infty}^\infty f(x) \cos \alpha x \, dx, \qquad B(\alpha) = \frac{1}{\pi} \int_{-\infty}^\infty f(x) \sin \alpha x \, dx.$$

Expressions (8) and (9) bear a resemblance to Fourier series representations on $-\pi < x < \pi$ and formulas for the coefficients a_n and b_n, derived in Sec. 15.

52. AN INTEGRATION FORMULA

Just as we prefaced our Fourier theorem in Sec. 19 with some preliminary theory, we include here and in Sec. 53 background that is essential to our proof of a *Fourier integral theorem*, which gives conditions under which representation (7) in Sec. 51 is valid.

This section is devoted to the evaluation of an improper integral that is prominent in applied mathematics. We show here that[†]

$$(1) \qquad \int_0^\infty \frac{\sin x}{x} \, dx = \frac{\pi}{2}.$$

Our method of evaluation requires us to first show that the integral actually converges. We note that the integrand $(\sin x)/x$ is piecewise continuous on any bounded interval; and since

$$(2) \qquad \int_0^c \frac{\sin x}{x} \, dx = \int_0^1 \frac{\sin x}{x} \, dx + \int_1^c \frac{\sin x}{x} \, dx,$$

where c is any positive number, it suffices to show that the integral

$$(3) \qquad \int_1^\infty \frac{\sin x}{x} \, dx = \lim_{c \to \infty} \int_1^c \frac{\sin x}{x}$$

converges. To accomplish this, we use the method of integration by parts and write

$$(4) \qquad \int_1^c \frac{\sin x}{x} \, dx = \cos 1 - \frac{\cos c}{c} - \int_1^c \frac{\cos x}{x^2} \, dx.$$

Since $|(\cos c)/c| \leqslant 1/c$, the second term on the right-hand side here tends to

---

[†] For another approach that is fairly standard, see, for instance, the book by Buck (1978, pp. 293–295). A method of evaluation involving complex variables is indicated in the authors' book (1990, pp. 197–198). Both books are listed in the Bibliography.

zero as c tends to infinity. Also, since $|(\cos x)/x^2| \leq 1/x^2$, the integral

$$
(5) \qquad \int_1^\infty \frac{\cos x}{x^2}\, dx
$$

is (absolutely) convergent. The limit of the left-hand side of equation (4) as c tends to infinity therefore exists; that is, integral (3) converges.

Now that we have established that integral (1) converges to some number L, or that

$$
\lim_{c \to \infty} \int_0^c \frac{\sin x}{x}\, = L,
$$

we note that, in particular,

$$
(6) \qquad \lim_{N \to \infty} \int_0^{(2N+1)\pi/2} \frac{\sin x}{x}\, dx = L
$$

as N passes through positive integers. That is,

$$
(7) \qquad \lim_{N \to \infty} \int_0^\pi \frac{\sin\left[(2N+1)u/2\right]}{u}\, du = L,
$$

where the substitution $x = (2N + 1)u/2$ has been made for the variable of integration. Observe that equation (7) can be written

$$
(8) \qquad \lim_{N \to \infty} \int_0^\pi g(u) D_N(u)\, du = L,
$$

where

$$
(9) \qquad g(u) = \frac{\sin(u/2)}{(u/2)}
$$

and where $D_N(u)$ is the Dirichlet kernel (Sec. 18)

$$
(10) \qquad D_N(u) = \frac{\sin\left[(2N+1)u/2\right]}{2\sin(u/2)}.
$$

The function $g(u)$, moreover, satisfies the conditions in Lemma 2, Sec. 18 (see Problem 1, Sec. 54), and $g(0 +\,) = 1$. So, by that lemma, limit (8) has the value $\pi/2$; and, by uniqueness of limits, $L = \pi/2$. Integration formula (1) is now established.

53. TWO LEMMAS

The two lemmas in this section are analogues of the ones in Sec. 18, leading up to a convergence theorem for Fourier series. The statement and proof of the corresponding theorem for Fourier integrals follow in Sec. 54.

Lemma 1. *If a function $G(u)$ is piecewise continuous on the interval $0 < x < c$, then*

(1)
$$\lim_{r \to \infty} \int_0^c G(u) \sin ru \, du = 0.$$

This is the general statement of the *Riemann-Lebesgue lemma* involving a sine function. Lemma 1 in Sec. 18 is a special case of it, where $c = \pi$ and r tends to infinity through the half integers $r = (2N + 1)/2$ $(N = 1, 2, \ldots)$, rather than continuously as it does here. This lemma also holds when $\sin ru$ is replaced by $\cos ru$; and the proof is similar to the one below involving $\sin ru$.

To verify limit (1), it is sufficient to show that if $G(u)$ is continuous at each point of an interval $a \leqslant u \leqslant b$, then

(2)
$$\lim_{r \to \infty} \int_a^b G(u) \sin ru \, du = 0.$$

For, in view of the discussion of integrals of piecewise continuous functions in Sec. 10, the integral in limit (1) can be expressed as the sum of a finite number of integrals of the type appearing in limit (2).

Assuming, then, that $G(u)$ is continuous on the closed bounded interval $a \leqslant u \leqslant b$, we note that it must also be *uniformly* continuous there. That is, for each positive number ε, there exists a positive number δ such that

$$|G(u) - G(v)| < \varepsilon$$

whenever u and v lie in the interval and satisfy the inequality $|u - v| < \delta$.[†]
Writing

$$\varepsilon = \frac{\varepsilon_0}{2(b - a)},$$

where ε_0 is an arbitrary positive number, we are thus assured that there is a positive number δ such that

(3) $|G(u) - G(v)| < \dfrac{\varepsilon_0}{2(b - a)}$ whenever $|u - v| < \delta.$

To obtain the limit (2), divide the interval $a \leqslant u \leqslant b$ into N subintervals of equal length $(b - a)/N$ by means of the points $a = u_0, u_1, u_2, \ldots, u_N = b$, where $u_0 < u_1 < u_2 < \cdots < u_N$, and let N be so large that the length of each

---

[†] See, for instance, the book by Taylor and Mann (1983, pp. 529–531), listed in the Bibliography.

subinterval is less than the number δ in condition (3). Then write

$$\int_a^b G(u) \sin ru \, du = \sum_{n=1}^{N} \int_{u_{n-1}}^{u_n} G(u) \sin ru \, du$$

$$= \sum_{n=1}^{N} \int_{u_{n-1}}^{u_n} [G(u) - G(u_n)] \sin ru \, du$$

$$+ \sum_{n=1}^{N} G(u_n) \int_{u_{n-1}}^{u_n} \sin ru \, du,$$

from which it follows that

(4) $$\left| \int_a^b G(u) \sin ru \, du \right|$$

$$\leqslant \sum_{n=1}^{N} \int_{u_{n-1}}^{u_n} |G(u) - G(u_n)| \, |\sin ru| \, du + \sum_{n=1}^{N} |G(u_n)| \left| \int_{u_{n-1}}^{u_n} \sin ru \, du \right|.$$

In view of condition (3) and the fact that $|\sin ru| \leqslant 1$, it is easy to see that

$$\int_{u_{n-1}}^{u_n} |G(u) - G(u_n)| \, |\sin ru| \, du < \frac{\varepsilon_0}{2(b-a)} \cdot \frac{b-a}{N} = \frac{\varepsilon_0}{2N}$$

$$(n = 1, 2, \ldots, N).$$

Also, since $G(u)$ is continuous on the closed interval $a \leqslant u \leqslant b$, it is bounded there; that is, there is a positive number M such that $|G(u)| \leqslant M$ for all u in that interval. Furthermore,

$$\left| \int_{u_{n-1}}^{u_n} \sin ru \, du \right| \leqslant \frac{|\cos ru_n| + |\cos ru_{n-1}|}{r} \leqslant \frac{2}{r} \qquad (n = 1, 2, \ldots, N),$$

where it is understood that $r > 0$. With these observations, we find that inequality (4) yields the statement

$$\left| \int_a^b G(u) \sin ru \, du \right| < \frac{\varepsilon_0}{2} + \frac{2MN}{r}.$$

Now write $R = 4MN/\varepsilon_0$ and observe that if $r > R$, then $2MN/r < \varepsilon_0/2$. Consequently,

$$\left| \int_a^b G(u) \sin ru \, du \right| < \frac{\varepsilon_0}{2} + \frac{\varepsilon_0}{2} = \varepsilon_0 \qquad \text{whenever } r > R;$$

and limit (2) is established.

Our second lemma makes direct use of the first one.

Lemma 2. *Suppose that a function $g(u)$ is piecewise continuous on every bounded interval of the positive u axis and that the right-hand derivative $g_R'(0)$*

exists. If the improper integral

$$(5) \qquad\qquad \int_0^\infty |g(u)|\, du$$

converges, then

$$(6) \qquad\qquad \lim_{r\to\infty} \int_0^\infty g(u)\, \frac{\sin ru}{u}\, du = \frac{\pi}{2} g(0+).$$

Observe that the integrand appearing in equation (6) is piecewise continuous on the same intervals as $g(u)$ and that when $u \geq 1$,

$$\left| g(u)\, \frac{\sin ru}{u} \right| \leq |g(u)|.$$

Thus the convergence of integral (5) ensures the existence of the integral in equation (6).

We begin the proof of the lemma by demonstrating its validity when the range of integration is replaced by any bounded interval $0 < x < c$. That is, we first show that if a function $g(u)$ is piecewise continuous on a bounded interval $0 < x < c$ and $g'_R(0)$ exists, then

$$(7) \qquad\qquad \lim_{r\to\infty} \int_0^c g(u)\, \frac{\sin ru}{u}\, du = \frac{\pi}{2} g(0+).$$

To prove this, we write

$$\int_0^c g(u)\, \frac{\sin ru}{u}\, du = I(r) + J(r),$$

where

$$I(r) = \int_0^c \frac{g(u) - g(0+)}{u}\, \sin ru\, du \qquad \text{and} \qquad J(r) = \int_0^c g(0+)\, \frac{\sin ru}{u}\, du.$$

Since the function $G(u) = [g(u) - g(0+)]/u$ is piecewise continuous on the interval $0 < x < c$, where $G(0+) = g'_R(0)$, we need only refer to Lemma 1 to see that

$$(8) \qquad\qquad \lim_{r\to\infty} I(r) = 0.$$

On the other hand, if we substitute $x = ru$ in the integral representing $J(r)$, the integration formula in Sec. 52 tells us that

$$(9) \qquad\qquad \lim_{r\to\infty} J(r) = g(0+) \lim_{r\to\infty} \int_0^{cr} \frac{\sin x}{x}\, dx = \frac{\pi}{2} g(0+).$$

Limit (7) is evidently now a consequence of limits (8) and (9).

To actually obtain limit (6), we note that

$$\left| \int_c^\infty g(u)\, \frac{\sin ru}{u}\, du \right| \leq \int_c^\infty |g(u)|\, du,$$

where we assume that $c \geqslant 1$. We then write

(10)
$$\left| \int_0^\infty g(u) \frac{\sin ru}{u} \, du - \frac{\pi}{2} g(0+) \right|$$

$$\leqslant \left| \int_0^c g(u) \frac{\sin ru}{u} \, du - \frac{\pi}{2} g(0+) \right| + \int_c^\infty |g(u)| \, du,$$

choosing c to be so large that the value of the last integral on the right, which is the remainder of integral (5), is less than $\varepsilon/2$, where ε is an arbitrary positive number independent of the value of r. In view of limit (7), there exists a positive number R such that whenever $r > R$, the first absolute value on the right-hand side of inequality (10) is also less than $\varepsilon/2$. It then follows that

$$\left| \int_0^\infty g(u) \frac{\sin ru}{u} \, du - \frac{\pi}{2} g(0+) \right| < \frac{\varepsilon}{2} + \frac{\varepsilon}{2} = \varepsilon$$

whenever $r > R$, and this is the same as statement (6).

54.   A FOURIER INTEGRAL THEOREM

The following theorem gives conditions under which the Fourier integral representation (7), Sec. 51, is valid.[†]

   Theorem. Let f denote a function which is piecewise continuous on every bounded interval of the x axis, and suppose that it is absolutely integrable over that axis; that is, the improper integral

$$\int_{-\infty}^\infty |f(x)| \, dx$$

converges. Then the Fourier integral

(1)
$$\frac{1}{\pi} \int_0^\infty \int_{-\infty}^\infty f(s) \cos \alpha(s - x) \, ds \, d\alpha$$

converges to the mean value

(2)
$$\frac{f(x+) + f(x-)}{2}$$

of the one-sided limits of f at each point x $(-\infty < x < \infty)$ where both of the one-sided derivatives $f_R'(x)$ and $f_L'(x)$ exist.

[†] For other conditions, see the books by Carslaw (1952, pp. 315ff) and Titchmarsh (1986, pp. 13ff), both listed in the Bibliography.

We begin our proof with the observation that integral (1) represents the limit as r tends to infinity of the integral

(3) $\qquad \dfrac{1}{\pi} \displaystyle\int_0^r \int_{-\infty}^{\infty} f(s) \cos \alpha(s - x)\, ds\, d\alpha = \dfrac{1}{\pi}[I(r, x) + J(r, x)],$

where

$$I(r, x) = \int_0^r \int_x^{\infty} f(s) \cos \alpha(s - x)\, ds\, d\alpha$$

and

$$J(r, x) = \int_0^r \int_{-\infty}^{x} f(s) \cos \alpha(s - x)\, ds\, d\alpha.$$

We now show that the individual integrals $I(r, x)$ and $J(r, x)$ exist; and, assuming that $f_R'(x)$ and $f_L'(x)$ exist, we examine the behavior of these integrals as r tends to infinity.

Turning to $I(r, x)$ first, we introduce the new variable of integration $u = s - x$ and write that integral in the form

(4) $\qquad\qquad I(r, x) = \displaystyle\int_0^r \int_0^{\infty} f(x + u) \cos \alpha u\, du\, d\alpha.$

Since

$$|f(x + u) \cos \alpha u| \leqslant |f(x + u)|$$

and because

$$\int_0^{\infty} |f(x + u)|\, du = \int_x^{\infty} |f(s)|\, ds \leqslant \int_{-\infty}^{\infty} |f(s)|\, ds,$$

the Weierstrass M-test for improper integrals applies to show that the integral

$$\int_0^{\infty} f(x + u) \cos \alpha u\, du$$

converges uniformly with respect to the variable α. Consequently, not only does the iterated integral (4) exist, but also the order of integration there can be reversed:[†]

$$I(r, x) = \int_0^{\infty} \int_0^r f(x + u) \cos \alpha u\, d\alpha\, du = \int_0^{\infty} f(x + u) \frac{\sin ru}{u}\, du.$$

Now the function $g(u) = f(x + u)$ satisfies all the conditions in Lemma 2, Sec. 53 (compare Sec. 19). Hence, if we apply that lemma to this last integral, we find

---

[†] Theorems on improper integrals used here are developed in the book by Buck (1978, sec. 6.4), listed in the Bibliography, as well as in most other texts on advanced calculus. The theorems are usually given for integrals with continuous integrands, but they are also valid when the integrands are piecewise continuous.

that

(5)
$$\lim_{r \to \infty} I(r, x) = \frac{\pi}{2} f(x+).$$

The limit of $J(r, x)$ as r tends to infinity is treated similarly. Here we make the substitution $u = x - s$ and write

$$J(r, x) = \int_0^r \int_0^\infty f(x - u) \cos \alpha u \, du \, d\alpha = \int_0^\infty f(x - u) \frac{\sin ru}{u} \, du.$$

When $g(u) = f(x - u)$, the limit

(6)
$$\lim_{r \to \infty} J(r, x) = \frac{\pi}{2} f(x-)$$

also follows from Lemma 2 in Sec. 53.

Finally, in view of limits (5) and (6), we see that the limit of the left-hand side of equation (3) as r tends to infinity has the value (2), which is, then, the value of integral (1) at any point where the one-sided derivatives of f exist.

Note that since the integrals in expressions (9), Sec. 51, for the coefficients $A(\alpha)$ and $B(\alpha)$ exist when f satisfies the conditions stated in the theorem, the form (8), Sec. 51, of the Fourier integral formula is also justified.

PROBLEMS

1. Show that the function

$$g(u) = \frac{\sin(u/2)}{(u/2)},$$

used in equation (8), Sec. 52, satisfies the conditions in Lemma 2, Sec. 18. To be precise, show that g is piecewise continuous on the interval $0 < x < \pi$ and that $g_R'(0)$ exists.

Suggestion: To obtain $g_R'(0)$, show that

$$g_R'(0) = \lim_{\substack{u \to 0 \\ u > 0}} \frac{2\sin(u/2) - u}{u^2}.$$

Then apply l'Hospital's rule twice.

2. Verify that all the conditions in the theorem in Sec. 54 are satisfied by the function f that is defined by the equations

$$f(x) = \begin{cases} 1 & \text{when } |x| < 1, \\ 0 & \text{when } |x| > 1, \end{cases}$$

and $f(\pm 1) = \frac{1}{2}$. Thus show that, for every x $(-\infty < x < \infty)$,

$$f(x) = \frac{1}{\pi} \int_0^\infty \frac{\sin \alpha(1 + x) + \sin \alpha(1 - x)}{\alpha} \, d\alpha = \frac{2}{\pi} \int_0^\infty \frac{\sin \alpha \cos \alpha x}{\alpha} \, d\alpha.$$

3. Show that the function defined by the equations

$$f(x) = \begin{cases} 0 & \text{when } x < 0, \\ \exp(-x) & \text{when } x > 0, \end{cases}$$

and $f(0) = \frac{1}{2}$ satisfies the conditions in the theorem in Sec. 54 and hence that

$$f(x) = \frac{1}{\pi} \int_0^\infty \frac{\cos \alpha x + \alpha \sin \alpha x}{1 + \alpha^2} \, d\alpha \qquad (-\infty < x < \infty).$$

Verify this representation directly at the point $x = 0$.

4. Show how it follows from the result in Problem 3 that

$$\exp(-|x|) = \frac{2}{\pi} \int_0^\infty \frac{\cos \alpha x}{1 + \alpha^2} \, d\alpha \qquad (-\infty < x < \infty).$$

5. Use the theorem in Sec. 54 to show that if

$$f(x) = \begin{cases} 0 & \text{when } x < 0 \text{ or } x > \pi, \\ \sin x & \text{when } \quad 0 \leqslant x \leqslant \pi, \end{cases}$$

then

$$f(x) = \frac{1}{\pi} \int_0^\infty \frac{\cos \alpha x + \cos \alpha(\pi - x)}{1 - \alpha^2} \, d\alpha \qquad (-\infty < x < \infty).$$

In particular, write $x = \pi/2$ to show that

$$\int_0^\infty \frac{\cos(\alpha\pi/2)}{1 - \alpha^2} \, d\alpha = \frac{\pi}{2}.$$

6. Show why the Fourier integral formula fails to represent the function $f(x) = 1$ $(-\infty < x < \infty)$. Also, point out which condition in the theorem in Sec. 54 is not satisfied by that function.

7. Give details showing that the integral $J(r, x)$ in Sec. 54 actually exists and that limit (6) in that section holds.

8. Let f be a nonzero function that is periodic, with period $2c$. Point out why the integrals

$$\int_{-\infty}^\infty f(x) \, dx \qquad \text{and} \qquad \int_{-\infty}^\infty |f(x)| \, dx$$

fail to exist.

9. Prove Lemma 1 in Sec. 53 when $\sin ru$ is replaced by $\cos ru$ in integral (1) there.

10. Assume that a function $f(x)$ has the Fourier integral representation (8), Sec. 51, which can be written

$$f(x) = \lim_{c \to \infty} \int_0^c [A(\alpha) \cos \alpha x + B(\alpha) \sin \alpha x] \, d\alpha.$$

Use the exponential forms (compare Problem 15, Sec. 21)

$$\cos \theta = \frac{e^{i\theta} + e^{-i\theta}}{2}, \qquad \sin \theta = \frac{e^{i\theta} - e^{-i\theta}}{2i}$$

of the cosine and sine functions to show formally that

$$f(x) = \lim_{c \to \infty} \int_{-c}^{c} a(\alpha)e^{i\alpha x} \, d\alpha,$$

where

$$a(\alpha) = \frac{A(\alpha) - iB(\alpha)}{2}, \qquad a(-\alpha) = \frac{A(\alpha) + iB(\alpha)}{2} \qquad (\alpha > 0).$$

Then use expressions (9), Sec. 51, for $A(\alpha)$ and $B(\alpha)$ to obtain the single formula[†]

$$a(\alpha) = \frac{1}{2\pi} \int_{-\infty}^{\infty} f(x)e^{-i\alpha x} \, dx \qquad (-\infty < \alpha < \infty).$$

55. THE COSINE AND SINE INTEGRALS

Let f denote a function satisfying the conditions stated in the theorem in Sec. 54. As noted at the end of the proof of that theorem, the Fourier integral representation of $f(x)$ remains valid when written

$$(1) \qquad f(x) = \int_{0}^{\infty} [A(\alpha) \cos \alpha x + B(\alpha) \sin \alpha x] \, d\alpha,$$

where

$$(2) \qquad A(\alpha) = \frac{1}{\pi} \int_{-\infty}^{\infty} f(x) \cos \alpha x \, dx, \qquad B(\alpha) = \frac{1}{\pi} \int_{-\infty}^{\infty} f(x) \sin \alpha x \, dx.$$

Also, in view of the theorem in Sec. 17, representation (1) is valid for any function f that is absolutely integrable over the entire x axis and *piecewise smooth* on every bounded interval of it.

Observe that if the function f is *even*, then $f(x) \sin \alpha x$ is odd in the variable x. The graph of $y = f(x) \sin \alpha x$ is, therefore, symmetric with respect to the origin. Hence $B(\alpha) = 0$, and representation (1) reduces to

$$(3) \qquad f(x) = \int_{0}^{\infty} A(\alpha) \cos \alpha x \, d\alpha.$$

The function $f(x) \cos \alpha x$ is even in x, and so the graph of $y = f(x) \cos \alpha x$ is symmetric with respect to the y axis. Consequently,

$$(4) \qquad A(\alpha) = \frac{2}{\pi} \int_{0}^{\infty} f(x) \cos \alpha x \, dx.$$

[†] The function $a(\alpha)$ is known as the *exponential Fourier transform* of $f(x)$ and is of particular importance in electrical engineering. For a development of this and other types of Fourier transforms, see, for example, the book by Churchill (1972) that is listed in the Bibliography.

The *Fourier cosine integral formula* (3) can, of course, be written

$$(5) \qquad f(x) = \frac{2}{\pi} \int_0^\infty \cos \alpha x \int_0^\infty f(s) \cos \alpha s \, ds \, d\alpha.$$

If, on the other hand, f is an *odd* function, then $A(\alpha) = 0$; and representation (1) becomes

$$(6) \qquad f(x) = \int_0^\infty B(\alpha) \sin \alpha x \, d\alpha,$$

where

$$(7) \qquad B(\alpha) = \frac{2}{\pi} \int_0^\infty f(x) \sin \alpha x \, dx.$$

Integral (6) is known as the *Fourier sine integral formula*. The compact form is

$$(8) \qquad f(x) = \frac{2}{\pi} \int_0^\infty \sin \alpha x \int_0^\infty f(s) \sin \alpha s \, ds \, d\alpha.$$

Suppose now that f is defined only when $x > 0$ and that it has the following properties:

(a) f is absolutely integrable over the positive x axis and piecewise smooth on every bounded interval of it;
(b) $f(x)$ at each point of discontinuity of f is the mean value of the one-sided limits $f(x +)$ and $f(x -)$.

When the even extension of f is made, integral (3) represents that extension for every nonzero x and equals $f(0 +)$ when $x = 0$. Likewise, integral (6) represents the odd extension of f for every nonzero x and has value zero when $x = 0$. The Fourier integral theorem in Sec. 54 thus provides us with a theorem that will be especially useful in the applications.

Theorem. *Let f denote a function that is defined on the positive x axis and satisfies conditions (a) and (b). Then the Fourier cosine integral representation (3), where the coefficient $A(\alpha)$ is defined by equation (4), is valid for each x ($x > 0$); and the same is true of the Fourier sine integral representation (6), where the coefficient $B(\alpha)$ is given by equation (7).*

Representation (3) is needed in the applications because of the solutions of the eigenvalue problem

$$(9) \qquad X''(x) + \lambda X(x) = 0, \qquad X'(0) = 0, \qquad |X(x)| < M \qquad (x > 0),$$

where M is a positive constant. This problem is *singular* (Sec. 42) because its fundamental interval $x > 0$ is unbounded. If $\lambda = 0$, $X(x)$ is any constant multiple of unity. If λ is a real number such that $\lambda > 0$ and we write $\lambda = \alpha^2$ ($\alpha > 0$), we readily find that, except for constant factors, the eigenfunctions are

$X(x) = \cos \alpha x$, where α takes on all positive values. The eigenvalues $\lambda = \alpha^2$ are continuous rather than discrete. If $\lambda < 0$, or $\lambda = -\alpha^2$ $(\alpha > 0)$, the solution of the differential equation and the boundary condition at $x = 0$ is $X(x) = 2C_1 \cosh \alpha x$. This is, however, unbounded on the half line $x > 0$ unless $C_1 = 0$. So the case $\lambda < 0$ yields no eigenfunctions. Cases other than those in which λ is real need not be considered since they yield unbounded solutions of the differential equation (see Problem 7). Although the eigenfunctions $X(x) = \cos \alpha x$ $(\alpha \geqslant 0)$ have no orthogonality property, the Fourier cosine integral formula (3) gives representations of functions $f(x)$ on the interval $x > 0$ which are generalized linear combinations of those eigenfunctions.

Likewise, $\lambda = \alpha^2$ and $X(x) = \sin \alpha x$ $(\alpha > 0)$ are the eigenvalues and eigenfunctions of the singular problem

$$(10) \qquad X''(x) + \lambda X(x) = 0, \qquad X(0) = 0, \qquad |X(x)| < M \quad (x > 0);$$

and formula (6) represents functions $f(x)$ in terms of $\sin \alpha x$.

PROBLEMS

1. By applying the Fourier sine integral formula and the theorem in Sec. 55 to the function defined by the equations

$$f(x) = \begin{cases} 1 & \text{when } 0 < x < b, \\ 0 & \text{when } \quad x > b, \end{cases}$$

and $f(b) = \frac{1}{2}$, obtain the representation

$$f(x) = \frac{2}{\pi} \int_0^\infty \frac{1 - \cos b\alpha}{\alpha} \sin \alpha x \, d\alpha \qquad (x > 0).$$

2. Verify that the function $\exp(-bx)$, where b is a positive constant, satisfies the conditions in the theorem in Sec. 55, and show that the coefficient $B(\alpha)$ in the Fourier sine integral representation of that function is

$$B(\alpha) = \frac{2}{\pi} \int_0^\infty e^{-bx} \sin \alpha x \, dx = \frac{2}{\pi} \cdot \frac{\alpha}{\alpha^2 + b^2}.$$

Thus prove that

$$e^{-bx} = \frac{2}{\pi} \int_0^\infty \frac{\alpha \sin \alpha x}{\alpha^2 + b^2} \, d\alpha \qquad (b > 0, \, x > 0).$$

3. Verify the Fourier sine integral representation

$$\frac{x}{x^2 + b^2} = \frac{2}{\pi} \int_0^\infty \sin \alpha x \int_0^\infty \frac{s \sin \alpha s}{s^2 + b^2} \, ds \, d\alpha \qquad (b > 0, \, x \geqslant 0)$$

by first observing that, according to the final result in Problem 2, the value of the inner integral here is $(\pi/2) \exp(-b\alpha)$. Then, by referring to the expression for $B(\alpha)$ in Problem 2, complete the verification. Show that the function $x/(x^2 + b^2)$ is *not*, however, absolutely integrable over the positive x axis.

4. As already verified in Problem 2, the function $\exp(-bx)$, where b is a positive constant, satisfies the conditions in the theorem in Sec. 55. Show that the coefficient $A(\alpha)$ in the Fourier cosine integral representation of that function is

$$A(\alpha) = \frac{2}{\pi} \int_0^\infty e^{-bx} \cos \alpha x \, dx = \frac{2}{\pi} \cdot \frac{b}{\alpha^2 + b^2}.$$

Thus prove that

$$e^{-bx} = \frac{2b}{\pi} \int_0^\infty \frac{\cos \alpha x}{\alpha^2 + b^2} \, d\alpha \qquad (b > 0, \, x \geqslant 0).$$

5. By regarding the positive constant b in the final equation obtained in Problem 4 as a variable and then differentiating each side of that equation with respect to b, show *formally* that

$$(1 + x)e^{-x} = \frac{4}{\pi} \int_0^\infty \frac{\cos \alpha x}{(\alpha^2 + 1)^2} \, d\alpha \qquad (x \geqslant 0).$$

6. Verify that the function $e^{-x} \cos x$ satisfies the conditions in the theorem in Sec. 55, and show that the coefficient $A(\alpha)$ in the Fourier cosine integral representation of that function can be written

$$A(\alpha) = \frac{1}{\pi} \int_0^\infty e^{-x} \cos(\alpha + 1)x \, dx + \frac{1}{\pi} \int_0^\infty e^{-x} \cos(\alpha - 1)x \, dx.$$

Then use the expression for the corresponding coefficients in Problem 4 to prove that

$$e^{-x} \cos x = \frac{2}{\pi} \int_0^\infty \frac{\alpha^2 + 2}{\alpha^4 + 4} \cos \alpha x \, d\alpha \qquad (x \geqslant 0).$$

7. Let λ denote a complex number that is not real, so that the square roots of $-\lambda$ are of the form $\pm(\alpha + i\beta)$, where α and β are real numbers and $\alpha \neq 0$. Use the identities[†]

$$|\cosh(x + iy)|^2 = \sinh^2 x + \cos^2 y, \qquad |\sinh(x + iy)|^2 = \sinh^2 x + \sin^2 y,$$

where x and y are real numbers, to show that such a λ cannot be an eigenvalue of
(*a*) the singular eigenvalue problem (9), Sec. 55;
(*b*) the singular eigenvalue problem (10), Sec. 55.

8. Show that the eigenvalues of the singular eigenvalue problem

$$X''(x) + \lambda X(x) = 0, \qquad |X(x)| < M \qquad (-\infty < x < \infty),$$

where M is a positive constant, are $\lambda = \alpha^2$ ($\alpha \geqslant 0$) and that the corresponding eigenfunctions are constant multiples of unity when $\alpha = 0$ and arbitrary linear combinations of $\cos \alpha x$ and $\sin \alpha x$ when $\alpha > 0$. (Use the method in Problem 7 to show that the eigenvalues must be real.)

[†] For results from the theory of functions of a complex variable used here, see the authors' book (1990, secs. 7 and 25), listed in the Bibliography.

9. Let $A(\alpha)$ and $B(\alpha)$ denote the coefficients (9), Sec. 51, in the Fourier integral representation (8) in that section for a function $f(x)$ $(-\infty < x < \infty)$ that satisfies the conditions in the theorem in Sec. 54.

(a) By considering even and odd functions of α, point out why

$$\int_{-\infty}^{\infty} [A(\alpha) \cos \alpha x + B(\alpha) \sin \alpha x] \, d\alpha = 2f(x)$$

and

$$\int_{-\infty}^{\infty} [B(\alpha) \cos \alpha x + A(\alpha) \sin \alpha x] \, d\alpha = 0.$$

(b) By adding corresponding sides of the equations in part (a), obtain the following symmetric form of the Fourier integral formula:[†]

$$f(x) = \frac{1}{\sqrt{2\pi}} \int_{-\infty}^{\infty} g(\alpha)(\cos \alpha x + \sin \alpha x) \, d\alpha \quad (-\infty < x < \infty),$$

where

$$g(\alpha) = \frac{1}{\sqrt{2\pi}} \int_{-\infty}^{\infty} f(x)(\cos \alpha x + \sin \alpha x) \, dx.$$

56. MORE ON SUPERPOSITION OF SOLUTIONS

In Sec. 26 we showed that linear combinations of solutions of linear homogeneous differential equations and boundary conditions are also solutions. In the same section we also extended that result to include infinite series of solutions, thus providing the basis of the technique for solving boundary value problems that was used in Chaps. 4 and 5. Another useful extension is illustrated by the following example, where superposition consists of integration with respect to a parameter α instead of summation with respect to an index n. It will enable us to solve certain boundary value problems in which Fourier integrals, rather than Fourier series, are required.

EXAMPLE. Consider the set of functions $\exp(-\alpha y) \sin \alpha x$, where each function corresponds to a value of the parameter α $(\alpha > 0)$ and where α is independent of x and y. Each function satisfies Laplace's equation

$$(1) \qquad\qquad\qquad u_{xx}(x, y) + u_{yy}(x, y) = 0 \qquad\qquad (x > 0, y > 0)$$

and the boundary condition

$$(2) \qquad\qquad\qquad u(0, y) = 0 \qquad\qquad\qquad (y > 0).$$

These functions are bounded in the domain $x > 0$, $y > 0$ (Fig. 59) and are

---

[†] This form is useful in certain types of transmission problems. See R. V. L. Hartley, *Proc. Inst. Radio Engrs.*, vol. 30, no. 3, pp. 144–150, 1942.

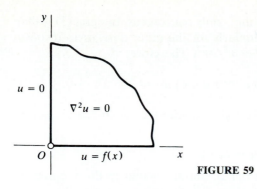

FIGURE 59

obtained from conditions (1) and (2) by the method of separation of variables when that boundedness condition is included (Problem 1, Sec. 58).

We now show that their combination of the type

$$(3) \qquad\qquad u(x, y) = \int_0^\infty B(\alpha)e^{-\alpha y} \sin \alpha x\, d\alpha \qquad\qquad (x > 0,\, y > 0)$$

also represents a solution of the homogeneous conditions (1) and (2) which is bounded in the domain $x > 0$, $y > 0$ for each function $B(\alpha)$ that is bounded and continuous on the half line $\alpha > 0$ and absolutely integrable over it.

To accomplish this, we use tests for improper integrals that are analogous to those for infinite series.[†] The integral in equation (3) converges absolutely and uniformly with respect to x and y because

$$(4) \qquad\qquad |B(\alpha)e^{-\alpha y} \sin \alpha x| \leqslant |B(\alpha)| \qquad\qquad (x \geqslant 0,\, y \geqslant 0)$$

and $B(\alpha)$ is independent of x and y and absolutely integrable from zero to infinity with respect to α. Moreover, since

$$(5) \qquad |u(x, y)| \leqslant \int_0^\infty |B(\alpha)e^{-\alpha y} \sin \alpha x|\, d\alpha \leqslant \int_0^\infty |B(\alpha)|\, d\alpha,$$

u is bounded. It is also a continuous function of x and y ($x \geqslant 0$, $y \geqslant 0$) because of the uniform convergence of the integral in equation (3) and the continuity of the integrand. Clearly, $u = 0$ when $x = 0$.

When $y > 0$,

$$(6) \qquad \frac{\partial u}{\partial x} = \frac{\partial}{\partial x} \int_0^\infty B(\alpha)e^{-\alpha y} \sin \alpha x\, d\alpha = \int_0^\infty \frac{\partial}{\partial x}[B(\alpha)e^{-\alpha y} \sin \alpha x]\, d\alpha;$$

for if $|B(\alpha)| \leqslant B_0$ and $y \geqslant y_0$, where y_0 is some small positive number, then the absolute value of the integrand of the integral on the far right does not exceed $B_0\alpha \exp(-\alpha y_0)$, which is independent of x and y and integrable from

---

[†] See the book by Kaplan (1991, pp. 471ff) or Taylor and Mann (1983, pp. 682ff), listed in the Bibliography.

$\alpha = 0$ to $\alpha = \infty$. Hence that integral is uniformly convergent. Integral (3) is then differentiable with respect to x, and similarly for the other derivatives involved in the laplacian operator $\nabla^2 = \partial^2/\partial x^2 + \partial^2/\partial y^2$. Therefore,

$$(7) \qquad \nabla^2 u = \int_0^\infty B(\alpha)\nabla^2(e^{-\alpha y}\sin\alpha x)\,d\alpha = 0 \qquad (x>0,\ y>0).$$

Suppose now that the function (3) is also required to satisfy the nonhomogeneous boundary condition

$$(8) \qquad u(x,0) = f(x) \qquad (x>0),$$

where f is a given function satisfying the conditions stated in the theorem in Sec. 55. We need to determine the function $B(\alpha)$ in equation (3) so that

$$(9) \qquad f(x) = \int_0^\infty B(\alpha)\sin\alpha x\,d\alpha \qquad (x>0).$$

This is easily done since representation (9) is the Fourier sine integral formula (6), Sec. 55, when

$$(10) \qquad B(\alpha) = \frac{2}{\pi}\int_0^\infty f(x)\sin\alpha x\,dx \qquad (\alpha>0).$$

We have shown here that the function (3), with $B(\alpha)$ given by equation (10), is a solution of the boundary value problem consisting of equations (1), (2), and (8), together with the requirement that u be bounded.

57. TEMPERATURES IN A SEMI-INFINITE SOLID

The face $x = 0$ of a semi-infinite solid $x \geqslant 0$ is kept at temperature zero (Fig. 60). Let us find the temperatures $u(x,t)$ in the solid when the initial temperature distribution is $f(x)$, assuming at present that f is piecewise smooth on each bounded interval of the positive x axis and that f is bounded and absolutely integrable from $x = 0$ to $x = \infty$.

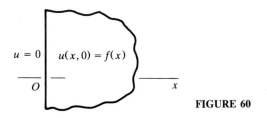

FIGURE 60

If the solid is considered as a limiting case of a slab $0 \leqslant x \leqslant c$ as c increases, some condition corresponding to a thermal condition on the face $x = c$ seems to be needed. Otherwise, the temperatures on that face may be increased in any manner as c increases. We require that our function u be bounded; that condition also implies that there is no instantaneous source of heat on the face $x = 0$ at the instant $t = 0$. Then

$$(1) \qquad\qquad u_t(x, t) = ku_{xx}(x, t) \qquad\qquad (x > 0, t > 0),$$

$$(2) \qquad\qquad u(0, t) = 0 \qquad\qquad (t > 0),$$

$$(3) \qquad\qquad u(x, 0) = f(x) \qquad\qquad (x > 0),$$

and $|u(x, t)| < M$, where M is some positive constant.

Linear combinations of functions $u = X(x)T(t)$ will not ordinarily be bounded unless X and T are themselves bounded. Upon separating variables, we thus have the conditions

$$(4) \qquad X''(x) + \lambda X(x) = 0, \qquad X(0) = 0, \qquad |X(x)| < M_1 \qquad (x > 0)$$

and

$$(5) \qquad\qquad T'(t) + \lambda kT(t) = 0, \qquad |T(t)| < M_2 \qquad (t > 0),$$

where M_1 and M_2 are positive constants. As pointed out at the end of Sec. 55, the singular eigenvalue problem (4) has continuous eigenvalues $\lambda = \alpha^2$, where α represents *all positive real numbers*; $X(x) = \sin \alpha x$ are the eigenfunctions. In this case, the corresponding functions $T(t) = \exp(-\alpha^2 kt)$ are bounded. The generalized linear combination of the functions $X(x)T(t)$ for all positive α,

$$(6) \qquad\qquad u(x, t) = \int_0^\infty B(\alpha)e^{-\alpha^2 kt} \sin \alpha x \, d\alpha,$$

will formally satisfy all the conditions in the boundary value problem if the function $B(\alpha)$ can be determined so that

$$(7) \qquad\qquad f(x) = \int_0^\infty B(\alpha) \sin \alpha x \, d\alpha \qquad\qquad (x > 0).$$

As in Sec. 56, we note that representation (7) is the Fourier sine integral formula (6), Sec. 55, for $f(x)$ if

$$(8) \qquad\qquad B(\alpha) = \frac{2}{\pi} \int_0^\infty f(x) \sin \alpha x \, dx \qquad\qquad (\alpha > 0).$$

Our formal solution (6), with $B(\alpha)$ defined by equation (8), can also be written

$$(9) \qquad\qquad u(x, t) = \frac{2}{\pi} \int_0^\infty e^{-\alpha^2 kt} \sin \alpha x \int_0^\infty f(s) \sin \alpha s \, ds \, d\alpha.$$

We can simplify this result by formally reversing the order of integration, replacing $2 \sin \alpha s \sin \alpha x$ by $\cos \alpha(s - x) - \cos \alpha(s + x)$, and then applying the integration formula (Problem 19, Sec. 58)

$$(10) \qquad\qquad \int_0^\infty e^{-\alpha^2 a} \cos \alpha b \, d\alpha = \frac{1}{2}\sqrt{\frac{\pi}{a}} \, \exp\left(-\frac{b^2}{4a}\right) \qquad\qquad (a > 0).$$

Equation (9) then becomes

(11) $u(x,t) = \dfrac{1}{2\sqrt{\pi k t}} \displaystyle\int_0^\infty f(s)\left\{\exp\left[-\dfrac{(s-x)^2}{4kt}\right] - \exp\left[-\dfrac{(s+x)^2}{4kt}\right]\right\} ds$

when $t > 0$. An alternative form of equation (11), obtained by introducing new variables of integration, is

(12) $u(x,t) = \dfrac{1}{\sqrt{\pi}} \displaystyle\int_{-x/(2\sqrt{kt})}^\infty f(x + 2\sigma\sqrt{kt})e^{-\sigma^2}\,d\sigma$

$- \dfrac{1}{\sqrt{\pi}} \displaystyle\int_{x/(2\sqrt{kt})}^\infty f(-x + 2\sigma\sqrt{kt})e^{-\sigma^2}\,d\sigma.$

Our use of the Fourier sine integral formula in obtaining solution (9) suggests that we apply the theorem in Sec. 55 in verifying that solution. The forms (11) and (12) suggest, however, that the condition in the theorem that $|f(x)|$ be integrable from zero to infinity can be relaxed in the verification. More precisely, when s is kept fixed and $t > 0$, the functions

$$\frac{1}{\sqrt{t}}\exp\left[-\frac{(s\pm x)^2}{4kt}\right]$$

satisfy the heat equation (1). Then, under the assumption that $f(x)$ is continuous and bounded when $x \geqslant 0$, it is possible to show that the function (11) is bounded and satisfies the heat equation when $x_0 < x < x_1$ and $t_0 < t < t_1$, where x_0, x_1, t_0, and t_1 are any positive numbers. Conditions (2) and (3) can be verified by using expression (12). By adding step functions to f (see Problem 4, Sec. 58), we can permit f to have a finite number of jumps on the half line $x > 0$. Except for special cases, details in the verification of formal solutions of this problem are, however, tedious.

When $f(x) = 1$, it follows from equation (12) that

(13) $u(x,t) = \dfrac{1}{\sqrt{\pi}}\left(\displaystyle\int_{-x/(2\sqrt{kt})}^\infty e^{-\sigma^2}\,d\sigma - \int_{x/(2\sqrt{kt})}^\infty e^{-\sigma^2}\,d\sigma\right).$

In terms of the *error function*

(14) $\mathrm{erf}(x) = \dfrac{2}{\sqrt{\pi}}\displaystyle\int_0^x e^{-\sigma^2}\,d\sigma,$

where $\mathrm{erf}(x)$ tends to unity as x tends to infinity (see Problem 18, Sec. 58), expression (13) can be written

(15) $u(x,t) = \mathrm{erf}\left(\dfrac{x}{2\sqrt{kt}}\right).$

The full verification of that result is not difficult.

58. TEMPERATURES IN AN UNLIMITED MEDIUM

For an application of the general Fourier integral formula, we now derive expressions for the temperatures $u(x, t)$ in a medium that occupies all space, where the initial temperature distribution is $f(x)$. We assume that $f(x)$ is bounded and, for the present, that it satisfies conditions under which it is represented by its Fourier integral formula. The boundary value problem consists of a boundedness condition $|u(x, t)| < M$ and the conditions

$$(1) \qquad\qquad u_t(x, t) = k u_{xx}(x, t) \qquad\qquad (-\infty < x < \infty, t > 0),$$

$$(2) \qquad\qquad u(x, 0) = f(x) \qquad\qquad (-\infty < x < \infty).$$

Separation of variables leads to the singular eigenvalue problem

$$X''(x) + \lambda X(x) = 0, \qquad |X(x)| < M_1 \qquad (-\infty < x < \infty),$$

whose eigenvalues are $\lambda = \alpha^2$, where $\alpha \geq 0$, and to the two linearly independent eigenfunctions $\cos \alpha x$ and $\sin \alpha x$ corresponding to each nonzero value of α (Problem 8, Sec. 55).

Our generalized linear combination of functions $X(x)T(t)$ is

$$(3) \qquad u(x, t) = \int_0^\infty \exp(-\alpha^2 kt)\, [A(\alpha) \cos \alpha x + B(\alpha) \sin \alpha x]\, d\alpha.$$

The coefficients $A(\alpha)$ and $B(\alpha)$ are to be determined so that the integral here represents $f(x)$ $(-\infty < x < \infty)$ when $t = 0$. According to equations (8) and (9) in Sec. 51 and our Fourier integral theorem (Sec. 54), the representation is valid if

$$A(\alpha) = \frac{1}{\pi} \int_{-\infty}^\infty f(x) \cos \alpha x\, dx, \qquad B(\alpha) = \frac{1}{\pi} \int_{-\infty}^\infty f(x) \sin \alpha x\, dx.$$

Thus

$$(4) \qquad u(x, t) = \frac{1}{\pi} \int_0^\infty e^{-\alpha^2 kt} \int_{-\infty}^\infty f(s) \cos \alpha(s - x)\, ds\, d\alpha.$$

If we formally reverse the order of integration here, the integration formula (10) in Sec. 57 can be used to write equation (4) in the form

$$(5) \qquad u(x, t) = \frac{1}{2\sqrt{\pi kt}} \int_{-\infty}^\infty f(s) \exp\left[-\frac{(s - x)^2}{4kt} \right] ds \qquad (t > 0).$$

An alternative form of this is

$$(6) \qquad u(x, t) = \frac{1}{\sqrt{\pi}} \int_{-\infty}^\infty f(x + 2\sigma\sqrt{kt})e^{-\sigma^2}\, d\sigma.$$

Forms (5) and (6) can be verified by assuming only that f is piecewise continuous over some bounded interval $|x| < c$ and continuous and bounded over the rest of the x axis, or when $|x| \geq c$. If f is an odd function, $u(x,t)$ becomes the function found in Sec. 57 for positive values of x.

PROBLEMS

1. Give details showing how the functions $\exp(-\alpha y)\sin \alpha x$ $(\alpha > 0)$ arise by means of separation of variables from conditions (1) and (2), Sec. 56, and the condition that the function $u(x, y)$ there be bounded when $x > 0$, $y > 0$.

2. (a) Substitute expression (10), Sec. 56, for the function $B(\alpha)$ into equation (3) of that section. Then, by formally reversing the order of integration, show that the solution of the boundary value problem treated in Sec. 56 can be written

$$u(x, y) = \frac{y}{\pi} \int_0^\infty f(s) \left[\frac{1}{(s - x)^2 + y^2} - \frac{1}{(s + x)^2 + y^2} \right] ds.$$

(b) Show that when $f(x) = 1$, the form of the solution obtained in part (a) can be written in terms of the inverse tangent function as

$$u(x, y) = \frac{2}{\pi} \tan^{-1} \frac{x}{y}.$$

3. Verify that the function $u = \mathrm{erf}[x/(2\sqrt{kt})]$ in Sec. 57 satisfies the heat equation $u_t = ku_{xx}$ when $x > 0$, $t > 0$ as well as the conditions

$$u(0 + , t) = 0 \qquad\qquad (t > 0),$$
$$u(x, 0 +) = 1 \qquad\qquad (x > 0),$$
$$|u(x, t)| < 1 \qquad\qquad (x > 0, t > 0).$$

4. Show that if

$$f(x) = \begin{cases} 0 & \text{when } 0 < x < c, \\ 1 & \text{when } \quad x > c, \end{cases}$$

expression (12), Sec. 57, reduces to

$$u(x, t) = \frac{1}{2} \, \mathrm{erf}\left(\frac{c + x}{2\sqrt{kt}} \right) - \frac{1}{2} \, \mathrm{erf}\left(\frac{c - x}{2\sqrt{kt}} \right).$$

Verify this solution of the boundary value problem in Sec. 57 when f is this function.

5. The face $x = 0$ of a semi-infinite solid $x \geq 0$ is kept at a constant temperature u_0 after its interior $x > 0$ is initially at temperature zero throughout. Obtain an expression for the temperatures $u(x, t)$ in the body.

Answer: $u(x, t) = u_0 \left[1 - \mathrm{erf}\left(\dfrac{x}{2\sqrt{kt}} \right) \right].$

6. (a) The face $x = 0$ of a semi-infinite solid $x \geq 0$ is insulated, and the initial temperature distribution is $f(x)$. Derive the temperature formula

$$u(x, t) = \frac{1}{\sqrt{\pi}} \int_{-x/(2\sqrt{kt})}^\infty f(x + 2\sigma\sqrt{kt}) e^{-\sigma^2} d\sigma$$

$$+ \frac{1}{\sqrt{\pi}} \int_{x/(2\sqrt{kt})}^\infty f(-x + 2\sigma\sqrt{kt}) e^{-\sigma^2} d\sigma.$$

(b) Show that if the function f in part (a) is defined by the equations

$$f(x) = \begin{cases} 1 & \text{when } 0 < x < c, \\ 0 & \text{when } \quad x > c, \end{cases}$$

then

$$u(x,t) = \frac{1}{2}\,\text{erf}\!\left(\frac{c+x}{2\sqrt{kt}}\right) + \frac{1}{2}\,\text{erf}\!\left(\frac{c-x}{2\sqrt{kt}}\right).$$

7. Let the initial temperature distribution $f(x)$ in the unlimited medium in Sec. 58 be defined by the equations

$$f(x) = \begin{cases} 0 & \text{when } x < 0, \\ 1 & \text{when } x > 0. \end{cases}$$

Show that

$$u(x,t) = \frac{1}{2} + \frac{1}{2}\,\text{erf}\!\left(\frac{x}{2\sqrt{kt}}\right).$$

Verify this solution of the boundary value problem in Sec. 58 when f is this function.

8. Derive this solution of the wave equation $y_{tt} = a^2 y_{xx}$ $(-\infty < x < \infty,\ t > 0)$, which satisfies the conditions $y(x,0) = f(x)$, $y_t(x,0) = 0$ when $-\infty < x < \infty$:

$$y(x,t) = \frac{1}{\pi}\int_0^\infty \cos \alpha at \int_{-\infty}^\infty f(s)\cos \alpha(s-x)\,ds\,d\alpha.$$

Also, reduce the solution to the form obtained in Example 2, Sec. 8:

$$y(x,t) = \frac{1}{2}[f(x+at) + f(x-at)].$$

9. A semi-infinite string, with one end fixed at the origin, is stretched along the positive half of the x axis and released at rest from a position $y = f(x)$ $(x \geqslant 0)$. Derive the expression

$$y(x,t) = \frac{2}{\pi}\int_0^\infty \cos \alpha at \sin \alpha x \int_0^\infty f(s)\sin \alpha s\,ds\,d\alpha$$

for the transverse displacements. Let $F(x)$ $(-\infty < x < \infty)$ denote the odd extension of $f(x)$, and show how this result reduces to the form

$$y(x,t) = \frac{1}{2}[F(x+at) + F(x-at)].$$

[Compare solution (10), Sec. 30, of the string problem treated in that section.]

10. Find the bounded harmonic function $u(x,y)$ in the horizontal semi-infinite strip $x > 0,\ 0 < y < 1$ that satisfies the boundary conditions

$$u_x(0,y) = 0, \qquad u_y(x,0) = 0, \qquad u(x,1) = e^{-x}.$$

Answer: $u(x,y) = \dfrac{2}{\pi}\displaystyle\int_0^\infty \dfrac{\cos \alpha x \cosh \alpha y}{(1+\alpha^2)\cosh \alpha}\,d\alpha.$

11. Find $u(x,y)$ when the boundary conditions in Problem 10 are replaced by the conditions

$$u_x(0,y) = 0, \qquad u_y(x,1) = -u(x,1), \qquad u(x,0) = f(x),$$

where

$$f(x) = \begin{cases} 1 & \text{when } 0 < x < 1, \\ 0 & \text{when } \quad x > 1. \end{cases}$$

Interpret this problem physically.

$$\textit{Answer: } u(x, y) = \frac{2}{\pi} \int_0^\infty \frac{\alpha \cosh \alpha (1 - y) + \sinh \alpha (1 - y)}{\alpha^2 \cosh \alpha + \alpha \sinh \alpha} \sin \alpha \cos \alpha x \, d\alpha.$$

12. Find the bounded harmonic function $u(x, y)$ in the semi-infinite strip $0 < x < 1$, $y > 0$ that satisfies the conditions

$$u_y(x, 0) = 0, \qquad u(0, y) = 0, \qquad u_x(1, y) = f(y).$$

$$\textit{Answer: } u(x, y) = \frac{2}{\pi} \int_0^\infty \frac{\sinh \alpha x \cos \alpha y}{\alpha \cosh \alpha} \int_0^\infty f(s) \cos \alpha s \, ds \, d\alpha.$$

13. Find the bounded harmonic function $u(x, y)$ in the strip $-\infty < x < \infty$, $0 < y < b$ such that $u(x, 0) = 0$ and $u(x, b) = f(x)$ $(-\infty < x < \infty)$, where f is bounded and represented by its Fourier integral.

$$\textit{Answer: } u(x, y) = \frac{1}{\pi} \int_0^\infty \frac{\sinh \alpha y}{\sinh \alpha b} \int_{-\infty}^\infty f(s) \cos \alpha (s - x) \, ds \, d\alpha.$$

14. Let a semi-infinite solid $x \geqslant 0$, which is initially at a uniform temperature, be cooled or heated by keeping its boundary at a uniform constant temperature (Sec. 57). Show that the times required for two interior points to reach the same temperature are proportional to the squares of the distances of those points from the boundary plane.

15. Solve the following boundary value problem for steady temperatures $u(x, y)$ in a thin plate in the shape of a semi-infinite strip when heat transfer to the surroundings at temperature zero takes place at the faces of the plate:

$$u_{xx}(x, y) + u_{yy}(x, y) - bu(x, y) = 0 \qquad\qquad (x > 0, 0 < y < 1),$$

$$u_x(0, y) = 0 \qquad\qquad (0 < y < 1),$$

$$u(x, 0) = 0, \qquad u(x, 1) = f(x) \qquad\qquad (x > 0),$$

where b is a positive constant and

$$f(x) = \begin{cases} 1 & \text{when } 0 < x < c, \\ 0 & \text{when } \quad x > c. \end{cases}$$

$$\textit{Answer: } u(x, y) = \frac{2}{\pi} \int_0^\infty \frac{\sin \alpha c \cos \alpha x \sinh \left(y \sqrt{\alpha^2 + b} \right)}{\alpha \sinh \sqrt{\alpha^2 + b}} \, d\alpha.$$

16. Verify that, for any constant C, the function

$$v(x, t) = Cxt^{-3/2} \exp \left(\frac{-x^2}{4kt} \right)$$

satisfies the heat equation $v_t = kv_{xx}$ when $x > 0$ and $t > 0$. Also verify that $v(0 + , t) = 0$ when $t > 0$ and that $v(x, 0 +) = 0$ when $x > 0$. Thus v can be added to the function u found in Sec. 57 to form other solutions of the problem there if the temperature function is not required to be bounded. But note that v is unbounded as x and t tend to zero, as can be seen by letting x vanish while $t = x^2$.

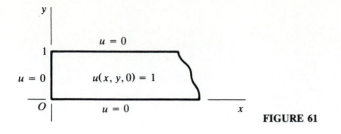

FIGURE 61

17. Let $u = u(x, y, t)$ denote the bounded solution of the two-dimensional temperature problem indicated in Fig. 61, where

$$u_t = k(u_{xx} + u_{yy}) \qquad (x > 0, 0 < y < 1, t > 0),$$

and let $v = v(x, t)$ and $w = w(y, t)$ denote the bounded solutions of the following one-dimensional temperature problems:

$$v_t = kv_{xx}, \qquad v(0, t) = 0, \qquad v(x, 0) = 1 \qquad\qquad (x > 0, t > 0),$$

$$w_t = kw_{yy}, \qquad w(0, t) = w(1, t) = 0, \qquad w(y, 0) = 1 \quad (0 < y < 1, t > 0).$$

(*a*) With the aid of the result obtained in Problem 3, Sec. 40, show that $u = vw$.

(*b*) By referring to the solution (15), Sec. 57, of the temperature problem there and to the temperature function found in Problem 4(*b*), Sec. 32, write explicit expressions for v and w. Then use the result in part (*a*) to show that

$$u(x, y, t) = \frac{4}{\pi} \operatorname{erf}\left(\frac{x}{2\sqrt{kt}}\right) \sum_{n=1}^{\infty} \frac{\sin(2n-1)\pi y}{2n-1} \exp\left[-(2n-1)^2\pi^2 kt\right].$$

18. Let I denote the integral of $\exp(-x^2)$ from zero to infinity, and write

$$I^2 = \int_0^{\infty} e^{-x^2}\, dx \int_0^{\infty} e^{-y^2}\, dy = \int_0^{\infty}\int_0^{\infty} e^{-(x^2+y^2)}\, dx\, dy.$$

Evaluate this iterated integral by using polar coordinates, and show that $I = \sqrt{\pi}/2$. Thus verify that $\operatorname{erf}(x)$, defined in equation (14), Sec. 57, tends to unity as x tends to infinity.

19. Derive the integration formula (10), Sec. 57, by first writing

$$y(x) = \int_0^{\infty} e^{-\alpha^2 a} \cos \alpha x\, d\alpha \qquad\qquad (a > 0)$$

and differentiating the integral to find $y'(x)$. Then integrate the new integral by parts to show that $2ay'(x) = -xy(x)$, point out why

$$y(0) = \frac{1}{2}\sqrt{\frac{\pi}{a}}$$

(see Problem 18), and solve for $y(x)$. The desired result is the value of y when $x = b$.[†]

[†] Another derivation is indicated in the authors' book (1990, p. 199), listed in the Bibliography.

CHAPTER

7

BESSEL FUNCTIONS AND APPLICATIONS

In boundary value problems that involve the laplacian $\nabla^2 u$ expressed in cylindrical or polar coordinates, the process of separating variables often produces a differential equation of the form

$$(1) \qquad \rho^2 \frac{d^2 y}{d\rho^2} + \rho \frac{dy}{d\rho} + (\lambda \rho^2 - \nu^2) y = 0,$$

where y is a function of the coordinate ρ. In such a problem, $-\lambda$ is a separation constant; and the values of λ are the eigenvalues of a Sturm-Liouville problem involving equation (1). The parameter ν is a nonnegative number determined by other aspects of the boundary value problem. Usually, ν is either zero or a positive integer.

In our applications, it turns out that $\lambda \geqslant 0$; and, when $\lambda > 0$, the substitution $x = \sqrt{\lambda}\,\rho$ can be used to transform equation (1) into a form that is free of λ:

$$(2) \qquad x^2 y''(x) + xy'(x) + (x^2 - \nu^2) y(x) = 0.$$

This differential equation is known as *Bessel's equation*. Its solutions are called *Bessel functions*, or sometimes cylindrical functions.

Equation (2) is an ordinary differential equation of the second order that is linear and homogeneous; and, upon comparing it with the standard form

$$y''(x) + A(x)y'(x) + B(x)y(x) = 0$$

242

of such equations, we see that $A(x) = 1/x$ and $B(x) = 1 - (\nu/x)^2$. These coefficients are continuous except at the origin, which is a singular point of Bessel's equation. The lemma in Sec. 44, regarding the existence and unique-ness of solutions, applies to that equation on any closed bounded interval that does not include the origin. But for boundary value problems in regions $\rho \leqslant c$, bounded by cylinders or circles, the origin $x = 0$ corresponds to the axis or center $\rho = 0$, which is interior to the region. The interval for the variable x then has zero as an end point.

   We limit our attention primarily to the cases $\nu = n$, where $n = 0, 1, 2, \ldots$. For such a case, we shall discover a solution of Bessel's equation that is represented by a power series which, together with all its derivatives, converges for every value of x, including $x = 0$. That solution, denoted by $J_n(x)$, and its derivatives of all orders are, therefore, everywhere continuous functions. In referring to any power series, we shall always mean a Maclaurin series, or a Taylor series about the origin.

59. BESSEL FUNCTIONS J_n

We let n denote any fixed nonnegative integer and seek a solution of Bessel's equation

$$(1) \qquad x^2 y''(x) + x y'(x) + (x^2 - n^2) y(x) = 0 \qquad (n = 0, 1, 2, \ldots)$$

in the form of a power series multiplied by x^c, where the first term in that series is nonzero and c is some constant. That is, we propose to determine c and the coefficients a_j so that the function

$$(2) \qquad y = x^c \sum_{j=0}^{\infty} a_j x^j = \sum_{j=0}^{\infty} a_j x^{c+j} \qquad (a_0 \neq 0)$$

satisfies equation (1).[†]

   Assume for the present that the series is differentiable. Then, upon substituting the function (2) and its derivatives into equation (1), we obtain the equation

$$\sum_{j=0}^{\infty} \left[(c+j)(c+j-1) + (c+j) - n^2 \right] a_j x^{c+j} + \sum_{j=0}^{\infty} a_j x^{c+j+2} = 0.$$

But $(c+j)(c+j-1) + (c+j) = (c+j)^2$, and the second summation here can be written

$$\sum_{j=2}^{\infty} a_{j-2} x^{c+j}.$$

[†] The series method used here to solve equation (1) is often referred to as the *method of Frobenius* and is treated in introductory texts on ordinary differential equations, such as the one by Boyce and DiPrima (1992) or the one by Rainville and Bedient (1989). Both are listed in the Bibliography.

Hence

$$\sum_{j=0}^{\infty} \left[(c+j)^2 - n^2 \right] a_j x^{c+j} + \sum_{j=2}^{\infty} a_{j-2} x^{c+j} = 0.$$

Multiplying through this equation by x^{-c} and writing out the $j = 0$ and $j = 1$ terms of the first series separately, we have the equation

$$(3) \qquad (c-n)(c+n)a_0 + (c+1-n)(c+1+n)a_1 x$$

$$+ \sum_{j=2}^{\infty} \left[(c+j-n)(c+j+n)a_j + a_{j-2} \right] x^j = 0.$$

Equation (3) is an identity in x if the coefficient of each power of x vanishes. Thus $c = n$ or $c = -n$ if the constant term is to vanish; and, in either case, $a_1 = 0$. Furthermore,

$$(c+j-n)(c+j+n)a_j + a_{j-2} = 0 \qquad (j = 2, 3, \dots).$$

We make the choice $c = n$. Then the *recurrence relation*

$$(4) \qquad a_j = \frac{-1}{j(2n+j)} a_{j-2} \qquad (j = 2, 3, \dots)$$

is obtained, giving each coefficient a_j $(j = 2, 3, \dots)$ in terms of the second coefficient preceding it in the series. Note that when n is positive, the choice $c = -n$ does not lead to a well-defined relation of the type (4), where the denominator on the right is never zero.

Since $a_1 = 0$, relation (4) requires that $a_3 = 0$; then $a_5 = 0$, etc. That is,

$$(5) \qquad a_{2k+1} = 0 \qquad (k = 0, 1, 2, \dots).$$

To obtain the remaining coefficients, we let k denote any positive integer and use relation (4) to write the following k equations:

$$a_2 = \frac{-1}{1(n+1)2^2} a_0,$$

$$a_4 = \frac{-1}{2(n+2)2^2} a_2,$$

$$\vdots$$

$$a_{2k} = \frac{-1}{k(n+k)2^2} a_{2k-2}.$$

Upon equating the product of the left-hand sides of these equations to the product of their right-hand sides, and then canceling the common factors $a_2, a_4, \dots, a_{2k-2}$ on each side of the resulting equation, we arrive at the expression

$$(6) \qquad a_{2k} = \frac{(-1)^k}{k!(n+1)(n+2)\cdots(n+k)2^{2k}} a_0 \qquad (k = 1, 2, \dots).$$

In view of identity (5) and since $c = n$, series (2) now takes the form

$$(7) \qquad y = a_0 x^n + \sum_{k=1}^{\infty} a_{2k} x^{n+2k},$$

where the coefficients a_{2k} ($k = 1, 2, \dots$) are those in expression (6). This series is absolutely convergent for all x, according to the ratio test:

$$\lim_{k \to \infty} \left| \frac{a_{2(k+1)} x^{n+2(k+1)}}{a_{2k} x^{n+2k}} \right| = \lim_{k \to \infty} \frac{1}{(k+1)(n+k+1)} \left(\frac{|x|}{2} \right)^2 = 0.$$

Hence it represents a continuous function and is differentiable with respect to x any number of times. Since it is differentiable and its coefficients satisfy the recurrence relation needed to make its sum satisfy Bessel's equation (1), series (7) is, indeed, a solution of that equation.

The coefficient a_0 in series (7) may have any nonzero value. If we substitute expression (6) into that series and write

$$y = a_0 x^n \left[1 + \sum_{k=1}^{\infty} \frac{(-1)^k}{k!(n+1)(n+2)\cdots(n+k)} \left(\frac{x}{2} \right)^{2k} \right],$$

we see that the choice

$$a_0 = \frac{1}{n!2^n}$$

simplifies our solution of Bessel's equation to $y = J_0(x)$, where

$$(8) \qquad J_n(x) = \frac{1}{n!} \left(\frac{x}{2} \right)^n + \sum_{k=1}^{\infty} \frac{(-1)^k}{k!(n+k)!} \left(\frac{x}{2} \right)^{n+2k}.$$

This function $J_n(x)$ is known as the *Bessel function of the first kind of order n* ($n = 0, 1, 2, \dots$). With the convention that $0! = 1$, it is written more compactly as

$$(9) \qquad J_n(x) = \sum_{k=0}^{\infty} \frac{(-1)^k}{k!(n+k)!} \left(\frac{x}{2} \right)^{n+2k}.$$

From expression (9), we note that

$$(10) \qquad J_n(-x) = (-1)^n J_n(x) \qquad (n = 0, 1, 2, \dots);$$

that is, J_n is an even function if $n = 0, 2, 4, \dots$ but odd if $n = 1, 3, 5, \dots$. Also, it is clear from expression (8) that $J_n(0) = 0$ when $n = 1, 2, \dots$ but that $J_0(0) = 1$.

The case in which $n = 0$ will be of special interest to us in the applications. Bessel's equation (1) then becomes

(11) $$xy''(x) + y'(x) + xy(x) = 0;$$

and expression (9) reduces to

(12) $$J_0(x) = \sum_{k=0}^{\infty} \frac{(-1)^k}{(k!)^2}\left(\frac{x}{2}\right)^{2k}.$$

Since

$$(k!)^2 2^{2k} = \left[(1)(2)(3)\cdots(k)2^k\right]^2 = 2^2 4^2 6^2 \cdots (2k)^2$$

when $k \geqslant 1$, another form is

(13) $$J_0(x) = 1 - \frac{x^2}{2^2} + \frac{x^4}{2^2 4^2} - \frac{x^6}{2^2 4^2 6^2} + \cdots .$$

Expressions (12) and (13) bear some resemblance to the power series for cos x. There is also a similarity between the power series representations of the odd functions $J_1(x)$ and sin x. Similarities between the properties of those functions include, as we shall see, the differentiation formula $J_0'(x) = -J_1(x)$, corresponding to the formula for the derivative of cos x. Graphs of $y = J_0(x)$ and $y = J_1(x)$ are shown in Fig. 62. More details regarding these graphs, especially the nature of the zeros of $J_0(x)$ and $J_1(x)$, will be developed later in the chapter.

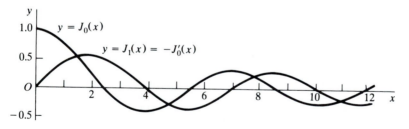

FIGURE 62

60.   GENERAL SOLUTIONS OF BESSEL'S EQUATION

A function linearly independent of J_n that satisfies Bessel's equation

(1) $$x^2 y''(x) + xy'(x) + (x^2 - n^2)y(x) = 0 \qquad\qquad (n = 0, 1, 2, \ldots)$$

can be obtained by various methods of a fairly elementary nature.

The singular point $x = 0$ of equation (1) is of a special type and is known as a *regular* singular point. The power series procedure, extended so as to give general solutions near regular singular points, applies to Bessel's equation. We do not give further details here but only state the results.

When $n = 0$, the general solution is found to be

$$(2) \quad y = AJ_0(x) + B\left[J_0(x) \ln x \right.$$

$$\left. + \frac{x^2}{2^2} - \frac{x^4}{2^2 4^2}\left(1 + \frac{1}{2}\right) + \frac{x^6}{2^2 4^2 6^2}\left(1 + \frac{1}{2} + \frac{1}{3}\right) - \cdots \right],$$

where A and B are arbitrary constants and $x > 0$. Observe that, as long as $B \neq 0$, any choice of A and B yields a solution which is unbounded as x tends to zero through positive values. Such a solution cannot, therefore, be expressed as a constant times $J_0(x)$, which tends to unity as x tends to zero. So $J_0(x)$ and the solution (2) are linearly independent when $B \neq 0$. It is most common to use *Euler's constant* $\gamma = 0.5772\ldots$, which is defined as the limit of the sequence

$$(3) \qquad\qquad s_n = 1 + \frac{1}{2} + \frac{1}{3} + \cdots + \frac{1}{n} - \ln n \qquad (n = 1, 2, \ldots),$$

and to write

$$A = \frac{2}{\pi}(\gamma - \ln 2) \qquad \text{and} \qquad B = \frac{2}{\pi}.$$

When A and B are assigned those values, the second solution that arises is *Weber's Bessel function of the second kind of order zero:*[†]

$$(4) \quad Y_0(x) = \frac{2}{\pi}\left[\left(\ln \frac{x}{2} + \gamma\right)J_0(x) \right.$$

$$\left. + \frac{x^2}{2^2} - \frac{x^4}{2^2 4^2}\left(1 + \frac{1}{2}\right) + \frac{x^6}{2^2 4^2 6^2}\left(1 + \frac{1}{2} + \frac{1}{3}\right) - \cdots \right].$$

More generally, when n has any one of the values $n = 0, 1, 2, \ldots$, equation (1) has a solution $Y_n(x)$ that is valid when $x > 0$ and is unbounded as x tends to zero. Since $J_n(x)$ is continuous at $x = 0$, then, $J_n(x)$ and $Y_n(x)$ are linearly independent; and when $x > 0$, the general solution of equation (1) can be written

$$(5) \qquad\qquad y = C_1 J_n(x) + C_2 Y_n(x) \qquad (n = 0, 1, 2, \ldots),$$

[†] There are other Bessel functions, and the notation varies widely throughout the literature. The treatise by Watson (1952) that is listed in the Bibliography is, however, usually regarded as the standard reference.

where C_1 and C_2 are arbitrary constants. The theory of the second solution $Y_n(x)$ is considerably more involved than that of $J_n(x)$, and we shall limit our applications to problems in which it is only necessary to know that $Y_n(x)$ is discontinuous at $x = 0$.

To write the general solution of Bessel's equation

$$(6) \qquad x^2 y''(x) + x y'(x) + (x^2 - \nu^2) y(x) = 0 \qquad (\nu > 0; \nu \neq 1, 2, \dots),$$

where ν is any positive number other than $1, 2, \dots$, we mention here some elementary properties of the *gamma function*, defined when $\nu > 0$ by means of the equation[†]

$$(7) \qquad \Gamma(\nu) = \int_0^\infty e^{-t} t^{\nu - 1} \, dt \qquad (\nu > 0).$$

An integration by parts shows that

$$(8) \qquad \Gamma(\nu + 1) = \nu \Gamma(\nu)$$

when $\nu > 0$. This property is *assigned* to the function when $\nu < 0$, so that $\Gamma(\nu) = \Gamma(\nu + 1)/\nu$ when $-1 < \nu < 0$, $-2 < \nu < -1$, etc. Thus equations (7) and (8), together, define $\Gamma(\nu)$ for all ν except $\nu = 0, -1, -2, \dots$ (Fig. 63). We find from equation (7) that $\Gamma(1) = 1$; also, it can be shown that $\Gamma(\nu)$ is continuous and positive when $\nu > 0$. It then follows from the identity $\Gamma(\nu) = \Gamma(\nu + 1)/\nu$ that $\Gamma(0+) = \infty$ and, furthermore, that $|\Gamma(\nu)|$ becomes infinite as $\nu \to -n$ $(n = 0, 1, 2, \dots)$. This means that $1/\Gamma(\nu)$ tends to zero as ν tends to $-n$ $(n = 0, 1, 2, \dots)$; and, for brevity, we write $1/\Gamma(-n) = 0$ when $n = 0, 1, 2, \dots$. Note that the reciprocal $1/\Gamma(\nu)$ is then continuous for all ν.

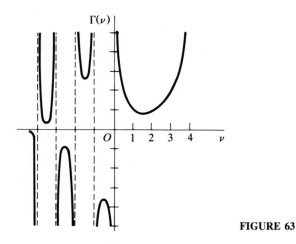

FIGURE 63

[†] Thorough developments of the gamma function appear in the books by Lebedev (1972, chap. 1) and Rainville (1971, chap. 2) that are listed in the Bibliography.

When $v = 1, 2, 3, \ldots, \Gamma(v)$ reduces to a factorial:

(9) $\cdot$ $$\Gamma(n + 1) = n! \qquad\qquad (n = 0, 1, 2, \ldots).$$

The verification of property (9) and the further property that

(10) $$\Gamma\left(\frac{1}{2}\right) = \sqrt{\pi}$$

is left to the problems.

The *Bessel function of the first kind of order* v $(v \geq 0)$ is defined as

(11) $$J_v(x) = \sum_{k=0}^{\infty} \frac{(-1)^k}{k!\,\Gamma(v + k + 1)}\left(\frac{x}{2}\right)^{v+2k}.$$

We note that this becomes expression (9), Sec. 59, for $J_n(x)$ when $v = n = 0, 1, 2, \ldots$. The Bessel function $J_{-v}(x)$ $(v > 0)$ is also well defined when v is replaced by $-v$ in equation (11). If $v = n = 1, 2, \ldots$, however, the summation in the resulting series starts from $k = n$ since $1/\Gamma(-n + k + 1)$ is zero when $0 \leq k \leq n - 1$. It is not difficult to verify by direct substitution that J_v and J_{-v} are solutions of equation (6). Those solutions are arrived at by a modification, involving property (8) of the gamma function, of the procedure used in Sec. 59.

When $v > 0$ and $v \neq 1, 2, \ldots$, the Bessel function $J_{-v}(x)$ is the product of $1/x^v$ and a power series in x whose initial term $(k = 0)$ is nonzero; hence $J_{-v}(x)$ is unbounded as $x \to 0$. Since $J_v(x)$ tends to zero as $x \to 0$, it is evident that J_v and J_{-v} are linearly independent functions. The general solution of Bessel's equation (6) is, therefore,

(12) $$y = C_1 J_v(x) + C_2 J_{-v}(x) \qquad\qquad (v > 0;\, v \neq 1, 2, \ldots),$$

where C_1 and C_2 are arbitrary constants. [Contrast this with solution (5) of equation (1).]

It can be shown that J_n and J_{-n} are linearly dependent because

(13) $$J_{-n}(x) = (-1)^n J_n(x) \qquad\qquad (n = 0, 1, 2, \ldots)$$

(see Problem 1, Sec. 61). So if $v = n = 0, 1, 2, \ldots$, solution (12) cannot be the *general* solution of equation (6).

61. RECURRENCE RELATIONS

Starting with the equation

$$x^{-n} J_n(x) = \frac{1}{2^n} \sum_{k=0}^{\infty} \frac{(-1)^k}{k!(n + k)!}\left(\frac{x}{2}\right)^{2k} \qquad (n = 0, 1, 2, \ldots),$$

we write

$$\frac{d}{dx}\left[x^{-n} J_n(x)\right] = \frac{1}{2^n} \sum_{k=1}^{\infty} \frac{k(-1)^k}{k(k-1)!(n+k)!}\left(\frac{x}{2}\right)^{2k-1}.$$

If we replace k by $k + 1$ here, so that k runs from zero to infinity again, it follows that

$$\frac{d}{dx}[x^{-n}J_n(x)] = x^{-n}\left(\frac{x}{2}\right)^n \sum_{k=0}^{\infty} \frac{(-1)^{k+1}}{k!(n+k+1)!}\left(\frac{x}{2}\right)^{2k+1}$$

$$= -x^{-n} \sum_{k=0}^{\infty} \frac{(-1)^k}{k!(n+1+k)!}\left(\frac{x}{2}\right)^{n+1+2k},$$

or

(1) $$\qquad\qquad \frac{d}{dx}[x^{-n}J_n(x)] = -x^{-n}J_{n+1}(x) \qquad (n = 0,1,2,\dots).$$

The special case

(2) $$\qquad\qquad J_0'(x) = -J_1(x)$$

was mentioned at the end of Sec. 59.

Similarly, from the power series representation of $x^n J_n(x)$, one can show that

(3) $$\qquad\qquad \frac{d}{dx}[x^n J_n(x)] = x^n J_{n-1}(x) \qquad (n = 1,2,\dots).$$

Relations (1) and (3), which are called *recurrence relations*, can be written

$$xJ_n'(x) = nJ_n(x) - xJ_{n+1}(x),$$

$$xJ_n'(x) = -nJ_n(x) + xJ_{n-1}(x).$$

Eliminating $J_n'(x)$ from these equations, we find that

(4) $$\qquad\qquad xJ_{n+1}(x) = 2nJ_n(x) - xJ_{n-1}(x) \qquad (n = 1,2,\dots).$$

This recurrence relation expresses J_{n+1} in terms of the functions J_n and J_{n-1} of lower orders.

From equation (3), we have the integration formula

(5) $$\qquad\qquad \int_0^x s^n J_{n-1}(s)\,ds = x^n J_n(x) \qquad (n = 1,2,\dots).$$

An important special case is

(6) $$\qquad\qquad \int_0^x s J_0(s)\,ds = x J_1(x).$$

Relations (1), (3), and (4) are valid when n is replaced by the unrestricted parameter ν. Modifications of the derivations simply consist of writing

$$\Gamma(\nu + k + 1) \qquad \text{or} \qquad (\nu + k)\Gamma(\nu + k)$$

in place of $(n + k)!$.

PROBLEMS

1. Using series (11), Sec. 60, and recalling that certain terms are to be dropped, show that

$$J_{-n}(x) = (-1)^n J_n(x) \qquad\qquad (n = 1, 2, \dots)$$

and hence that the functions J_n and J_{-n} are linearly dependent.

2. Derive the recurrence relation (3), Sec. 61.

3. Establish the differentiation formula

$$x^2 J_n''(x) = (n^2 - n - x^2) J_n(x) + x J_{n+1}(x) \qquad\qquad (n = 0, 1, 2, \dots).$$

4. (*a*) Derive the *reduction formula*

$$\int_0^x s^n J_0(s)\, ds = x^n J_1(x) + (n-1) x^{n-1} J_0(x) - (n-1)^2 \int_0^x s^{n-2} J_0(s)\, ds$$

$$(n = 2, 3, \dots)$$

by applying integration by parts twice and using the relations (Sec. 61)

$$\frac{d}{ds}[s J_1(s)] = s J_0(s), \qquad \frac{d}{ds} J_0(s) = -J_1(s)$$

in the first and second of those integrations, respectively.

(*b*) Note that, in view of equation (6), Sec. 61, the identity obtained in part (*a*) can be applied successively to evaluate the integral on the left-hand side of that identity when the integer n is *odd*.[†] Illustrate this by showing that

$$\int_0^x s^5 J_0(s)\, ds = x(x^2 - 8)\big[4x J_0(x) + (x^2 - 8) J_1(x)\big].$$

5. Let y be any solution of Bessel's equation of order zero, and let $\mathscr{L}$ denote the self-adjoint (Sec. 41) differential operator defined by the equation

$$\mathscr{L}[X(x)] = [x X'(x)]' + x X(x).$$

(*a*) By writing $X = J_0$ and $Y = y$ in Lagrange's identity [Problem 3(*b*), Sec. 43]

$$X \mathscr{L}[Y] - Y \mathscr{L}[X] = \frac{d}{dx}[x(XY' - X'Y)]$$

for that operator, show that there is a constant B such that

$$\frac{d}{dx}\left[\frac{y(x)}{J_0(x)}\right] = \frac{B}{x[J_0(x)]^2}.$$

[†] Note, too, that when n is *even*, the reduction formula can be used to transform the problem of evaluating $\int_0^x s^n J_0(s)\, ds$ into that of evaluating $\int_0^x J_0(s)\, ds$, which is tabulated for various values of x in, for example, the book edited by Abramowitz and Stegun (1972, pp. 492–493) that is listed in the Bibliography. Further references are given on pp. 490–491 of that book.

(b) Assuming that the function $1/[J_0(x)]^2$ has a Maclaurin series expansion of the form[†]

$$\frac{1}{[J_0(x)]^2} = 1 + \sum_{k=1}^{\infty} c_k x^{2k}$$

and that the expansion obtained by multiplying each side of this by $1/x$ can be integrated term by term, use the result in part (a) to show formally that y can be written in the form

$$y = AJ_0(x) + B\left[J_0(x) \ln x + \sum_{k=1}^{\infty} d_k x^{2k} \right],$$

where A, B, and d_k ($k = 1, 2, \ldots$) are constants. [Compare equation (2), Sec. 60.]

6. Let s_n ($n = 1, 2, \ldots$) be the sequence defined in equation (3), Sec. 60. Show that $s_n > 0$ and $s_n - s_{n+1} > 0$ for each n. Thus show that the sequence is bounded and decreasing and that it therefore converges to some number γ. Also, point out how it follows that $0 \leqslant \gamma < 1$.

Suggestion: Observe from the graph of the function $y = 1/x$ that

$$\sum_{k=1}^{n-1} \frac{1}{k} > \int_1^n \frac{dx}{x} = \ln n \qquad\qquad (n \geqslant 2)$$

and

$$\frac{1}{n+1} < \int_n^{n+1} \frac{dx}{x} = \ln(n+1) - \ln n \qquad\qquad (n \geqslant 1).$$

7. (a) Derive the property $\Gamma(\nu + 1) = \nu\Gamma(\nu)$ of the gamma function, as stated in Sec. 60.

(b) Show that $\Gamma(1) = 1$ and, using mathematical induction, verify that $\Gamma(n + 1) = n!$ when $n = 0, 1, 2, \ldots$.

8. Verify that the function J_ν ($\nu \geqslant 0$), defined by equation (11), Sec. 60, satisfies Bessel's equation (6) in that section. Point out how it follows that $J_{-\nu}$ is also a solution.

9. Derive the differentiation formula

$$\frac{d}{dx}[x^{-\nu}J_\nu(x)] = -x^{-\nu}J_{\nu+1}(x),$$

where $\nu \geqslant 0$, and point out why it is also valid when ν is replaced by $-\nu$ ($\nu > 0$). [Compare relation (1), Sec. 61.]

10. Refer to the result obtained in Problem 18, Sec. 58, and show that

$$\Gamma\left(\frac{1}{2}\right) = 2\int_0^{\infty} e^{-x^2}\, dx = \sqrt{\pi}\,.$$

[†] This valid assumption is easily justified by methods from the theory of functions of a complex variable. See the author's book (1990, chap. 5), listed in the Bibliography.

11. With the aid of mathematical induction, verify that

$$\Gamma\left(k + \frac{1}{2}\right) = \frac{(2k)!}{k!2^{2k}}\sqrt{\pi} \qquad (k = 0, 1, 2, \dots).$$

12. Use the series representation (11), Sec. 60, for $J_\nu(x)$ and the identity in Problem 11 to show that

$$(a) \quad J_{1/2}(x) = \sqrt{\frac{2}{\pi x}} \sin x; \qquad (b) \quad J_{-1/2}(x) = \sqrt{\frac{2}{\pi x}} \cos x.$$

13. Use results in Problems 9 and 12 to show that

$$J_{3/2}(x) = \sqrt{\frac{2}{\pi x}}\left(\frac{\sin x}{x} - \cos x\right).$$

14. Show that if y is a differentiable function of x and if $s = \alpha x$, where α is a nonzero constant, then

$$\frac{dy}{dx} = \alpha\frac{dy}{ds} \qquad \text{and} \qquad \frac{d^2y}{dx^2} = \alpha^2\frac{d^2y}{ds^2}.$$

Thus show that the substitution $s = \alpha x$ transforms the differential equation

$$x^2\frac{d^2y}{dx^2} + x\frac{dy}{dx} + (\alpha^2x^2 - n^2)y = 0 \qquad (n = 0, 1, 2, \dots)$$

into Bessel's equation

$$s^2\frac{d^2y}{ds^2} + s\frac{dy}{ds} + (s^2 - n^2)y = 0 \qquad (n = 0, 1, 2, \dots),$$

which is free of α. Conclude that the general solution of the first differential equation here is

$$y = C_1 J_n(\alpha x) + C_2 Y_n(\alpha x).$$

15. From the series representation (9), Sec. 59, for $J_n(x)$, show that

$$i^{-n}J_n(ix) = \sum_{k=0}^{\infty} \frac{1}{k!(n+k)!}\left(\frac{x}{2}\right)^{n+2k} \qquad (n = 0, 1, 2, \dots).$$

The function $I_n(x) = i^{-n}J_n(ix)$ is the *modified Bessel function of the first kind of order* n. Show that the series here converges for all x, that $I_n(x) > 0$ when $x > 0$, and that $I_n(-x) = (-1)^n I_n(x)$. Also, by referring to the result in Problem 14, point out why $I_n(x)$ is a solution of the modified Bessel equation

$$x^2y''(x) + xy'(x) - (x^2 + n^2)y(x) = 0.$$

62. BESSEL'S INTEGRAL FORM OF $J_n(x)$

We now derive a useful integral representation for $J_n(x)$. To do this, we first note that the series in the expansions

$$(1) \qquad \exp\left(\frac{xt}{2}\right) = \sum_{j=0}^{\infty} \frac{x^j}{j!2^j}t^j, \qquad \exp\left(-\frac{x}{2t}\right) = \sum_{k=0}^{\infty} \frac{(-1)^k x^k}{k!2^k}t^{-k}$$

are absolutely convergent when x is any number and $t \neq 0$. Hence the product of these exponential functions is itself represented by a series formed by multiplying each term in one series by every term in the other and then summing the resulting terms *in any order.*[†] Clearly, the variable t occurs in each of those resulting terms as a factor t^n $(n = 0, 1, 2, \ldots)$ or t^{-n} $(n = 1, 2, \ldots)$; and the terms involving any particular power of t may be collected as a sum.

In the case of t^n $(n = 0, 1, 2, \ldots)$, that sum is obtained by multiplying the kth term in the second series by the term in the first series whose index is $j = n + k$ and then summing from $k = 0$ to $k = \infty$. The result is

$$\sum_{k=0}^{\infty} \frac{(-1)^k}{k!(n+k)!} \left(\frac{x}{2}\right)^{n+2k} t^n = J_n(x)t^n.$$

Similarly, the sum of the terms involving t^{-n} $(n = 1, 2, \ldots)$ is found by multiplying the jth term in the first series by the term in the second series with index $k = n + j$ and summing from $j = 0$ to $j = \infty$. That sum may be written

$$(-1)^n \sum_{j=0}^{\infty} \frac{(-1)^j}{j!(n+j)!} \left(\frac{x}{2}\right)^{n+2j} t^{-n} = (-1)^n J_n(x)t^{-n}.$$

A series representation for the product of the exponential functions (1) is, therefore,

$$(2) \qquad \exp\left[\frac{x}{2}\left(t - \frac{1}{t}\right)\right] = J_0(x) + \sum_{n=1}^{\infty} \left[J_n(x)t^n + (-1)^n J_n(x)t^{-n}\right].$$

Let us write $t = e^{i\phi}$ in equation (2). In view of Euler's formula

$$e^{i\phi} = \cos\phi + i\sin\phi,$$

we know that

$$e^{i\phi} - e^{-i\phi} = 2i\sin\phi$$

and

$$e^{in\phi} = \cos n\phi + i\sin n\phi, \qquad e^{-in\phi} = \cos n\phi - i\sin n\phi.$$

It thus follows from equation (2) that

$$(3) \qquad \exp(ix\sin\phi) = J_0(x) + \sum_{n=1}^{\infty} \left[1 + (-1)^n\right] J_n(x)\cos n\phi$$

$$+ i\sum_{n=1}^{\infty} \left[1 - (-1)^n\right] J_n(x)\sin n\phi.$$

[†] For a justification of this procedure, see, for example, the book by Taylor and Mann (1983, pp. 601–602) that is listed in the Bibliography.

Now, again by Euler's formula,

$$\exp(ix\sin\phi) = \cos(x\sin\phi) + i\sin(x\sin\phi);$$

and if we equate the real parts on each side of equation (3), we find that

$$\cos(x\sin\phi) = J_0(x) + \sum_{n=1}^{\infty}\left[1 + (-1)^n\right]J_n(x)\cos n\phi.$$

Holding x fixed and regarding this equation as a Fourier cosine series representation of the function $\cos(x\sin\phi)$ on the interval $0 < \phi < \pi$, we need only recall the formula for the coefficients in such a series to write

$$(4) \qquad \left[1 + (-1)^n\right]J_n(x) = \frac{2}{\pi}\int_0^{\pi}\cos(x\sin\phi)\cos n\phi\, d\phi \qquad (n = 0, 1, 2, \ldots).$$

If, on the other hand, we equate the imaginary parts on each side of equation (3), we obtain the Fourier sine series representation

$$\sin(x\sin\phi) = \sum_{n=1}^{\infty}\left[1 - (-1)^n\right]J_n(x)\sin n\phi$$

for $\sin(x\sin\phi)$ on the same interval. Consequently,

$$(5) \qquad \left[1 - (-1)^n\right]J_n(x) = \frac{2}{\pi}\int_0^{\pi}\sin(x\sin\phi)\sin n\phi\, d\phi \qquad (n = 1, 2, \ldots).$$

According to expressions (4) and (5), then,

$$(6) \qquad J_{2n}(x) = \frac{1}{\pi}\int_0^{\pi}\cos(x\sin\phi)\cos 2n\phi\, d\phi \qquad (n = 0, 1, 2, \ldots)$$

and

$$(7) \qquad J_{2n-1}(x) = \frac{1}{\pi}\int_0^{\pi}\sin(x\sin\phi)\sin(2n - 1)\phi\, d\phi \qquad (n = 1, 2, \ldots).$$

A single expression for $J_n(x)$ can be obtained by adding corresponding sides of equations (4) and (5) and writing

$$2J_n(x) = \frac{2}{\pi}\int_0^{\pi}\left[\cos n\phi\cos(x\sin\phi) + \sin n\phi\sin(x\sin\phi)\right]d\phi.$$

That is,

$$(8) \qquad J_n(x) = \frac{1}{\pi}\int_0^{\pi}\cos(n\phi - x\sin\phi)\, d\phi \qquad (n = 0, 1, 2, \ldots).$$

This is known as *Bessel's integral form* of $J_n(x)$, and expressions (6) and (7) are special cases of it.

63.   CONSEQUENCES OF THE INTEGRAL REPRESENTATIONS

From the integral representation

(1) $$J_n(x) = \frac{1}{\pi} \int_0^\pi \cos(n\phi - x \sin \phi) \, d\phi \qquad (n = 0, 1, 2, \dots)$$

just obtained, it follows that

$$J_n'(x) = \frac{1}{\pi} \int_0^\pi \sin(n\phi - x \sin \phi) \sin \phi \, d\phi.$$

Continued differentiation yields integral representations for $J_n''(x)$, etc. In each case, the absolute value of the integrand that arises does not exceed unity. The following *boundedness properties* are, then, consequences of Bessel's integral form (1):

(2) $$|J_n(x)| \le 1, \qquad \left| \frac{d^k}{dx^k} J_n(x) \right| \le 1 \qquad (k = 1, 2, \dots).$$

The first of these inequalities, for example, is obtained by writing

$$|J_n(x)| \le \frac{1}{\pi} \int_0^\pi |\cos(n\phi - x \sin \phi)| \, d\phi \le \frac{1}{\pi} \int_0^\pi d\phi = 1.$$

Sometimes it is useful to write the integral representations

$$J_{2n}(x) = \frac{1}{\pi} \int_0^\pi \cos(x \sin \phi) \cos 2n\phi \, d\phi \qquad (n = 0, 1, 2, \dots)$$

and

$$J_{2n-1}(x) = \frac{1}{\pi} \int_0^\pi \sin(x \sin \phi) \sin(2n - 1)\phi \, d\phi \qquad (n = 1, 2, \dots),$$

obtained in Sec. 62, as

(3) $$J_{2n}(x) = \frac{2}{\pi} \int_0^{\pi/2} \cos(x \sin \phi) \cos 2n\phi \, d\phi \qquad (n = 0, 1, 2, \dots)$$

and

(4) $$J_{2n-1}(x) = \frac{2}{\pi} \int_0^{\pi/2} \sin(x \sin \phi) \sin(2n - 1)\phi \, d\phi \qquad (n = 1, 2, \dots).$$

Expressions (3) and (4) follow from the fact that, when x is fixed, the graphs of the integrands

$$y = g(\phi) = \cos(x \sin \phi) \cos 2n\phi,$$
$$y = h(\phi) = \sin(x \sin \phi) \sin(2n - 1)\phi$$

are symmetric with respect to the line $\phi = \pi/2$:

$$g(\pi - \phi) = g(\phi), \qquad h(\pi - \phi) = h(\phi).$$

We note the special case

(5) $$J_0(x) = \frac{2}{\pi} \int_0^{\pi/2} \cos(x \sin \phi) \, d\phi$$

of representation (3), which can also be written

(6) $$J_0(x) = \frac{2}{\pi} \int_0^{\pi/2} \cos(x \cos \theta) \, d\theta$$

by means of the substitution $\theta = (\pi/2) - \phi$.

Representations (3) and (4) may be used to verify that, for each fixed n ($n = 0, 1, 2, \ldots$),

(7) $$\lim_{x \to \infty} J_n(x) = 0.$$

To give the details when $n = 0$, we substitute $u = \sin \phi$ in equation (5) to write

$$\frac{\pi}{2} J_0(x) = \int_0^c \frac{\cos xu}{\sqrt{1 - u^2}} \, du + \int_c^1 \frac{\cos xu}{\sqrt{1 - u^2}} \, du,$$

where $0 < c < 1$. The second integral here is improper but uniformly convergent with respect to x. Corresponding to any positive number ε, the absolute value of that integral can be made less than $\varepsilon/2$, uniformly for all x, by selecting c so that $1 - c$ is sufficiently small and positive. The Riemann-Lebesgue lemma involving a cosine function (Sec. 53) then applies to the first integral with that value of c. That is, there is a number x_ε such that the absolute value of the first integral is less than $\varepsilon/2$ whenever $x > x_\varepsilon$. Therefore,

$$\frac{\pi}{2}|J_0(x)| < \varepsilon \qquad\qquad \text{whenever } x > x_\varepsilon;$$

and this establishes property (7) when $n = 0$. Verification when n is a positive integer is left to the problems.

It is interesting to contrast limit (7) with the limit

(8) $$\lim_{n \to \infty} J_n(x) = 0,$$

which is valid for each fixed x $(-\infty < x < \infty)$. This limit follows from the fact that the coefficients

$$a_{2n} = 2J_{2n}(x) \qquad \text{and} \qquad b_{2n-1} = 2J_{2n-1}(x)$$

in the Fourier cosine and sine series for certain functions of ϕ in Sec. 62 must tend to zero as n tends to infinity, according to Sec. 16. Limit (8) can also be obtained by applying the Riemann-Lebesgue lemma to the integral representations for $J_{2n}(x)$ and $J_{2n-1}(x)$.

PROBLEMS

1. Use integral representations for $J_n(x)$ to verify that
   (a) $J_0(0) = 1$;   (b) $J_n(0) = 0$ $(n = 1, 2, \dots)$;   (c) $J_0'(x) = -J_1(x)$.
2. Derive representation (3), Sec. 63, for $J_{2n}(x)$ by writing the Fourier cosine series for $\cos(x \sin \phi)$ in Sec. 62 as

$$\cos(x \sin \phi) = J_0(x) + 2 \sum_{n=1}^{\infty} J_{2n}(x) \cos 2n\phi$$

   and then interpreting it as a Fourier cosine series on the interval $0 < \phi < \pi/2$.
3. Deduce from expression (3), Sec. 63, that

$$J_{2n}(x) = (-1)^n \frac{2}{\pi} \int_0^{\pi/2} \cos(x \cos \theta) \cos 2n\theta \, d\theta \qquad (n = 0, 1, 2, \dots).$$

4. Deduce from expression (4), Sec. 63, that

$$J_{2n-1}(x) = (-1)^{n+1} \frac{2}{\pi} \int_0^{\pi/2} \sin(x \cos \theta) \cos(2n-1)\theta \, d\theta \qquad (n = 1, 2, \dots).$$

5. Complete the verification of property (7), Sec. 63, that

$$\lim_{x \to \infty} J_n(x) = 0$$

   for each fixed n $(n = 0, 1, 2, \dots)$.
6. Apply integration by parts to representations (3) and (4) in Sec. 63 and then use the Riemann-Lebesgue lemma (Sec. 53) to show that

$$\lim_{n \to \infty} nJ_n(x) = 0$$

   for each fixed x.
7. Verify directly from the representation (Sec. 63)

$$J_0(x) = \frac{2}{\pi} \int_0^{\pi/2} \cos(x \sin \phi) \, d\phi$$

   that $J_0(x)$ satisfies Bessel's equation

$$xy''(x) + y'(x) + xy(x) = 0.$$

8. According to Sec. 24, if a function f and its derivative f' are continuous on the interval $-\pi \leqslant x \leqslant \pi$ and if $f(-\pi) = f(\pi)$, then Parseval's equation

$$\frac{1}{\pi} \int_{-\pi}^{\pi} [f(x)]^2 \, dx = \frac{a_0^2}{2} + \sum_{n=1}^{\infty} (a_n^2 + b_n^2)$$

   holds, where the numbers a_n $(n = 0, 1, 2, \dots)$ and b_n $(n = 1, 2, \dots)$ are the Fourier coefficients

$$a_n = \frac{1}{\pi} \int_{-\pi}^{\pi} f(x) \cos nx \, dx, \qquad b_n = \frac{1}{\pi} \int_{-\pi}^{\pi} f(x) \sin nx \, dx.$$

(a) By applying that result to $f(\phi) = \cos(x \sin \phi)$, an even function of ϕ, and referring to the Fourier (cosine) series for $f(\phi)$ in Sec. 62, show that

$$\frac{1}{\pi} \int_0^\pi \cos^2(x \sin \phi) \, d\phi = [J_0(x)]^2 + 2 \sum_{n=1}^\infty [J_{2n}(x)]^2 \qquad (-\infty < x < \infty).$$

(b) Similarly, by writing $f(\phi) = \sin(x \sin \phi)$ and referring to the Fourier (sine) series expansion in Sec. 62, show that

$$\frac{1}{\pi} \int_0^\pi \sin^2(x \sin \phi) \, d\phi = 2 \sum_{n=1}^\infty [J_{2n-1}(x)]^2 \qquad (-\infty < x < \infty).$$

(c) Combine the results in parts (a) and (b) to show that

$$[J_0(x)]^2 + 2 \sum_{n=1}^\infty [J_n(x)]^2 = 1 \qquad (-\infty < x < \infty),$$

and point out how it follows from this identity that

$$|J_0(x)| \leqslant 1 \qquad \text{and} \qquad |J_n(x)| \leqslant \frac{1}{\sqrt{2}} \qquad (n = 1, 2, \ldots)$$

for all x.

9. By writing $t = i$ in the series representation (2), Sec. 62, derive the expansions

$$\cos x = J_0(x) + 2 \sum_{n=1}^\infty (-1)^n J_{2n}(x)$$

and

$$\sin x = 2 \sum_{n=1}^\infty (-1)^{n+1} J_{2n-1}(x),$$

which are valid for all x.

10. Show that series representation (2), Sec. 62, can be written in the form

$$\exp\left[\frac{x}{2}\left(t - \frac{1}{t}\right)\right] = \lim_{N \to \infty} \sum_{n=-N}^N J_n(x) t^n \qquad (t \neq 0).$$

This exponential function is, then, a *generating function* for the Bessel functions $J_n(x)$ $(n = 0, \pm 1, \pm 2, \ldots)$.

64. THE ZEROS OF $J_0(x)$

A modified form of Bessel's equation

$$(1) \qquad x^2 y''(x) + xy'(x) + (x^2 - \nu^2) y(x) = 0$$

in which the term containing the first derivative is absent is sometimes useful. That form is easily found (Problem 1, Sec. 66) by making the substitution $y(x) = x^c u(x)$ in equation (1) and observing that the coefficient of $u'(x)$ in the resulting differential equation

$$(2) \qquad x^2 u''(x) + (1 + 2c) xu'(x) + (x^2 - \nu^2 + c^2) u(x) = 0$$

is zero if $c = -\frac{1}{2}$. The desired modified form of equation (1) is, then,

$$(3) \qquad x^2 u''(x) + \left(x^2 - \nu^2 + \frac{1}{4}\right) u(x) = 0;$$

and the function $u = \sqrt{x}\, J_\nu(x)$ is evidently a solution of it. In particular, when $\nu = 0$, we see that the function $u = \sqrt{x}\, J_0(x)$ satisfies the differential equation

$$(4) \qquad u''(x) + \left(1 + \frac{1}{4x^2}\right) u(x) = 0.$$

We shall now use equation (4) to show that *the positive zeros of $J_0(x)$, or roots of the equation $J_0(x) = 0$, form an increasing sequence of numbers x_j $(j = 1, 2, \ldots)$ such that $x_j \to \infty$ as $j \to \infty$.*[†]

We start with the observation that the differential operator $\mathcal{L} = d^2/dx^2$ is self-adjoint (Sec. 41) and that Lagrange's identity [Problem 3(b), Sec. 43] for this operator is

$$(5) \qquad U(x)V''(x) - V(x)U''(x) = \frac{d}{dx}[U(x)V'(x) - U'(x)V(x)],$$

where $U(x)$ and $V(x)$ are any functions that are twice-differentiable. We write

$$(6) \qquad U(x) = \sqrt{x}\, J_0(x) \qquad \text{and} \qquad V(x) = \sin x.$$

From equation (4), we know that

$$U''(x) + U(x) = \frac{U(x)}{4x^2};$$

furthermore,

$$V''(x) + V(x) = 0.$$

The left-hand side of identity (5) then becomes $U(x)V(x)/(4x^2)$, and it follows that

$$(7) \qquad \int_a^b \frac{U(x)V(x)}{4x^2}\, dx = [U(x)V'(x) - U'(x)V(x)]_a^b,$$

where $0 < a < b$.

It is now easy to show that our function $U(x) = \sqrt{x}\, J_0(x)$, and hence $J_0(x)$, has at least one zero in each interval

$$2k\pi \leqslant x \leqslant 2k\pi + \pi \qquad\qquad (k = 1, 2, \ldots).$$

[†] Our method is a modification of the one used by A. Czarnecki, *Amer. Math. Monthly*, vol. 71, no. 4, pp. 403–404, 1964, who considers Bessel functions $J_\nu(x)$, where $-\frac{1}{2} \leqslant \nu \leqslant \frac{1}{2}$.

We do this by assuming that $U(x) \neq 0$ anywhere in an interval

$$2k\pi \leqslant x \leqslant 2k\pi + \pi$$

and obtaining a contradiction. According to that assumption, either $U(x) > 0$ for all x in the interval or $U(x) < 0$ for all such x, since $U(x)$ is continuous and thus cannot change sign without having a zero value at some point in the interval.

Suppose that $U(x) > 0$ when $2k\pi \leqslant x \leqslant 2k\pi + \pi$. In identity (7), write $a = 2k\pi$ and $b = 2k\pi + \pi$. Since $V(a) = V(b) = 0$, $V'(a) = 1$, and $V'(b) = -1$, that identity becomes

(8)
$$\int_{2k\pi}^{2k\pi+\pi} U(x) \frac{\sin x}{4x^2} \, dx = -[U(2k\pi) + U(2k\pi + \pi)].$$

The integrand here is positive when $2k\pi < x < 2k\pi + \pi$. Hence the left-hand side of this equation has a positive value while the right-hand side is negative, giving a contradiction.

If, on the other hand, $U(x) < 0$ when $2k\pi \leqslant x \leqslant 2k\pi + \pi$, those two sides of equation (8) are negative and positive, respectively. This is again a contradiction. Thus $J_0(x)$ has at least one zero in each interval

$$2k\pi \leqslant x \leqslant 2k\pi + \pi \qquad\qquad (k = 1, 2, \dots).$$

Actually, $J_0(x)$ can have *at most* a finite number of zeros in any closed bounded interval $a \leqslant x \leqslant b$. To see that this is so, we assume that the interval $a \leqslant x \leqslant b$ does contain an infinite number of zeros. From advanced calculus, we know that if a given infinite set of points lies in a closed bounded interval, there is always a sequence of distinct points in that set which converges to a point in the interval.[†] In particular, then, our assumption that the interval $a \leqslant x \leqslant b$ contains an infinite number of zeros of $J_0(x)$ implies that there exists a sequence x_m $(m = 1, 2, \dots)$ of distinct zeros such that $x_m \to c$ as $m \to \infty$, where c is a point which also lies in the interval. Since the function $J_0(x)$ is continuous, $J_0(c) = 0$; and, by the definition of the limit of a sequence, every interval centered about c contains other zeros of $J_0(x)$. But the fact that $J_0(x)$ is not identically zero and has a Maclaurin series representation which is valid for all x means that there exists some interval centered at c which contains no other zeros.[‡] Since this is contrary to what has just been shown, the number of zeros in the interval $a \leqslant x \leqslant b$ cannot, then, be infinite.

It is now evident that the positive zeros of $J_0(x)$ can, in fact, be arranged as an increasing sequence of numbers tending to infinity. The table below gives the values, correct to four significant figures, of the first five zeros of $J_0(x)$ and the corresponding values of $J_1(x)$ [see Fig. 62 (Sec. 59)]. Extensive tables of

---

[†] See, for example, the book by Taylor and Mann (1983, pp. 515–519), listed in the Bibliography.

[‡] That is, the zeros of such a function are *isolated*. An argument for this is given in a somewhat more general setting in the authors' book (1990, p. 181), listed in the Bibliography.

numerical values of Bessel and related functions, together with their zeros, will be found in books listed in the Bibliography.[†]

$$J_0(x_j) = 0$$

j	1	2	3	4	5
x_j	2.405	5.520	8.654	11.79	14.93
$J_1(x_j)$	0.5191	−0.3403	0.2715	−0.2325	0.2065

65. ZEROS OF RELATED FUNCTIONS

If $J_n(a) = 0$ and $J_n(b) = 0$ for two distinct positive numbers a and b, then $x^{-n}J_n(x)$ also vanishes when $x = a$ and when $x = b$. It thus follows from Rolle's theorem that the derivative of $x^{-n}J_n(x)$ vanishes for at least one value of x between a and b. But (Sec. 61)

$$\frac{d}{dx}[x^{-n}J_n(x)] = -x^{-n}J_{n+1}(x) \qquad (n = 0, 1, 2, \ldots);$$

and so, when $n = 0, 1, 2, \ldots$, there is at least one zero of $J_{n+1}(x)$ between any two positive zeros of $J_n(x)$. Also, just as in the case of $J_0(x)$ (Sec. 64), the function $J_{n+1}(x)$ can have at most a finite number of zeros in each bounded interval.

We have already shown that the positive zeros of $J_0(x)$ form an un-bounded increasing sequence of numbers. It now follows that the zeros of $J_1(x)$ must form such a set. The same is then true for $J_2(x)$, etc. That is, for each fixed nonnegative integer n, the set of all positive roots of the equation $J_n(x) = 0$ forms an increasing sequence $x = x_{nj}$ ($j = 1, 2, \ldots$), where $x_{nj} \to \infty$ as $j \to \infty$.

The function $y = J_n(x)$ satisfies Bessel's equation, which is a linear homo-geneous differential equation of the second order with the origin as a singular point. According to the lemma in Sec. 44 on the uniqueness of solutions of second-order linear differential equations, there is just one solution that satisfies the conditions $y(c) = y'(c) = 0$, where $c > 0$; that solution is identically zero. Consequently, there is no positive number c such that $J_n(c) = J_n'(c) = 0$. That is, $J_n'(x)$ cannot vanish at a positive zero of $J_n(x)$; thus $J_n(x)$ must change its sign at that point.

Let a and b ($0 < a < b$) be two consecutive zeros of $J_n(x)$. If $J_n'(a) > 0$, then $J_n(x) > 0$ when $a < x < b$ and $J_n(x)$ is decreasing at its zero b; that is, $J_n'(b) < 0$. Similarly, if $J_n'(a) < 0$, then $J_n'(b) > 0$. So the values of J_n' alternate in sign at consecutive positive zeros of J_n.

We now consider the function $hJ_n(x) + xJ_n'(x)$, where h is a nonnegative constant. The zeros of this function will also arise in certain boundary value problems. If a and b are consecutive positive zeros of $J_n(x)$, it follows that $hJ_n(x) + xJ_n'(x)$ must have the values $aJ_n'(a)$ and $bJ_n'(b)$ at the points $x = a$ and

---

[†] See, especially, the book edited by Abramowitz and Stegun (1972) and the ones by Jahnke, Emde, and Lösch (1960), Gray and Mathews (1966), and Watson (1952).

$x = b$, respectively. Since one of those values is positive and the other negative, the function vanishes at some point, or at some finite number of points, between a and b. It therefore has an increasing sequence of positive zeros tending to infinity.[†] We collect our principal results as follows.

Theorem. *For each fixed* n ($n = 0, 1, 2, \ldots$), *the positive roots of the equation*

$$(1) \qquad\qquad\qquad J_n(x) = 0$$

form an increasing sequence $x = x_{nj}$ ($j = 1, 2, \ldots$) *such that* $x_{nj} \to \infty$ *as* $j \to \infty$; *also, the positive roots of the equation*

$$(2) \qquad\qquad hJ_n(x) + xJ_n'(x) = 0 \qquad\qquad (h \geqslant 0),$$

where h *is a constant, always form a sequence of that type.*

Observe that $x = 0$ is a root of both equations (1) and (2) if n is a positive integer, since $J_n(0) = 0$ ($n = 1, 2, \ldots$). It is also a root of equation (2) when $h = n = 0$.

If $x = c$ is a root of equation (1), then $x = -c$ is also a root since $J_n(-c) = (-1)^n J_n(c)$. That statement is true of equation (2) as well; for, in view of the recurrence relation (Sec. 61)

$$xJ_n'(x) = nJ_n(x) - xJ_{n+1}(x),$$

equation (2) can be written

$$(3) \qquad\qquad (h + n)J_n(x) - xJ_{n+1}(x) = 0,$$

and we note that

$$(h + n)J_n(-c) - (-c)J_{n+1}(-c) = (-1)^n\big[(h + n)J_n(c) - cJ_{n+1}(c)\big].$$

Finally, although our discussion leading up to the theorem need not have excluded the possibility that h be negative, those values of h will not arise in our applications.

66. ORTHOGONAL SETS OF BESSEL FUNCTIONS

As indicated at the beginning of the chapter, where somewhat different notation was used, the physical applications in this chapter will involve solutions of the

---

[†] In the important special case $n = 0$, the first few zeros are tabulated for various positive values of h in, for example, the book on heat conduction by Carslaw and Jaeger (1959, p. 493) that is listed in the Bibliography.

differential equation

(1)
$$x^2 \frac{d^2X}{dx^2} + x\frac{dX}{dx} + (\lambda x^2 - n^2)X = 0 \quad (n = 0, 1, 2, \dots),$$

whose self-adjoint form (Sec. 41) is

(2)
$$\frac{d}{dx}\left(x\frac{dX}{dx}\right) + \left(-\frac{n^2}{x} + \lambda x\right)X = 0 \quad (n = 0, 1, 2, \dots).$$

More specifically, we shall need to solve a singular (Sec. 42) Sturm-Liouville problem, on an interval $0 \leqslant x \leqslant c$, consisting of the differential equation (1) and a boundary condition of the type

(3)
$$b_1 X(c) + b_2 X'(c) = 0.$$

The constants b_1 and b_2 are real and not both zero, and X and X' are to be *continuous on the entire interval* $0 \leqslant x \leqslant c$.

In the important special case when $b_2 = 0$, the boundary condition (3) is

(4)
$$X(c) = 0.$$

When $b_2 \neq 0$, we may multiply through condition (3) by c/b_2 and write it as

(5)
$$hX(c) + cX'(c) = 0,$$

where $h = cb_1/b_2$. In solving our Sturm-Liouville problem, we shall find it convenient to use the boundary condition (3) in its separate forms (4) and (5); and, when using condition (5), *we shall always assume that* $h \geqslant 0$.

The corollary in Sec. 43, applied to case (a) of the theorem in that section, ensures that any eigenvalue of our singular Sturm-Liouville problem must be a real number. We now consider the three possibilities of λ being zero, positive, or negative.

When $\lambda = 0$, equation (1) is a Cauchy-Euler equation (see Problem 3, Sec. 35):

(6)
$$x^2 \frac{d^2X}{dx^2} + x\frac{dX}{dx} - n^2X = 0.$$

To solve it, we write $x = \exp s$ and put it in the form

(7)
$$\frac{d^2X}{ds^2} - n^2X = 0.$$

If n is positive, it follows that $X = Ae^{ns} + Be^{-ns}$, where A and B are constants; that is, $X(x) = Ax^n + Bx^{-n}$. Since our solution must be continuous, and therefore bounded, on the interval $0 \leqslant x \leqslant c$, we require that $B = 0$. Hence $X(x) = Ax^n$. It is now easy to see that $A = 0$ if either condition (4) or (5) is to be satisfied, and we arrive at only the trivial solution $X(x) \equiv 0$. Thus zero is not an eigenvalue if n is positive.

If, on the other hand, $n = 0$, equation (7) has the general solution $X = As + B$; and the general solution of equation (6) when $n = 0$ is, therefore,

$X(x) = A \ln x + B$. According to the continuity requirements, then, $X(x) = B$. When condition (4) is imposed, $B = 0$; the same is true of condition (5) when $h > 0$. But when $h = 0$, condition (5) becomes simply $X'(c) = 0$, and B can remain arbitrary. So *if $n = 0$ and condition (5) is used when $h = 0$, we have the eigenfunction*

$$(8) \qquad\qquad X(x) = 1, \qquad corresponding\ to \qquad \lambda = 0.$$

This is the only case in which $\lambda = 0$ is an eigenvalue, and any eigenfunction corresponding to that eigenvalue is a constant multiple of the function (8).

We consider next the case in which $\lambda > 0$ and write $\lambda = \alpha^2$ ($\alpha > 0$). Equation (1) is then

$$(9) \qquad\qquad x^2 \frac{d^2X}{dx^2} + x\frac{dX}{dx} + (\alpha^2x^2 - n^2)X = 0,$$

and we know from Problem 14, Sec. 61, that its general solution is

$$(10) \qquad\qquad X(x) = C_1 J_n(\alpha x) + C_2 Y_n(\alpha x).$$

Our continuity requirements imply that $C_2 = 0$, since $Y_n(\alpha x)$ is discontinuous at $x = 0$ (see Sec. 60). Hence any nontrivial solution of equation (9) that meets those requirements must be a constant multiple of the function $X(x) = J_n(\alpha x)$.

In applying the boundary condition at $x = c$, we emphasize that *the symbol $J_n'(\alpha x)$ stands for the derivative of $J_n(s)$ with respect to s, evaluated at $s = \alpha x$.* Then $d/dx\,[J_n(\alpha x)] = \alpha J_n'(\alpha x)$; and conditions (4) and (5) require that

$$(11) \qquad\qquad J_n(\alpha c) = 0$$

and

$$(12) \qquad\qquad h J_n(\alpha c) + (\alpha c) J_n'(\alpha c) = 0,$$

respectively. Note that since equation (2), Sec. 65, can be written in the form (3) in that section, equation (12) can also be written as

$$(h + n)J_n(\alpha c) - (\alpha c)J_{n+1}(\alpha c) = 0.$$

According to the theorem in Sec. 65, each of equations (11) and (12) has an infinite number of positive roots

$$(13) \qquad\qquad \alpha_j = \frac{x_{nj}}{c} \qquad\qquad (j = 1, 2, \ldots),$$

where x_{nj} ($j = 1, 2, \ldots$) is the unbounded increasing sequence in the statement of that theorem. The numbers α_j here depend, of course, on the value of n and also on the value of h in the case of equation (12). Our Sturm-Liouville problem thus has eigenvalues $\lambda_j = \alpha_j^2$ ($j = 1, 2, \ldots$), and the corresponding eigenfunctions are

$$(14) \qquad\qquad X_j(x) = J_n(\alpha_j x) \qquad\qquad (j = 1, 2, \ldots).$$

We note that if the numbers α_j are the positive roots of equation (12) when $n = h = 0$, which is the only case where $\lambda = 0$ was found to be an eigenvalue, that equation can be written as

$$(15) \qquad J_1(\alpha c) = 0.$$

The numbers α_j are then given more directly as the positive roots of equation (15). Also, making a minor exception in our notation, we let the subscript j range over the values $j = 2, 3, \ldots$, instead of starting from unity. The subscript $j = 1$ is reserved for writing $\alpha_1 = 0$ and $\lambda_1 = \alpha_1^2 = 0$. This allows us to include the eigenvalue $\lambda_1 = 0$ and the eigenfunction $X_1(x) = J_0(\alpha_1 x) = 1$, obtained earlier for the case $n = h = 0$. Note that it is also possible to describe the numbers α_j $(j = 1, 2, \ldots)$ here as the *nonnegative* roots of equation (15).

Finally, we consider the case in which $\lambda < 0$, or $\lambda = -\alpha^2$ $(\alpha > 0)$, and write equation (1) as

$$(16) \qquad x^2 \frac{d^2 X}{dx^2} + x \frac{dX}{dx} - (\alpha^2 x^2 + n^2) X = 0.$$

The substitution $s = \alpha x$ can be used here to put equation (16) in the form (compare Problem 14, Sec. 61)

$$(17) \qquad s^2 \frac{d^2 X}{ds^2} + s \frac{dX}{ds} - (s^2 + n^2) X = 0.$$

From Problem 15, Sec. 61, we know that the modified Bessel function $X = I_n(s) = i^{-n} J_n(is)$ satisfies equation (17); and, since it has a power series representation that converges for all s, $X = I_n(\alpha x)$ satisfies the continuity requirements in our problem. As was the case with equation (9), equation (16) has a second solution which is *discontinuous* at $x = 0$, that solution being analogous to $Y_n(\alpha x)$.[†] Thus we know that, except for an arbitrary constant factor, $X(x) = I_n(\alpha x)$.

We now show that, for each positive value of α, the function $X(x) = I_n(\alpha x)$ fails to satisfy either of the boundary conditions (4) and (5). In each case, our proof rests on the fact that $I_n(x) > 0$ when $x > 0$, as demonstrated in Problem 15, Sec. 61.

Since $I_n(\alpha c) > 0$ when $\alpha > 0$, it is obvious that condition (4), which requires that $I_n(\alpha c) = 0$, fails to be satisfied for any positive number α. Also, in view of the alternative form (3), Sec. 65, of equation (2) in that section, condition (5), when applied to our function $X(x) = I_n(\alpha x) = i^{-n} J_n(i\alpha x)$, becomes

$$(h + n) i^{-n} J_n(i\alpha c) + \alpha c i^{-(n+1)} J_{n+1}(i\alpha c) = 0,$$

or

$$(18) \qquad (h + n) I_n(\alpha c) + \alpha c I_{n+1}(\alpha c) = 0.$$

[†] For a detailed discussion of this, see, for example, the book by Tranter (1969, pp. 16ff) that is listed in the Bibliography.

Since $\alpha > 0$, the left-hand side of this last equation is positive; and, once again, no positive values of α can occur as roots. We conclude, then, that there are no negative eigenvalues.

We have now completely solved our singular Sturm-Liouville problem consisting of equations (1) and (3), where, since we have assumed that the constant $h = cb_1/b_2$ is nonnegative, it is understood that the constants b_1 and b_2 in equation (3) have the same sign when neither is zero.

The eigenvalues are all represented by the numbers $\lambda_j = \alpha_j^2$ where the α_j are given by equation (13) and where $\lambda_j > 0$, except that $\lambda_1 = 0$ in the case $n = h = 0$. Since the numbers x_{nj} used in equation (13) form an unbounded increasing sequence, it is clear that the same is true of the eigenvalues $\lambda_j = \alpha_j^2$. That is, $\lambda_j < \lambda_{j+1}$ and $\lambda_j \to \infty$ as $j \to \infty$.

We summarize our results in the following theorem.

Theorem. *Let n have one of the values $n = 0, 1, 2, \ldots$. For the singular Sturm-Liouville problem consisting of the differential equation*

$$(19) \qquad x^2\frac{d^2X}{dx^2} + x\frac{dX}{dx} + (\lambda x^2 - n^2)X = 0 \qquad (0 < x < c),$$

which reduces to

$$x\frac{d^2X}{dx^2} + \frac{dX}{dx} + \lambda xX = 0 \qquad (0 < x < c)$$

when $n = 0$, and one of the boundary conditions

$$(20) \qquad\qquad X(c) = 0,$$

$$(21) \qquad\qquad hX(c) + cX'(c) = 0 \qquad (h \geqslant 0, h + n > 0),$$

$$(22) \qquad\qquad X'(c) = 0 \qquad (n = 0),$$

the eigenvalues and corresponding eigenfunctions are

$$\lambda_j = \alpha_j^2, \qquad X_j = J_n(\alpha_j x) \qquad (j = 1, 2, \ldots),$$

where the numbers α_j are defined as follows:

(a) When condition (20) is used, α_j ($j = 1, 2, \ldots$) are the positive roots of the equation

$$J_n(\alpha c) = 0.$$

(b) When condition (21) is used, α_j ($j = 1, 2, \ldots$) are the positive roots of the equation

$$hJ_n(\alpha c) + (\alpha c)J_n'(\alpha c) = 0,$$

which can also be written as $(h + n)J_n(\alpha c) - (\alpha c)J_{n+1}(\alpha c) = 0$.

(c) *When condition* (22) *is used,* $\alpha_1 = 0$ *and* α_j $(j = 2, 3, \ldots)$ *are the positive roots of the equation*

$$J_0'(\alpha c) = 0,$$

which can also be written as $J_1(\alpha c) = 0.$

Note that when n is positive ($n = 1, 2, \ldots$), the constant h in condition (21) can be zero. For that value of h, condition (21) is simply $X'(c) = 0$, and the condition in case (b) that is used to define the α_j becomes $J_n'(\alpha c) = 0$. Note, too, that condition (22) is condition (21) when $h = 0$ and $n = 0$. Since the first eigenvalue is then $\alpha_1 = 0$, as stated in case (c), the first eigenfunction is actually

$$X_1 = J_0(\alpha_1 x) = J_0(0) = 1.$$

For each of the cases in this theorem, the orthogonality property

$$(23) \qquad \int_0^c x J_n(\alpha_j x) J_n(\alpha_k x)\, dx = 0 \qquad\qquad (j \neq k)$$

follows from case (a) of the theorem in Sec. 43. Observe that this orthogonality of the eigenfunctions with respect to the weight function x, on the interval $0 < x < c$, is the same as ordinary orthogonality of the functions $\sqrt{x}\, J_n(\alpha_j x)$ on that same interval. Also, many orthogonal sets are represented here, depending on the values of c, n, and h. In the next two sections, we shall normalize these eigenfunctions and find formulas for the coefficients in generalized Fourier series expansions involving the normalized eigenfunctions.

PROBLEMS

1. By means of the substitution $y(x) = x^c u(x)$, transform Bessel's equation

$$x^2 y''(x) + xy'(x) + (x^2 - \nu^2) y(x) = 0$$

into the differential equation

$$x^2 u''(x) + (1 + 2c) xu'(x) + (x^2 - \nu^2 + c^2) u(x) = 0,$$

which becomes equation (3), Sec. 64, when $c = -\frac{1}{2}$.

2. Use equation (3), Sec. 64, to obtain a general solution of Bessel's equation when $\nu = \frac{1}{2}$. Then, using the expressions (Problem 12, Sec. 61)

$$J_{1/2}(x) = \sqrt{\frac{2}{\pi x}} \sin x, \qquad J_{-1/2}(x) = \sqrt{\frac{2}{\pi x}} \cos x,$$

point out how $J_{1/2}(x)$ and $J_{-1/2}(x)$ are special cases of that solution.

3. By referring to the theorem in Sec. 66, show that the eigenvalues of the singular Sturm-Liouville problem

$$xX'' + X' + \lambda x X = 0, \qquad X(2) = 0,$$

on the interval $0 \leqslant x \leqslant 2$, are the numbers $\lambda_j = \alpha_j^2$ $(j = 1, 2, \ldots)$, where α_j are the positive roots of the equation $J_0(2\alpha) = 0$, and that the corresponding eigenfunctions are $X_j = J_0(\alpha_j x)$ $(j = 1, 2, \ldots)$. With the aid of the table in Sec. 64, obtain the numerical values $\alpha_1 = 1.2$, $\alpha_2 = 2.8$, $\alpha_3 = 4.3$, valid to one decimal place.

4. Write $U(x) = \sqrt{x} J_n(\alpha x)$, where n has any one of the values $n = 0, 1, 2, \ldots$ and α is a positive constant.

 (a) Use equation (3), Sec. 64, to show that

$$U''(x) + \left(\alpha^2 + \frac{1 - 4n^2}{4x^2}\right) U(x) = 0.$$

 (b) Let c denote any fixed positive number and write $U_j(x) = \sqrt{x} J_n(\alpha_j x)$ $(j = 1, 2, \ldots)$, where α_j are the positive roots of the equation $J_n(\alpha c) = 0$. Use the result in part (a) to show that

$$\left(\alpha_j^2 - \alpha_k^2\right) U_j(x) U_k(x) = U_j U_k'' - U_k U_j''.$$

 (c) Use the result in part (b) and Lagrange's identity [Problem 3(b), Sec. 43] for the self-adjoint operator $\mathcal{L} = d^2/dx^2$ to show that the set $\{U_j(x)\}$ $(j = 1, 2, \ldots)$ in part (b) is orthogonal on the interval $0 < x < c$ with weight function unity. Thus give another proof that the set $\{J_n(\alpha_j x)\}$ $(j = 1, 2, \ldots)$ in case (a) of the theorem in Sec. 66 is orthogonal on that interval with weight function x.

5. Let n have any one of the fixed values $n = 0, 1, 2, \ldots$.

 (a) Suppose that $J_n(ib) = 0$ $(b \neq 0)$ and use results in Problem 15, Sec. 61, to reach a contradiction. Thus show that the function $J_n(z)$ has no pure imaginary zeros $z = ib$ $(b \neq 0)$.

 (b) Since our series representation of $J_n(x)$ (Sec. 59) converges when x is replaced by any complex number z and since the coefficients of the powers of z in that representation are all real, it follows that $J_n(\bar{z}) = \overline{J_n(z)}$, where $\bar{z}$ denotes the complex conjugate $x - iy$ of the number $z = x + iy$.[†] Also, the proof of orthogonality in Problem 4 above remains valid when α is a nonzero complex number and when the set of roots α_j there is allowed to include any nonzero complex roots of the equation $J_n(\alpha c) = 0$ that may occur. Use these facts to show that if the complex number $a + ib$ $(a \neq 0, b \neq 0)$ is a zero of $J_n(z)$, then $a - ib$ is also a zero and that

$$\int_0^1 x \left| J_n((a + ib)x) \right|^2 dx = \int_0^1 x J_n((a + ib)x) J_n((a - ib)x) \, dx = 0.$$

Point out why the value of the first integral here is actually positive and, with this contradiction, deduce that $J_n(z)$ has no zeros of the form $a + ib$ $(a \neq 0, b \neq 0)$. Conclude that if $z = x_j$ $(j = 1, 2, \ldots)$ are the positive zeros of $J_n(z)$, then the only other zeros, real or complex, are the numbers $z = -x_j$ $(j = 1, 2, \ldots)$, and also $z = 0$ when n is positive.

[†] For a discussion of power series representations in the complex plane, see the authors' book (1990), chap. 5, listed in the Bibliography.

67.   THE ORTHONORMAL FUNCTIONS

From Problem 14, Sec. 61, we know that if α is a positive constant, the function

$$X(x) = J_n(\alpha x)$$

satisfies the equation

(1) $$(xX')' + \left(\alpha^2 x - \frac{n^2}{x}\right)X = 0 \qquad (n = 0,1,2,\dots).$$

We multiply each side by $2xX'$ and write

$$\frac{d}{dx}(xX')^2 + (\alpha^2 x^2 - n^2)\frac{d}{dx}(X^2) = 0.$$

After integrating both terms here and using integration by parts in the second term, we find that

$$\left[(xX')^2 + (\alpha^2 x^2 - n^2)X^2\right]_0^c - 2\alpha^2\int_0^c xX^2\,dx = 0,$$

where c is any positive number. When $n = 0$, the quantity inside the brackets clearly vanishes at $x = 0$; and the same is true when $n = 1,2,\dots$, since $X(0) = J_n(0) = 0$ then. We thus arrive at the expression

(2) $$2\alpha^2\int_0^c x[J_n(\alpha x)]^2\,dx = \alpha^2 c^2[J_n'(\alpha c)]^2 + (\alpha^2 c^2 - n^2)[J_n(\alpha c)]^2,$$

which we now use to find the norms of all the eigenfunctions in the theorem in Sec. 66, except for the one corresponding to the zero eigenvalue in case (c) there. That norm is treated separately.

   (a) Suppose that α_j $(j = 1,2,\dots)$ are the positive roots of the equation

(3) $$J_n(\alpha c) = 0.$$

Expression (2) tells us that

$$2\int_0^c x[J_n(\alpha_j x)]^2\,dx = c^2[J_n'(\alpha_j c)]^2.$$

The integral here is the square of the norm of $J_n(\alpha_j x)$ on the interval $0 < x < c$, with weight function x. Also (Sec. 61)

$$xJ_n'(x) = nJ_n(x) - xJ_{n+1}(x),$$

and therefore $J_n'(\alpha_j c) = -J_{n+1}(\alpha_j c)$. Hence

(4) $$\|J_n(\alpha_j x)\|^2 = \frac{c^2}{2}[J_{n+1}(\alpha_j c)]^2 \qquad (j = 1,2,\dots).$$

   (b) Suppose that α_j $(j = 1,2,\dots)$ are the positive roots of the equation

(5) $$hJ_n(\alpha c) + (\alpha c)J_n'(\alpha c) = 0 \qquad (h \geqslant 0, h + n > 0).$$

We find from equation (2) that

(6)
$$\left\| J_n(\alpha_j x) \right\|^2 = \frac{\alpha_j^2 c^2 - n^2 + h^2}{2\alpha_j^2} \left[J_n(\alpha_j c) \right]^2 \qquad (j = 1, 2, \ldots).$$

(c) *Suppose that* $\alpha_1 = 0$ *and that* α_j $(j = 2, 3, \ldots)$ *are the positive roots of the equation*

(7)
$$J_0'(\alpha c) = 0.$$

Since $J_0(\alpha_1 x) = J_0(0) = 1$,

(8)
$$\left\| J_0(\alpha_1 x) \right\|^2 = \int_0^c x \, dx = \frac{c^2}{2}.$$

Expressions for $\| J_0(\alpha_j c) \|^2$ $(j = 2, 3, \ldots)$ are obtained by writing $n = h = 0$ in equation (6):

(9)
$$\left\| J_0(\alpha_j x) \right\|^2 = \frac{c^2}{2} \left[J_0(\alpha_j c) \right]^2 \qquad (j = 2, 3, \ldots).$$

For equation (7) is simply equation (5) when $n = h = 0$, and the restriction $h + n > 0$ is not actually needed in deriving expression (6).

The orthogonal eigenfunctions $X_j(x) = J_n(\alpha_j x)$ of our singular Sturm-Liouville problem can now be written in normalized form as

(10)
$$\phi_j(x) = \frac{J_n(\alpha_j x)}{\| J_n(\alpha_j x) \|} \qquad (j = 1, 2, \ldots).$$

The norms here are given for the eigenfunctions in cases (*a*) and (*b*) of the theorem in Sec. 66 by equations (4) and (6), respectively. In case (*c*), they are given by equations (8) and (9). The fact that the set (10) is orthonormal on the interval $0 < x < c$ with weight function x is, of course, expressed by the equations

$$\int_0^c x \phi_j(x) \phi_k(x) \, dx = \begin{cases} 0 & \text{when } j \neq k, \\ 1 & \text{when } j = k. \end{cases}$$

68. FOURIER-BESSEL SERIES

Let f be any piecewise continuous function defined on an interval $0 < x < c$. We consider here the generalized Fourier series

$$\sum_{j=1}^{\infty} c_j \phi_j(x)$$

for f with respect to the normalized eigenfunctions (10) in Sec. 67. According to Sec. 12, the coefficients c_j in that series are the numbers

$$c_j = (f, \phi_j) = \int_0^c xf(x)\phi_j(x)\,dx = \frac{1}{\|J_n(\alpha_j x)\|}\int_0^c xf(x)J_n(\alpha_j x)\,dx.$$

Writing $A_j = c_j/\|J_n(\alpha_j x)\|$, we thus have the correspondence

$$(1) \qquad\qquad f(x) \sim \sum_{j=1}^{\infty} A_j J_n(\alpha_j x) \qquad\qquad (0 < x < c),$$

where

$$(2) \qquad\qquad A_j = \frac{1}{\|J_n(\alpha_j x)\|^2}\int_0^c xf(x)J_n(\alpha_j x)\,dx \qquad (j = 1,2,\dots).$$

The norms found in Sec. 67 can now be used to adapt expression (2) to each of the three cases (a), (b), and (c) treated in that section and in the theorem in Sec. 66.

Theorem 1. *Let A_j be the coefficients in correspondence* (1).

(a) *If* α_j $(j = 1,2,\dots)$ *are the positive roots of the equation*
$$J_n(\alpha c) = 0,$$
then

$$(3) \qquad A_j = \frac{2}{c^2[J_{n+1}(\alpha_j c)]^2}\int_0^c xf(x)J_n(\alpha_j x)\,dx \qquad (j = 1,2,\dots).$$

(b) *If* α_j $(j = 1,2,\dots)$ *are the positive roots of the equation*
$$hJ_n(\alpha c) + (\alpha c)J_n'(\alpha c) = 0 \qquad (h \geqslant 0, h + n > 0),$$
which can also be written as $(h + n)J_n(\alpha c) - (\alpha c)J_{n+1}(\alpha c) = 0$, *then*

$$(4) \qquad A_j = \frac{2\alpha_j^2}{(\alpha_j^2 c^2 - n^2 + h^2)[J_n(\alpha_j c)]^2}\int_0^c xf(x)J_n(\alpha_j x)\,dx$$
$$(j = 1,2,\dots).$$

(c) *If* $n = 0$ *in series* (1) *and if* $\alpha_1 = 0$ *and* α_j $(j = 2,3,\dots)$ *are the positive roots of the equation*
$$J_0'(\alpha c) = 0,$$
which can also be written as $J_1(\alpha c) = 0$, *then*

$$(5) \qquad\qquad A_1 = \frac{2}{c^2}\int_0^c xf(x)\,dx$$

and

(6)
$$A_j = \frac{2}{c^2 [J_0(\alpha_j c)]^2} \int_0^c x f(x) J_0(\alpha_j x)\, dx \qquad (j = 2, 3, \dots).$$

Note that since $J_0(0) = 1$, expression (6) becomes expression (5) when $j = 1$. It is, however, more convenient in the applications to treat A_1 separately and to write correspondence (1) as

$$f(x) \sim A_1 + \sum_{j=2}^{\infty} A_j J_0(\alpha_j x) \qquad (0 < x < c)$$

when case (c) is to be used.

Proofs that correspondence (1) is actually an equality, under conditions similar to those used to ensure the representation of a function by its Fourier cosine or sine series, usually involve the theory of functions of a complex variable. We state, without proof, one form of such a representation theorem and refer the reader to the Bibliography.[†]

Theorem 2. *Let f denote a function that is piecewise smooth on an interval $0 < x < c$, and suppose that $f(x)$ at each point of discontinuity of f in that interval is defined as the mean value of the one-sided limits $f(x +)$ and $f(x -)$. Then*

(7)
$$f(x) = \sum_{j=1}^{\infty} A_j J_n(\alpha_j x) \qquad (0 < x < c),$$

where the coefficients A_j are defined by equation (3) or (4) or the pair of equations (5) and (6), depending on the particular equation that determines the numbers α_j.

Expansion (7) is called a *Fourier-Bessel series* representation of $f(x)$.

EXAMPLE 1. Let us expand the function $f(x) = 1$ $(0 < x < c)$ into a series of the type

$$\sum_{j=1}^{\infty} A_j J_0(\alpha_j x),$$

where α_j $(j = 1, 2, \dots)$ are the positive roots of the equation $J_0(\alpha c) = 0$. Case (a) in Theorem 1 is evidently applicable here, and expression (3) tells us that

(8)
$$A_j = \frac{2}{c^2 [J_1(\alpha_j c)]^2} \int_0^c x J_0(\alpha_j x)\, dx.$$

[†] This theorem is proved in the book by Watson (1952). Also see the work by Titchmarsh (1962), as well as the books by Gray and Mathews (1966) and Bowman (1958). These are all listed in the Bibliography.

This integral is readily evaluated by substituting $s = \alpha_j x$ and using the integration formula (Sec. 61)

(9)
$$\int_0^x sJ_0(s)\, ds = xJ_1(x).$$

To be specific,

$$\int_0^c xJ_0(\alpha_j x)\, dx = \frac{1}{\alpha_j^2}\int_0^{\alpha_j c} sJ_0(s)\, ds = \frac{c}{\alpha_j}J_1(\alpha_j c).$$

Consequently,

(10)
$$1 = \frac{2}{c}\sum_{j=1}^{\infty}\frac{J_0(\alpha_j x)}{\alpha_j J_1(\alpha_j c)}\qquad (0 < x < c),$$

where $J_0(\alpha_j c) = 0\ (\alpha_j > 0)$.

EXAMPLE 2. To represent the function $f(x) = x\ (0 < x < 1)$ in a series of the form

$$A_1 + \sum_{j=2}^{\infty} A_j J_0(\alpha_j x),$$

where $\alpha_j\ (j = 2, 3, \ldots)$ are the positive roots of the equation $J_1(\alpha) = 0$, we refer to case (c) in Theorem 1. According to expression (5),

$$A_1 = 2\int_0^1 x^2\, dx = \frac{2}{3};$$

and we find from expression (6) that

$$A_j = \frac{2}{[J_0(\alpha_j)]^2}\int_0^1 x^2 J_0(\alpha_j x)\, dx\qquad (j = 2, 3, \ldots).$$

This last integral can be evaluated by referring to the reduction formula (see Problem 4, Sec. 61, and the footnote with that problem)

$$\int_0^x s^2 J_0(s)\, ds = x^2 J_1(x) + xJ_0(x) - \int_0^x J_0(s)\, ds$$

and then recalling that $J_1(\alpha_j) = 0$:

$$\int_0^1 x^2 J_0(\alpha_j x)\, dx = \frac{1}{\alpha_j^3}\int_0^{\alpha_j} s^2 J_0(s)\, ds$$

$$= \frac{1}{\alpha_j^3}\left[\alpha_j J_0(\alpha_j) - \int_0^{\alpha_j} J_0(s)\, ds\right].$$

Thus

$$(11) \qquad x = \frac{2}{3} + 2 \sum_{j=2}^{\infty} \left[\alpha_1 J_0(\alpha_j) - \int_0^{\alpha_j} J_0(s)\, ds \right] \frac{J_0(\alpha_j x)}{\alpha_j^3 \left[J_0(\alpha_j) \right]^2} \qquad (0 < x < 1),$$

where $J_1(\alpha_j) = 0 \ (\alpha_j > 0)$.

The theorem in Sec. 65 and the two in this section are also valid when n is replaced by an arbitrary real number $\nu \ (\nu > -\frac{1}{2})$, although we have not developed properties of the function J_ν far enough to establish any such generalizations.

For functions f on the unbounded interval $x > 0$, there is an integral representation in terms of J_ν, analogous to the Fourier cosine and sine integral formulas. The representation, for a fixed $\nu \ (\nu > -\frac{1}{2})$, is

$$f(x) = \int_0^\infty \alpha J_\nu(\alpha x) \int_0^\infty s f(s) J_\nu(\alpha s)\, ds\, d\alpha \qquad (x > 0)$$

and is known as *Hankel's integral formula.*[†] It is valid if f is piecewise smooth on each bounded interval, if $\sqrt{x} f(x)$ is absolutely integrable from zero to infinity, and if $f(x)$ is defined as its mean value at each point of discontinuity.

If the interval $0 < x < c$ is replaced by some interval $a < x < b$, where $a > 0$, the Sturm-Liouville problem treated in Sec. 66 is no longer singular when the same differential equation is used and boundary conditions of type (3) in that section are applied at *each* end point. In general, the resulting eigenfunctions then involve both of the Bessel functions J_n and Y_n.

PROBLEMS

1. Show that

$$x = 2 \sum_{j=1}^{\infty} \left[1 - \frac{1}{\alpha_j^2 J_1(\alpha_j)} \int_0^{\alpha_j} J_0(s)\, ds \right] \frac{J_0(\alpha_j x)}{\alpha_j J_1(\alpha_j)} \qquad (0 < x < 1),$$

where $\alpha_j \ (j = 1, 2, \dots)$ are the positive roots of the equation $J_0(\alpha) = 0$.

2. Derive the representation

$$x^2 = \frac{c^2}{2} + 4 \sum_{j=2}^{\infty} \frac{J_0(\alpha_j x)}{\alpha_j^2 J_0(\alpha_j c)} \qquad (0 < x < c),$$

where $\alpha_j \ (j = 2, 3, \dots)$ are the positive roots of the equation $J_1(\alpha c) = 0$.

[†] See the book by Sneddon (1951, chap. 2) that is listed in the Bibliography. For a summary of representations in terms of Bessel functions, see the work edited by Erdélyi (1981, vol. 2, chap. 7) that is also listed.

3. Show that if

$$f(x) = \begin{cases} 1 & \text{when } 0 < x < 1, \\ 0 & \text{when } 1 < x < 2, \end{cases}$$

and $f(1) = \frac{1}{2}$, then

$$f(x) = \frac{1}{2} \sum_{j=1}^{\infty} \frac{J_1(\alpha_j)}{\alpha_j [J_1(2\alpha_j)]^2} J_0(\alpha_j x) \qquad (0 < x < 2),$$

where α_j $(j = 1, 2, \ldots)$ are the positive roots of the equation $J_0(2\alpha) = 0$.

4. Let α_j $(j = 1, 2, \ldots)$ denote the positive roots of the equation $J_0(\alpha c) = 0$, where c is a fixed positive number.

(a) Derive the expansion

$$x^2 = \frac{2}{c} \sum_{j=1}^{\infty} \frac{(\alpha_j c)^2 - 4}{\alpha_j^3 J_1(\alpha_j c)} J_0(\alpha_j x) \qquad (0 < x < c).$$

(b) Combine expansion (10) in Example 1, Sec. 68, with the one in part (a) to show that

$$c^2 - x^2 = \frac{8}{c} \sum_{j=1}^{\infty} \frac{J_0(\alpha_j x)}{\alpha_j^3 J_1(\alpha_j c)} \qquad (0 < x < c).$$

5. Find the coefficients A_j $(j = 1, 2, \ldots)$ in the expansion

$$1 = \sum_{j=1}^{\infty} A_j J_0(\alpha_j x) \qquad (0 < x < c)$$

when $\alpha_1 = 0$ and α_j $(j = 2, 3, \ldots)$ are the positive roots of the equation $J_0'(\alpha c) = 0$.
Answer: $A_1 = 1$, $A_j = 0$ $(j = 2, 3, \ldots)$.

6. (a) Obtain the representation

$$1 = 2c \sum_{j=1}^{\infty} \frac{\alpha_j J_1(\alpha_j c) J_0(\alpha_j x)}{(\alpha_j^2 c^2 + h^2)[J_0(\alpha_j c)]^2} \qquad (0 < x < c),$$

where α_j $(j = 1, 2, \ldots)$ are the positive roots of the equation

$$h J_0(\alpha c) + (\alpha c) J_0'(\alpha c) = 0 \qquad (h > 0).$$

[Contrast this representation with representation (10) in Example 1, Sec. 68, and the one in Problem 5.]

(b) Show how the result in part (a) can be written in the form

$$1 = \frac{2}{c} \sum_{j=1}^{\infty} \frac{1}{\alpha_j} \cdot \frac{J_1(\alpha_j c) J_0(\alpha_j x)}{[J_0(\alpha_j c)]^2 + [J_1(\alpha_j c)]^2} \qquad (0 < x < c).$$

7. Show that if

$$f(x) = \begin{cases} x & \text{when } 0 < x < 1, \\ 0 & \text{when } 1 < x < 2, \end{cases}$$

and $f(1) = \frac{1}{2}$, then

$$f(x) = 2 \sum_{j=1}^{\infty} \frac{\alpha_j J_2(\alpha_j)}{\left(4\alpha_j^2 - 1\right)\left[J_1(2\alpha_j)\right]^2} J_1(\alpha_j x) \qquad (0 < x < 2),$$

where α_j $(j = 1, 2, \ldots)$ are the positive roots of the equation $J_1'(2\alpha) = 0$.

Suggestion: Note that when $h = 0$ in case (b) of Theorem 1 in Sec. 68, the equation defining the α_j there becomes $J_n'(\alpha c) = 0$.

8. Let n have any one of the positive values $n = 1, 2, \ldots$. Show that

$$x^n = 2 \sum_{j=1}^{\infty} \frac{\alpha_j J_{n+1}(\alpha_j)}{\left(\alpha_j^2 - n^2\right)\left[J_n(\alpha_j)\right]^2} J_n(\alpha_j x) \qquad (0 < x < 1),$$

where α_j $(j = 1, 2, \ldots)$ are the positive roots of the equation $J_n'(\alpha) = 0$. (See the suggestion with Problem 7.)

9. Point out why the eigenvalues of the singular Sturm-Liouville problem

$$x^2 X'' + x X' + (\lambda x^2 - 1) X = 0, \qquad X(1) = 0,$$

on the interval $0 < x < 1$, are the numbers $\lambda_j = \alpha_j^2$ $(j = 1, 2, \ldots)$, where α_j are the positive roots of the equation $J_1(\alpha) = 0$, and why the corresponding eigenfunctions are $X_j = J_1(\alpha_j x)$ $(j = 1, 2, \ldots)$. Then obtain the representation

$$x = 2 \sum_{j=1}^{\infty} \frac{J_1(\alpha_j x)}{\alpha_j J_2(\alpha_j)} \qquad (0 < x < 1)$$

in terms of those eigenfunctions.

10. As indicated in Sec. 68, there exist conditions on f under which representation (7) there is valid when n is replaced by ν $(\nu > -\frac{1}{2})$, where ν is not necessarily an integer. In particular, suppose that

$$f(x) = \sum_{j=1}^{\infty} A_j J_{1/2}(\alpha_j x) \qquad (0 < x < c),$$

where $\sqrt{x} f(x)$ is piecewise smooth, where α_j are the positive roots of the equation $J_{1/2}(\alpha c) = 0$, and where [compare expression (3), Sec. 68]

$$A_j = \frac{2}{c^2 \left[J_{3/2}(\alpha_j c)\right]^2} \int_0^c x f(x) J_{1/2}(\alpha_j x)\, dx \qquad (j = 1, 2, \ldots).$$

Using the expressions [Problems 12(a) and 13, Sec. 61]

$$J_{1/2}(x) = \sqrt{\frac{2}{\pi x}} \sin x \qquad \text{and} \qquad J_{3/2}(x) = \sqrt{\frac{2}{\pi x}} \left(\frac{\sin x}{x} - \cos x\right)$$

to substitute for the Bessel functions involved, show that this Fourier-Bessel representation is actually the Fourier sine series representation of $\sqrt{x} f(x)$ on the interval $0 < x < c$.

69. TEMPERATURES IN A LONG CYLINDER

In both of the following examples, we shall use Bessel functions to find temperatures in an infinitely long circular cylinder $\rho \leqslant c$ whose lateral surface $\rho = c$ is kept at temperature zero (Fig. 64). Other thermal conditions will be such that the temperatures will depend only on the space variable ρ, which is the distance from the axis of the cylinder, and time t. We assume that the material of the solid is homogeneous.

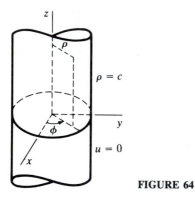

FIGURE 64

EXAMPLE 1. When the cylinder is as shown in Fig. 64 and the initial temperatures vary only with ρ, the temperatures $u = u(\rho, t)$ in the cylinder satisfy the special case (Sec. 4)

$$(1) \qquad u_t = k \left(u_{\rho\rho} + \frac{1}{\rho} u_\rho \right) \qquad (0 < \rho < c, t > 0)$$

of the heat equation in cylindrical coordinates and the boundary conditions

$$(2) \qquad\qquad u(c, t) = 0 \qquad\qquad (t > 0),$$
$$(3) \qquad\qquad u(\rho, 0) = f(\rho) \qquad\qquad (0 < \rho < c).$$

Also, when $t > 0$, the function u is to be continuous throughout the cylinder and, in particular, on the axis $\rho = 0$. We assume that f is piecewise smooth on the interval $0 < \rho < c$ and, for convenience, that f is defined as the mean value of its one-sided limits at each point in that interval where it is discontinuous.

Any solutions of the homogeneous equations (1) and (2) that are of the type $u = R(\rho)T(t)$ must satisfy the conditions

$$RT' = kT \left(R'' + \frac{1}{\rho} R' \right), \qquad R(c)T(t) = 0.$$

Separating variables in the first equation here, we have

$$\frac{T'}{kT} = \frac{1}{R}\left(R'' + \frac{1}{\rho}R'\right) = -\lambda,$$

where $-\lambda$ is the separation constant. Thus

(4) $\qquad\qquad \rho R''(\rho) + R'(\rho) + \lambda\rho R(\rho) = 0, \qquad R(c) = 0 \quad (0 < \rho < c),$

and

(5) $\qquad\qquad\qquad\qquad T'(t) + \lambda k T(t) = 0 \qquad\qquad\qquad (t > 0).$

  The differential equation in R is Bessel's equation, with the parameter λ, in which $n = 0$. Problem (4), together with continuity conditions on R and R' on the interval $0 \le \rho \le c$, is a special case of the singular Sturm-Liouville problem in the theorem in Sec. 66 when $n = 0$ and the boundary condition (20) there is taken. According to that theorem, the eigenvalues λ_j of problem (4) are the numbers $\lambda_j = \alpha_j^2 \ (j = 1, 2, \dots)$, where α_j are the positive roots of the equation

(6) $\qquad\qquad\qquad\qquad\qquad J_0(\alpha c) = 0;$

and $R_j = J_0(\alpha_j \rho)$ are the corresponding eigenfunctions.

  When $\lambda = \lambda_j$, equation (5) is satisfied by $T_j = \exp(-\alpha_j^2 kt)$. So the desired products are

$$u_j = R_j(\rho)T_j(t) = J_0(\alpha_j\rho)\exp\left(-\alpha_j^2 kt\right) \qquad\qquad (j = 1, 2, \dots);$$

and the generalized linear combination of these functions,

(7) $\qquad\qquad u(\rho, t) = \sum_{j=1}^{\infty} A_j J_0(\alpha_j\rho)\exp\left(-\alpha_j^2 kt\right),$

formally satisfies the homogeneous conditions (1) and (2) in our boundary value problem. It also satisfies the nonhomogeneous initial condition (3) when the coefficients A_j are such that

$$f(\rho) = \sum_{j=1}^{\infty} A_j J_0(\alpha_j\rho) \qquad\qquad (0 < \rho < c).$$

This is a valid Fourier-Bessel series representation (Theorem 2, Sec. 68) if the coefficients have the values

(8) $\qquad\qquad A_j = \frac{2}{c^2[J_1(\alpha_j c)]^2}\int_0^c \rho f(\rho)J_0(\alpha_j\rho)\,d\rho \qquad (j = 1, 2, \dots),$

obtained by writing $n = 0$ in equation (3), Sec. 68.

  The formal solution of the boundary value problem is, therefore, given by equation (7) with the coefficients (8), where α_j are the positive roots of equation

(6). Thus our temperature formula can be written

$$(9) \qquad u(\rho,t) = \frac{2}{c^2} \sum_{j=1}^{\infty} \frac{J_0(\alpha_j\rho)}{\left[J_1(\alpha_jc)\right]^2} \exp\left(-\alpha_j^2 kt\right) \int_0^c sf(s)J_0(\alpha_js)\,ds.$$

EXAMPLE 2. Suppose now that heat is generated in the cylinder in Example 1 at a constant rate per unit volume and that the surface and initial temperatures are both zero. The temperatures $u = u(\rho,t)$ must satisfy the conditions

$$(10) \qquad u_t = k\left(u_{\rho\rho} + \frac{1}{\rho}u_\rho\right) + q_0 \qquad (0 < \rho < c, t > 0),$$

where q_0 is a positive constant (Sec. 2), and

$$(11) \qquad u(c,t) = 0, \qquad u(\rho,0) = 0.$$

The function u is, of course, required to be continuous in the cylinder, as was the solution in Example 1.

The differential equation (10) is nonhomogeneous, because of the constant term q_0, and this suggests that we apply the method of variation of parameters, first used in Sec. 33. To be specific, we know from Example 1 above that, without the term q_0, the eigenfunctions $R_j = J_0(\alpha_j\rho)$, where $J_0(\alpha_jc) = 0$ $(\alpha_j > 0)$, arise. Hence we seek a solution of the present boundary value problem having the form

$$(12) \qquad u(\rho,t) = \sum_{j=1}^{\infty} A_j(t)J_0(\alpha_j\rho),$$

where the α_j are as just stated.

Substituting this series into equation (10) and noting how the representation

$$q_0 = \frac{2q_0}{c} \sum_{j=1}^{\infty} \frac{J_0(\alpha_j\rho)}{\alpha_j J_1(\alpha_jc)} \qquad (0 < \rho < c)$$

follows immediately from the one obtained in Example 1, Sec. 68, we find that if series (12) is to satisfy equation (10), then

$$\sum_{j=1}^{\infty} A_j'(t)J_0(\alpha_j\rho) = k\sum_{j=1}^{\infty} A_j(t)\left[\frac{d^2}{d\rho^2}J_0(\alpha_j\rho) + \frac{1}{\rho}\frac{d}{d\rho}J_0(\alpha_j\rho)\right]$$
$$+ \sum_{j=1}^{\infty} \frac{2q_0 J_0(\alpha_j\rho)}{c\alpha_j J_1(\alpha_jc)}.$$

But, according to the theorem in Sec. 66,

$$\frac{d^2}{d\rho^2}J_0(\alpha_j\rho) + \frac{1}{\rho}\frac{d}{d\rho}J_0(\alpha_j\rho) = -\alpha_j^2 J_0(\alpha_j\rho).$$

Thus

$$\sum_{j=1}^{\infty} \left[A'_j(t) + \alpha_j^2 k A_j(t) \right] J_0(\alpha_j\rho) = \sum_{j=1}^{\infty} \frac{2q_0}{c\alpha_j J_1(\alpha_j c)} J_0(\alpha_j\rho);$$

and, by equating coefficients on each side of this equation, we arrive at the differential equation

(13) $$A'_j(t) + \alpha_j^2 k A_j(t) = \frac{2q_0}{c\alpha_j J_1(\alpha_j c)} \qquad (j = 1, 2, \dots).$$

Furthermore, in view of the second of conditions (11),

$$\sum_{j=1}^{\infty} A_j(0) J_0(\alpha_j\rho) = 0 \qquad (0 < \rho < c).$$

Consequently,

(14) $$A_j(0) = 0 \qquad (j = 1, 2, \dots).$$

To solve the linear differential equation (13), we multiply each side by the integrating factor

$$\exp \int \alpha_j^2 k \, dt = \exp \alpha_j^2 kt.$$

This enables us to write the differential equation as

$$\frac{d}{dt} \left[e^{\alpha_j^2 kt} A_j(t) \right] = \frac{2q_0}{c\alpha_j J_1(\alpha_j c)} e^{\alpha_j^2 kt}.$$

After replacing t by τ here, we then integrate each side from $\tau = 0$ to $\tau = t$ and recall condition (14). The result is

$$e^{\alpha_j^2 kt} A_j(t) = \frac{2q_0}{ck\alpha_j^3 J_1(\alpha_j c)} (e^{\alpha_j^2 kt} - 1),$$

or

(15) $$A_j(t) = \frac{2q_0}{ck} \cdot \frac{1 - \exp\left(-\alpha_j^2 kt\right)}{\alpha_j^3 J_1(\alpha_j c)} \qquad (j = 1, 2, \dots).$$

Finally, by substituting this expression for the coefficients $A_j(t)$ into series (12), we arrive at the desired temperature formula:

(16) $$u(\rho, t) = \frac{2q_0}{ck} \sum_{j=1}^{\infty} \frac{1 - \exp\left(-\alpha_j^2 kt\right)}{\alpha_j^3 J_1(\alpha_j c)} J_0(\alpha_j\rho),$$

where $J_0(\alpha_j c) = 0 \ (\alpha_j > 0)$.

70. HEAT TRANSFER AT THE SURFACE OF THE CYLINDER

Let us replace the condition that the surface of the infinite cylinder in Example 1, Sec. 69, be at temperature zero by the condition that heat transfer take place there into surroundings at temperature zero. As in Sec. 3, where Newton's law for surface heat transfer was discussed, the flux through the surface is assumed to be proportional to the difference between the temperature of the surface and that of its surroundings. That is,

$$Ku_\rho(c,t) = -Hu(c,t) \qquad (K > 0,\ H > 0),$$

where K is the thermal conductivity of the material of the cylinder and H is its surface conductance.

The boundary value problem for the temperature function $u(\rho, t)$ is now

$$(1) \qquad u_t = k\left(u_{\rho\rho} + \frac{1}{\rho}u_\rho\right) \qquad (0 < \rho < c, t > 0),$$

$$(2) \qquad cu_\rho(c,t) = -hu(c,t) \qquad (t > 0),$$

$$(3) \qquad u(\rho,0) = f(\rho) \qquad (0 < \rho < c).$$

We have written $h = cH/K$; and, for convenience, we allow the possibility that the otherwise positive constant h be zero. In that case, condition (2) simply states that the surface $\rho = c$ is insulated.

When $u = R(\rho)T(t)$, separation of variables produces the eigenvalue problem

$$(4) \quad \rho R''(\rho) + R'(\rho) + \lambda\rho R(\rho) = 0, \qquad hR(c) + cR'(c) = 0 \quad (0 < \rho < c).$$

If $h > 0$, the eigenvalues are, according to the theorem in Sec. 66, $\lambda_j = \alpha_j^2$ $(j = 1, 2, \ldots)$, where α_j are the positive roots of the equation

$$(5) \qquad hJ_0(\alpha c) + (\alpha c)J_0'(\alpha c) = 0.$$

The corresponding eigenfunctions are $R_j = J_0(\alpha_j\rho)$ $(j = 1, 2, \ldots)$; and, since $T'(t) + \lambda kT(t) = 0$, we arrive at the products

$$u_j = R_j(\rho)T_j(t) = J_0(\alpha_j\rho)\exp\left(-\alpha_j^2 kt\right) \qquad (j = 1, 2, \ldots).$$

The formal solution of our problem is, then,

$$(6) \qquad u(\rho,t) = \sum_{j=1}^{\infty} A_j J_0(\alpha_j\rho)\exp\left(-\alpha_j^2 kt\right),$$

where, in view of expression (4), Sec. 68,

$$(7) \qquad A_j = \frac{2\alpha_j^2}{(\alpha_j^2 c^2 + h^2)[J_0(\alpha_j c)]^2}\int_0^c \rho f(\rho)J_0(\alpha_j\rho)\,d\rho \qquad (j = 1, 2, \ldots).$$

If $h = 0$, the boundary condition in our eigenvalue problem becomes $R'(c) = 0$. In that case, the theorem in Sec. 66 tells us that $\lambda_j = \alpha_j^2$ $(j = 1, 2, \ldots)$, where $\alpha_1 = 0$ and α_j $(j = 2, 3, \ldots)$ are the positive roots of the equation $J_0'(\alpha c) = 0$, or

(8) $$J_1(\alpha c) = 0.$$

The corresponding eigenfunctions are, moreover, $R_1 = 1$ and $R_j = J_0(\alpha_j \rho)$ $(j = 2, 3, \ldots)$. Thus

(9) $$u(\rho, t) = A_1 + \sum_{j=2}^{\infty} A_j J_0(\alpha_j \rho) \exp\left(-\alpha_j^2 kt\right).$$

By referring to expressions (5) and (6) in Sec. 68, we see that the coefficients here are

(10) $$A_1 = \frac{2}{c^2} \int_0^c \rho f(\rho) \, d\rho,$$

(11) $$A_j = \frac{2}{c^2 \left[J_0(\alpha_j c)\right]^2} \int_0^c \rho f(\rho) J_0(\alpha_j \rho) \, d\rho \qquad (j = 2, 3, \ldots).$$

A number of *steady-state* temperature problems in cylindrical coordinates, giving rise to Bessel functions, appear in the problems to follow. In those problems, the temperatures will continue to be independent of ϕ. The function $u = u(\rho, z)$ will, then, be harmonic and satisfy Laplace's equation $\nabla^2 u = 0$, where (see Sec. 4)

(12) $$\nabla^2 u = u_{\rho\rho} + \frac{1}{\rho} u_\rho + u_{zz}.$$

PROBLEMS

1. Let $u(\rho, t)$ denote the solution found in Example 1, Sec. 69, when $c = 1$ and $f(\rho) = u_0$, where u_0 is a constant. With the aid of the table at the end of Sec. 64, show that the first three terms in the series for $u(\rho, t)$ are, approximately, as follows:

$$u(\rho, t) = 2u_0 \Big[0.80 J_0(2.4\rho) e^{-5.8kt} - 0.53 J_0(5.5\rho) e^{-30kt}$$
$$+ 0.43 J_0(8.7\rho) e^{-76kt} - \cdots \Big].$$

2. Show that the solution of the temperature problem in Example 2, Sec. 69, can be written

$$u(\rho, t) = \frac{q_0}{4k} \left[c^2 - \rho^2 - \frac{8}{c} \sum_{j=1}^{\infty} \frac{J_0(\alpha_j \rho) \exp\left(-\alpha_j^2 kt\right)}{\alpha_j^3 J_1(\alpha_j c)} \right],$$

where $J_0(\alpha_j c) = 0$ $(\alpha_j > 0)$.
   Suggestion: Note that, according to Problem 4(b), Sec. 68,

$$\sum_{j=1}^{\infty} \frac{J_0(\alpha_j \rho)}{\alpha_j^3 J_1(\alpha_j c)} = \frac{c}{8}(c^2 - \rho^2) \qquad (0 < \rho < c).$$

3. In Example 2, Sec. 69, suppose that the rate per unit volume at which heat is internally generated is $q(t)$, rather than simply q_0. Derive the following generalization of the solution found in that example:

$$u(\rho, t) = \frac{2}{c} \sum_{j=1}^{\infty} \frac{J_0(\alpha_j \rho)}{\alpha_j J_1(\alpha_j c)} \int_0^t q(\tau) \exp\left[-\alpha_j^2 k(t - \tau)\right] d\tau,$$

where $J_0(\alpha_j c) = 0 \ (\alpha_j > 0)$.

4. Derive an expression for the steady temperatures $u(\rho, z)$ in the solid cylinder formed by the three surfaces $\rho = 1$, $z = 0$, and $z = 1$ when $u = 0$ on the side, the bottom is insulated, and $u = 1$ on the top.

Answer: $u(\rho, z) = 2 \sum_{j=1}^{\infty} \dfrac{J_0(\alpha_j \rho)}{\alpha_j J_1(\alpha_j)} \cdot \dfrac{\cosh \alpha_j z}{\cosh \alpha_j}$, where $J_0(\alpha_j) = 0 \ (\alpha_j > 0)$.

5. Find the bounded steady temperatures $u(\rho, z)$ in the semi-infinite cylinder $\rho \leqslant 1$, $z \geqslant 0$ when $u = 1$ on the base and there is heat transfer into surroundings at temperature zero, according to Newton's law (see Sec. 70), at the surface $\rho = 1$, $z > 0$.

Answer: $u(\rho, z) = 2h \sum_{j=1}^{\infty} \dfrac{J_0(\alpha_j \rho) \exp(-\alpha_j z)}{J_0(\alpha_j)(\alpha_j^2 + h^2)}$,

where $h J_0(\alpha_j) - \alpha_j J_1(\alpha_j) = 0 \ (\alpha_j > 0)$.

6. (a) A solid cylinder is formed by the three surfaces $\rho = 1$, $z = 0$, and $z = b \ (b > 0)$. The side is insulated, the bottom kept at temperature zero, and the top at temperatures $f(\rho)$. Derive this expression for the steady temperatures $u(\rho, z)$ in the cylinder:

$$u(\rho, z) = \frac{2z}{b} \int_0^1 sf(s) \, ds + 2 \sum_{j=2}^{\infty} \frac{J_0(\alpha_j \rho)}{[J_0(\alpha_j)]^2} \cdot \frac{\sinh \alpha_j z}{\sinh \alpha_j b} \int_0^1 sf(s) J_0(\alpha_j s) \, ds,$$

where $\alpha_2, \alpha_3, \ldots$ are the positive roots of the equation $J_1(\alpha) = 0$.

(b) Show that when $f(\rho) = 1 \ (0 < \rho < 1)$ in part (a), the solution there reduces to $u(\rho, z) = z/b$.

7. A function $u(\rho, z)$ is harmonic interior to the cylinder formed by the three surfaces $\rho = c$, $z = 0$, and $z = b \ (b > 0)$. Assuming that $u = 0$ on the first two of those surfaces and that $u(\rho, b) = f(\rho) \ (0 < \rho < c)$, derive the expression

$$u(\rho, z) = \sum_{j=1}^{\infty} A_j J_0(\alpha_j \rho) \frac{\sinh \alpha_j z}{\sinh \alpha_j b},$$

where α_j are the positive roots of the equation $J_0(\alpha c) = 0$ and the coefficients A_j are given by equation (8), Sec. 69.

8. Solve this Dirichlet problem (Sec. 7) for $u(\rho, z)$:

$$\nabla^2 u = 0 \qquad\qquad (0 < \rho < 1, z > 0),$$
$$u(1, z) = 0, \qquad u(\rho, 0) = 1,$$

and u is to be bounded in the domain $\rho < 1$, $z > 0$.

Answer: $u(\rho, z) = 2 \sum_{j=1}^{\infty} \dfrac{J_0(\alpha_j \rho)}{\alpha_j J_1(\alpha_j)} \exp(-\alpha_j z)$, where $J_0(\alpha_j) = 0 \ (\alpha_j > 0)$.

9. Solve the following problem for temperatures $u(\rho, t)$ in a thin circular plate with heat transfer from its faces into surroundings at temperature zero:

$$u_t = u_{\rho\rho} + \frac{1}{\rho}u_\rho - bu \qquad\qquad (0 < \rho < 1, t > 0),$$

$$u(1, t) = 0, \qquad u(\rho, 0) = 1,$$

where b is a positive constant.

Answer: $u(\rho, t) = 2\exp(-bt)\sum_{j=1}^{\infty} \frac{J_0(\alpha_j\rho)}{\alpha_j J_1(\alpha_j)}\exp(-\alpha_j^2 t),$

where $J_0(\alpha_j) = 0$ ($\alpha_j > 0$).

10. Solve Problem 9 after replacing the condition $u(1, t) = 0$ by this heat transfer condition at the edge:

$$u_\rho(1, t) = -hu(1, t) \qquad\qquad (h > 0).$$

11. Give a physical interpretation of the following boundary value problem for a function $u(\rho, t)$ (see Example 2, Sec. 69):

$$u_t = u_{\rho\rho} + \frac{1}{\rho}u_\rho + q_0 \qquad\qquad (0 < \rho < 1, t > 0),$$

$$u_\rho(1, t) = 0, \qquad u(\rho, 0) = a\rho^2,$$

where q_0 and a are positive constants. Then, after pointing out why it is reasonable to seek a solution of the form

$$u(\rho, t) = A_1(t) + \sum_{j=2}^{\infty} A_j(t)J_0(\alpha_j\rho),$$

where α_j $(j = 2, 3, \dots)$ are the positive roots of the equation $J_1(\alpha_j) = 0$, use the method of variation of parameters to actually find that solution.

Answer: $u(\rho, t) = \dfrac{a}{2} + q_0 t + 4a\sum_{j=2}^{\infty} \dfrac{J_0(\alpha_j\rho)\exp\left(-\alpha_j^2 t\right)}{\alpha_j^2 J_0(\alpha_j)},$

where the α_j are as stated above.

12. Interpret this boundary value problem as a temperature problem in a cylinder (see Sec. 3):

$$u_t = u_{\rho\rho} + \frac{1}{\rho}u_\rho \qquad\qquad (0 < \rho < 1, t > 0),$$

$$u_\rho(1, t) = B, \qquad u(\rho, 0) = 0,$$

where B is a positive constant. Then, after making the substitution

$$u(\rho, t) = U(\rho, t) + \frac{B}{2}\rho^2$$

to obtain a boundary value problem for $U(\rho, t)$, refer to the solution in Problem 11 to derive the temperature formula

$$u(\rho,t) = \frac{B}{4}\left[2\rho^2 + 8t - 1 - 8\sum_{j=2}^{\infty}\frac{J_0(\alpha_j\rho)\exp\left(-\alpha_j^2 t\right)}{\alpha_j^2 J_0(\alpha_j)}\right],$$

where α_j $(j = 2, 3, \ldots)$ are the positive roots of the equation $J_1(\alpha) = 0$. [Note that the substitution for $u(\rho, t)$ made here is suggested by the fact that $U_\rho(1, t) = 0$.]

13. Over a long solid cylinder $\rho \leqslant 1$, at uniform temperature A, there is tightly fitted a long hollow cylinder $1 \leqslant \rho \leqslant 2$ of the same material at temperature B. The outer surface $\rho = 2$ is then kept at temperature B. Let $u(\rho, t)$ denote the temperatures in the cylinder of radius 2 so formed, and set up the boundary value problem for those temperatures. Then, after making the substitution

$$u(\rho,t) = U(\rho,t) + B$$

to obtain a boundary value problem for $U(\rho, t)$, refer to the solution in Example 1, Sec. 69, to derive the temperature formula

$$u(\rho,t) = B + \frac{A - B}{2}\sum_{j=1}^{\infty}\frac{J_1(\alpha_j)}{\alpha_j\left[J_1(2\alpha_j)\right]^2}J_0(\alpha_j\rho)\exp\left(-\alpha_j^2 kt\right),$$

where α_j are the positive roots of the equation $J_0(2\alpha) = 0$. (This is a temperature problem in shrunken fittings.)

14. Solve this boundary value problem for $u(x, t)$:

$$xu_t = (xu_x)_x - \frac{n^2}{x}u \qquad\qquad (0 < x < c, t > 0),$$

$$u(c,t) = 0 \qquad\qquad (t > 0),$$

$$u(x,0) = f(x) \qquad\qquad (0 < x < c),$$

where u is continuous for $0 \leqslant x \leqslant c$, $t > 0$ and where n is a nonnegative integer.

   Answer: $u(x,t) = \sum_{j=1}^{\infty}A_j J_n(\alpha_j x)\exp\left(-\alpha_j^2 t\right)$, where α_j and A_j are the constants in case (a) of Theorem 1 in Sec. 68.

15. Let $u(\rho, z)$ denote a function which is harmonic interior to the cylinder formed by the three surfaces $\rho = c$, $z = 0$, and $z = b$ $(b > 0)$. Given that $u = 0$ on both the top and bottom of the cylinder and that $u(c, z) = f(z)$ $(0 < z < b)$, derive the expression

$$u(\rho,z) = \sum_{n=1}^{\infty}b_n\frac{I_0(n\pi\rho/b)}{I_0(n\pi c/b)}\sin\frac{n\pi z}{b},$$

where

$$b_n = \frac{2}{b}\int_0^b f(z)\sin\frac{n\pi z}{b}\,dz.$$

[See Problem 15, Sec. 61, as well as the comments immediately following equation (17), Sec. 66, regarding the solutions of that modified form of Bessel's equation.]

16. Let the steady temperatures $u(\rho, z)$ in a semi-infinite cylinder $\rho \leqslant 1$, $z \geqslant 0$, whose base is insulated, be such that $u(1, z) = f(z)$, where

$$f(z) = \begin{cases} 1 & \text{when } 0 < z < 1, \\ 0 & \text{when } z > 1. \end{cases}$$

With the aid of the Fourier cosine integral formula (Sec. 55), derive the expression

$$u(\rho, z) = \frac{2}{\pi} \int_0^\infty \frac{I_0(\alpha\rho)}{\alpha I_0(\alpha)} \cos \alpha z \sin \alpha \, d\alpha$$

for those temperatures. (See the remarks at the end of Problem 15.)

17. Given a function $f(z)$ that is represented by its Fourier integral formula (Sec. 51) for all real z, derive the following expression for the harmonic function $u(\rho, z)$ inside the infinite cylinder $\rho \leqslant c$, $-\infty < z < \infty$ such that $u(c, z) = f(z)$ $(-\infty < z < \infty)$:

$$u(\rho, z) = \frac{1}{\pi} \int_0^\infty \frac{I_0(\alpha\rho)}{I_0(\alpha c)} \int_{-\infty}^\infty f(s) \cos \alpha(s - z) \, ds \, d\alpha.$$

(See the remarks at the end of Problem 15.)

71. VIBRATION OF A CIRCULAR MEMBRANE

A membrane, stretched over a fixed circular frame $\rho = c$ in the plane $z = 0$, is given an initial displacement $z = f(\rho, \phi)$ and released at rest from that position. The transverse displacements $z(\rho, \phi, t)$, where ρ, ϕ, and z are cylindrical coordinates, are described by the continuous function that satisfies this boundary value problem:

$$(1) \qquad\qquad z_{tt} = a^2 \left(z_{\rho\rho} + \frac{1}{\rho} z_\rho + \frac{1}{\rho^2} z_{\phi\phi} \right),$$

$$(2) \qquad\qquad z(c, \phi, t) = 0 \qquad\qquad (-\pi \leqslant \phi \leqslant \pi, t \geqslant 0),$$

$$(3) \quad z(\rho, \phi, 0) = f(\rho, \phi), \qquad z_t(\rho, \phi, 0) = 0 \quad (0 \leqslant \rho \leqslant c, -\pi \leqslant \phi \leqslant \pi),$$

where the function $z(\rho, \phi, t)$ is periodic, with period 2π, in the variable ϕ.

A function $z = R(\rho)\Phi(\phi)T(t)$ satisfies equation (1) if

$$(4) \qquad\qquad \frac{T''}{a^2 T} = \frac{1}{R}\left(R'' + \frac{1}{\rho} R' \right) + \frac{1}{\rho^2} \frac{\Phi''}{\Phi} = -\lambda,$$

where $-\lambda$ is any constant. We separate variables again in the second of equations (4) and write $\Phi''/\Phi = -\mu$. Then we find that the function $R\Phi T$ satisfies the homogeneous conditions and has the necessary periodicity with respect to ϕ if R and Φ are eigenfunctions of the Sturm-Liouville problems

$$(5) \qquad \rho^2 R''(\rho) + \rho R'(\rho) + (\lambda\rho^2 - \mu)R(\rho) = 0, \qquad R(c) = 0,$$

$$(6) \quad \Phi''(\phi) + \mu\Phi(\phi) = 0, \qquad \Phi(-\pi) = \Phi(\pi), \qquad \Phi'(-\pi) = \Phi'(\pi)$$

and T is such that

$$T''(t) + \lambda a^2 T(t) = 0, \qquad T'(0) = 0.$$

If μ has one of the values

$$\mu = n^2 \qquad\qquad (n = 0, 1, 2, \dots),$$

the theorem in Sec. 66 can be applied to problem (5); and if we consider problem (6) first, we see that the constant μ must, in fact, have one of those values. For, according to Sec. 40, they are the eigenvalues of problem (6). To be precise, $\Phi(\phi) = \frac{1}{2}$ when $n = 0$; and when $n = 1, 2, \dots,$ $\Phi(\phi)$ can be any linear combination of $\cos n\phi$ and $\sin n\phi$. From the theorem in Sec. 66, we now see that the eigenvalues of problem (5) are the numbers $\lambda_{nj} = \alpha_{nj}^2$ $(j = 1, 2, \dots)$, where α_{nj} are the positive roots of the equation

$$(7) \qquad\qquad\qquad J_n(\alpha c) = 0 \qquad\qquad (n = 0, 1, 2, \dots),$$

the corresponding eigenfunctions being $R(\rho) = J_n(\alpha_{nj}\rho)$. Then $T(t) = \cos \alpha_{nj} at.$

The generalized linear combination of our functions $R\Phi T,$

$$(8) \quad z(\rho, \phi, t) = \frac{1}{2} \sum_{j=1}^{\infty} A_{0j} J_0(\alpha_{0j}\rho) \cos \alpha_{0j} at$$

$$+ \sum_{n=1}^{\infty} \sum_{j=1}^{\infty} J_n(\alpha_{nj}\rho)(A_{nj} \cos n\phi + B_{nj} \sin n\phi) \cos \alpha_{nj} at,$$

formally satisfies all the homogeneous conditions. It also satisfies the condition $z(\rho, \phi, 0) = f(\rho, \phi)$ if the coefficients A_{0j}, A_{nj}, and B_{nj} are such that

$$(9) \quad f(\rho, \phi) = \frac{1}{2} \sum_{j=1}^{\infty} A_{0j} J_0(\alpha_{0j}\rho)$$

$$+ \sum_{n=1}^{\infty} \left\{ \left[\sum_{j=1}^{\infty} A_{nj} J_n(\alpha_{nj}\rho) \right] \cos n\phi 2 + \left[\sum_{j=1}^{\infty} B_{nj} J_n(\alpha_{nj}\rho) \right] \sin n\phi \right\}$$

when $0 \leqslant \rho \leqslant c$, $-\pi \leqslant \phi \leqslant \pi$.

For each fixed value of ρ, series (9) is the Fourier series for $f(\rho, \phi)$ on the interval $-\pi \leqslant \phi \leqslant \pi$ if

$$\sum_{j=1}^{\infty} A_{nj} J_n(\alpha_{nj}\rho) = \frac{1}{\pi} \int_{-\pi}^{\pi} f(\rho, \phi) \cos n\phi \, d\phi \quad (n = 0, 1, 2, \dots),$$

$$\sum_{j=1}^{\infty} B_{nj} J_n(\alpha_{nj}\rho) = \frac{1}{\pi} \int_{-\pi}^{\pi} f(\rho, \phi) \sin n\phi \, d\phi \qquad (n = 1, 2, \dots).$$

For every fixed n, the series on the left-hand side of each equation here furnishes the Fourier-Bessel series representation, on the interval $0 < \rho < c$, of the corresponding function of ρ on the right-hand side, provided that (Sec. 68)

$$(10) \quad A_{nj} = \frac{2}{\pi c^2 \left[J_{n+1}(\alpha_{nj}c) \right]^2} \int_0^c \rho J_n(\alpha_{nj}\rho) \int_{-\pi}^{\pi} f(\rho, \phi) \cos n\phi \; d\phi \; d\rho,$$

$$(11) \quad B_{nj} = \frac{2}{\pi c^2 \left[J_{n+1}(\alpha_{nj}c) \right]^2} \int_0^c \rho J_n(\alpha_{nj}\rho) \int_{-\pi}^{\pi} f(\rho, \phi) \sin n\phi \; d\phi \; d\rho.$$

The displacements $z(\rho, \phi, t)$ are, then, given by equation (8) when the coefficients have the values (10) and (11). We assume, of course, that the function f is such that the series in expression (8) has adequate properties of convergence and differentiability.

PROBLEMS

1. Suppose that in Sec. 71 the initial displacement function $f(\rho, \phi)$ is a linear combination of a finite number of the functions $J_0(\alpha_{0j}\rho)$ and $J_n(\alpha_{nj}\rho) \cos n\phi$, $J_n(\alpha_{nj}\rho) \sin n\phi$ $(n = 1, 2, \dots)$. Point out why the iterated series in expression (8) of that section then contains only a finite number of terms and represents a rigorous solution of the boundary value problem.

2. Let the initial displacement of the membrane in Sec. 71 be $f(\rho)$, a function of ρ only, and derive the expression

$$z(\rho, t) = \frac{2}{c^2} \sum_{j=1}^{\infty} \frac{J_0(\alpha_j\rho) \cos \alpha_j at}{\left[J_1(\alpha_j c) \right]^2} \int_0^c sf(s)J_0(\alpha_j s) \; ds,$$

where α_j are the positive roots of the equation $J_0(\alpha c) = 0$, for the displacements when $t > 0$.

3. Show that if the initial displacement of the membrane in Sec. 71 is $AJ_0(\alpha_k\rho)$, where A is a constant and α_k is some positive root of the equation $J_0(\alpha c) = 0$, then the subsequent displacements are

$$z(\rho, t) = AJ_0(\alpha_k\rho) \cos \alpha_k at.$$

Observe that these displacements are all periodic in t with a common period; thus the membrane gives a musical note.

4. Replace the initial conditions (3), Sec. 71, by the conditions that $z = 0$ and $z_t = 1$ when $t = 0$. This is the case if the membrane and its frame are moving with unit velocity in the z direction and the frame is brought to rest at the instant $t = 0$. Derive the expression

$$z(\rho, t) = \frac{2}{ac} \sum_{j=1}^{\infty} \frac{\sin \alpha_j at}{\alpha_j^2 J_1(\alpha_j c)} J_0(\alpha_j \rho),$$

where α_j are the positive roots of the equation $J_0(\alpha c) = 0$, for the displacements when $t > 0$.

5. Suppose that the *damped* transverse displacements $z(\rho, t)$ in a membrane, stretched over a circular frame, satisfy the conditions

$$z_{tt} = z_{\rho\rho} + \frac{1}{\rho}z_\rho - 2bz_t \qquad (0 < \rho < 1, t > 0),$$

$$z(1, t) = 0, \qquad z(\rho, 0) = 0, \qquad z_t(\rho, 0) = v_0.$$

The constant coefficient of damping $2b$ is such that $0 < b < \alpha_1$, where α_1 is the smallest of the positive zeros of $J_0(\alpha)$. Derive the solution

$$z(\rho, t) = 2v_0 e^{-bt} \sum_{j=1}^{\infty} \frac{J_0(\alpha_j\rho)}{\alpha_j J_1(\alpha_j)} \cdot \frac{\sin\left(t\sqrt{\alpha_j^2 - b^2}\right)}{\sqrt{\alpha_j^2 - b^2}},$$

where $J_0(\alpha_j) = 0$ $(\alpha_j > 0)$, of this boundary value problem.

6. Derive the following expression for the temperatures $u(\rho, \phi, t)$ in an infinite cylinder $\rho \leqslant c$ when $u = 0$ on the surface $\rho = c$ and $u = f(\rho, \phi)$ at time $t = 0$:

$$u(\rho, \phi, t) = \frac{1}{2} \sum_{j=1}^{\infty} A_{0j} J_0(\alpha_{0j}\rho) \exp\left(-\alpha_{0j}^2 kt\right)$$

$$+ \sum_{n=1}^{\infty} \sum_{j=1}^{\infty} J_n(\alpha_{nj}\rho)(A_{nj}\cos n\phi + B_{nj}\sin n\phi) \exp\left(-\alpha_{nj}^2 kt\right),$$

where α_{nj}, A_{nj}, and B_{nj} are the numbers defined in Sec. 71.

7. Derive an expression for the temperatures $u(\rho, z, t)$ in a solid cylinder $\rho \leqslant c$, $0 \leqslant z \leqslant \pi$ whose entire surface is kept at temperature zero and whose initial temperature is a constant A. Show that it can be written as the product

$$u(\rho, z, t) = Av(z, t)w(\rho, t)$$

of A and the functions

$$v(z, t) = \frac{4}{\pi} \sum_{n=1}^{\infty} \frac{\sin(2n-1)z}{2n-1} \exp\left[-(2n-1)^2 kt\right]$$

and

$$w(\rho, t) = \frac{2}{c} \sum_{j=1}^{\infty} \frac{J_0(\alpha_j\rho)}{\alpha_j J_1(\alpha_j c)} \exp\left(-\alpha_j^2 kt\right),$$

where α_j are the positive roots of the equation $J_0(\alpha c) = 0$. Also, show that $v(z, t)$ represents temperatures in a slab $0 \leqslant z \leqslant \pi$ and $w(\rho, t)$ temperatures in an infinite cylinder $\rho \leqslant c$, both with zero boundary temperature and unit initial temperature (see Example 1, Sec. 32, and Example 1, Sec. 69).

8. Derive the following expression for temperatures $u(\rho, \phi, t)$ in the long right-angled cylindrical wedge formed by the surface $\rho = 1$ and the planes $\phi = 0$ and $\phi = \pi/2$ when $u = 0$ on its entire surface and $u = f(\rho, \phi)$ at time $t = 0$:

$$u(\rho, \phi, t) = \sum_{n=1}^{\infty} \sum_{j=1}^{\infty} B_{nj} J_{2n}(\alpha_{nj}\rho) \sin 2n\phi \exp\left(-\alpha_{nj}^2 kt\right),$$

where α_{nj} are the positive roots of the equation $J_{2n}(\alpha) = 0$ and

$$B_{nj}\left[J_{2n+1}(\alpha_{nj})\right]^2 = \frac{8}{\pi}\int_0^{\pi/2}\sin 2n\phi\int_0^1 \rho f(\rho,\phi)J_{2n}(\alpha_{nj}\rho)\,d\rho\,d\phi.$$

9. Show that if the plane $\phi = \pi/2$ in Problem 8 is replaced by a plane $\phi = \phi_0$, the expression for the temperatures in the wedge will, in general, involve Bessel functions J_ν of *nonintegral* orders.

10. Solve Problem 8 when the entire surface of the wedge is insulated, instead of being kept at temperature zero.

11. Solve the boundary value problem

$$u_{\rho\rho} + \frac{1}{\rho}u_\rho - \frac{n^2}{\rho^2}u + u_{zz} = 0 \qquad (0 < \rho < 1, z > 0),$$

$$u(1,z) = 0, \qquad u(\rho,0) = \rho^n,$$

where $u(\rho, z)$ is bounded and continuous for $0 \leqslant \rho < 1$, $z > 0$ and where n is a positive integer. (When $n = 0$, this problem becomes the Dirichlet problem that was solved in Problem 8, Sec. 70.)

Answer: $u(\rho, z) = 2\sum_{j=1}^{\infty} \dfrac{J_n(\alpha_j\rho)}{\alpha_j J_{n+1}(\alpha_j)}\exp(-\alpha_j z)$, where $J_n(\alpha_j) = 0$ $(\alpha_j > 0)$.

12. Let the function $u(\rho, \phi, z)$ satisfy Poisson's equation (Sec. 3) $\nabla^2 u + ay = 0$, where a is a constant, inside a semi-infinite half cylinder $0 < \rho < 1$, $0 < \phi < \pi$, $z \geqslant 0$, and suppose that $u = 0$ on the entire surface. The function u, which is assumed to be bounded and continuous for $0 \leqslant \rho < 1$, $0 < \phi < \pi$, $z > 0$, thus satisfies the boundary value problem

$$u_{\rho\rho} + \frac{1}{\rho}u_\rho + \frac{1}{\rho^2}u_{\phi\phi} + u_{zz} + a\rho\sin\phi = 0 \qquad (0 < \rho < 1, 0 < \phi < \pi, z > 0),$$

$$u(1,\phi,z) = 0, \qquad u(\rho,\phi,0) = 0, \qquad u(\rho,0,z) = u(\rho,\pi,z) = 0.$$

Use the following method to solve it.

(*a*) By writing $u(\rho, \phi, z) = a\sin\phi\, v(\rho, z)$, reduce the stated problem to the one

$$v_{\rho\rho} + \frac{1}{\rho}v_\rho - \frac{1}{\rho^2}v + v_{zz} + \rho = 0 \qquad (0 < \rho < 1, z > 0),$$

$$v(1,z) = 0, \qquad v(\rho,0) = 0$$

in $v(\rho, z)$, where v is bounded and continuous for $0 \leqslant \rho < 1$, $z > 0$.

(*b*) Note how, when $n = 1$, the solution in Problem 11 suggests that the method of variation of parameters (see Example 2, Sec. 69) be used to seek a solution of the form

$$v(\rho, z) = \sum_{j=1}^{\infty} A_j(z)J_1(\alpha_j\rho),$$

where $J_1(\alpha_j) = 0$ $(\alpha_j > 0)$, for the problem in part (*a*). Apply that method to obtain the initial value problem

$$A_j''(z) - \alpha_j^2 A_j(z) = -\frac{2}{\alpha_j J_2(\alpha_j)}, \qquad A_j(0) = 0$$

in ordinary differential equations. Then, by adding a particular solution of this differential equation, which is a constant that is readily found by inspection, to the general solution of the complementary equation $A_j''(z) - \alpha_j^2 A_j(z) = 0$ (compare Problem 13, Sec. 38), find $v(\rho, z)$. Thus arrive at the solution

$$u(\rho, \phi, z) = 2a \sin \phi \sum_{j=1}^{\infty} \frac{1 - \exp(-\alpha_j z)}{\alpha_j^3 J_2(\alpha_j)} J_1(\alpha_j \rho),$$

where $J_1(\alpha_j) = 0$ ($\alpha_j > 0$), of the original problem.

Suggestion: In obtaining the ordinary differential equation for $A_j(z)$ in part (b), one can write the needed Fourier-Bessel expansion for ρ by simply referring to the expansion already found in Problem 9, Sec. 68. Also, it is necessary to observe how the identity

$$\frac{d^2}{d\rho^2} J_1(\alpha_j \rho) + \frac{1}{\rho} \frac{d}{d\rho} J_1(\alpha_j \rho) - \frac{1}{\rho^2} J_1(\alpha_j \rho) = -\alpha_j^2 J_1(\alpha_j \rho)$$

follows immediately from that problem.

CHAPTER

8

LEGENDRE POLYNOMIALS AND APPLICATIONS

As we shall see later in this chapter (Secs. 77 and 78), an application of the method of separation of variables to Laplace's equation in the spherical coordinates r and θ leads, after the substitution $x = \cos\theta$ is made, to *Legendre's equation*

$$(1) \qquad \left[(1 - x^2)y'(x)\right]' + \lambda y(x) = 0,$$

where λ is the separation constant. The points $x = 1$ and $x = -1$ correspond to $\theta = 0$ and $\theta = \pi$, respectively, and we begin the chapter by using series to discover solutions of equation (1) that can be used when $-1 \leqslant x \leqslant 1$.

72. SOLUTIONS OF LEGENDRE'S EQUATION

To solve Legendre's equation, we write it as

$$(1) \qquad (1 - x^2)y''(x) - 2xy'(x) + \lambda y(x) = 0$$

and observe that, while $x = \pm 1$ are singular points, $x = 0$ is an *ordinary point*.

We thus seek a solution of the form[†]

$$(2) \qquad\qquad y = \sum_{j=0}^{\infty} a_j x^j.$$

Substitution of series (2) into equation (1) yields the identity

$$\sum_{j=0}^{\infty} j(j-1)a_j x^{j-2} - \sum_{j=0}^{\infty} [j(j-1) + 2j - \lambda]a_j x^j = 0.$$

Since the first two terms in the first series here are actually zero and since

$$j(j-1) + 2j = j(j+1)$$

in the second series, we may write

$$\sum_{j=2}^{\infty} j(j-1)a_j x^{j-2} - \sum_{j=0}^{\infty} [j(j+1) - \lambda]a_j x^j = 0.$$

Finally, by putting the first of these series in the form

$$\sum_{j=0}^{\infty} (j+2)(j+1)a_{j+2} x^j,$$

we arrive at the equation

$$(3) \qquad \sum_{j=0}^{\infty} \{(j+2)(j+1)a_{j+2} - [j(j+1) - \lambda]a_j\}x^j = 0,$$

involving a single series.

Equation (3) is an identity in x if the coefficients a_j satisfy the recurrence relation

$$(4) \qquad\qquad a_{j+2} = \frac{j(j+1) - \lambda}{(j+2)(j+1)} a_j \qquad (j = 0, 1, 2, \dots).$$

The power series (2) thus represents a solution of Legendre's equation within its interval of convergence if its coefficients satisfy relation (4). This leaves a_0 and a_1 as arbitrary constants.

If $a_1 = 0$, it follows from relation (4) that $a_3 = a_5 = \cdots = 0$. Thus one nontrivial solution of Legendre's equation, containing only even powers of x, is

$$(5) \qquad\qquad y_1 = a_0 + \sum_{k=1}^{\infty} a_{2k} x^{2k} \qquad (a_0 \neq 0),$$

where a_0 is an arbitrary nonzero constant and where the remaining coefficients $a_2, a_4, \dots$ are expressed in terms of a_0 by successive applications of relation (4).

[†] For a discussion of ordinary points and a justification for this substitution, see, for example, the books referred to earlier in the footnote in Sec. 59.

(See Problem 8, Sec. 75.) Another solution, containing only odd powers of x, is obtaining by writing $a_0 = 0$ and letting a_1 be arbitrary. More precisely, the series

$$(6) \qquad y_2 = a_1 x + \sum_{k=1}^{\infty} a_{2k+1} x^{2k+1} \qquad (a_1 \neq 0)$$

satisfies Legendre's equation for any nonzero value of a_1 when $a_3, a_5, \ldots$ are written in terms of a_1 in accordance with relation (4). These two solutions are, of course, linearly independent since they are not constant multiples of each other.

From relation (4), it is clear that the value of λ affects the values of all but the first coefficients in series (5) and (6). As we shall see in Sec. 73, there are certain values of λ that cause series (5) and (6) to terminate and become polynomials. Assuming for the moment that series (5) does not terminate, we note from relation (4), with $j = 2k$, that

$$\lim_{k \to \infty} \left| \frac{a_{2(k+1)} x^{2(k+1)}}{a_{2k} x^{2k}} \right| = \lim_{k \to \infty} \left| \frac{2k(2k+1) - \lambda}{(2k+2)(2k+1)} x^2 \right| = x^2.$$

So, according to the ratio and absolute convergence tests, series (5) converges when $x^2 < 1$ and diverges when $x^2 > 1$. Although it is somewhat more difficult to show, series (5) diverges when $x = \pm 1$.[†]

Similar arguments apply to series (6). In summary, then, *if λ is such that either of the series (5) or (6) does not terminate and become a polynomial, that series converges only when* $-1 < x < 1$.

73. LEGENDRE POLYNOMIALS

When Legendre's equation

$$(1 - x^2) y''(x) - 2xy'(x) + \lambda y(x) = 0$$

arises in the applications, it will be necessary to have a solution which, along with its derivative, is continuous on the closed interval $-1 \leqslant x \leqslant 1$. But we know from Sec. 72 that, unless it terminates, neither of the series solutions

$$(1) \qquad y_1 = a_0 + \sum_{k=1}^{\infty} a_{2k} x^{2k} \qquad (a_0 \neq 0),$$

$$(2) \qquad y_2 = a_1 x + \sum_{k=1}^{\infty} a_{2k+1} x^{2k+1} \qquad (a_1 \neq 0)$$

obtained there satisfies those continuity conditions.

[†] See, for instance, the book by Bell (1968, pp. 230–231), listed in the Bibliography.

Suppose now that the parameter λ in Legendre's equation has one of the integral values

(3) $$\lambda = n(n + 1) \qquad (n = 0, 1, 2, \ldots),$$

in which case the recurrence relation (4), Sec. 72, becomes

(4) $$a_{j+2} = \frac{j(j + 1) - n(n + 1)}{(j + 2)(j + 1)} a_j \qquad (j = 0, 1, 2, \ldots).$$

Since $a_{n+2} = 0$ and, consequently, $a_{n+4} = a_{n+6} = \cdots = 0$, it follows that one of the solutions y_1, y_2 is actually a polynomial.

Note that if $n = 0$, then $a_2 = a_4 = a_6 = \cdots = 0$; and series (1) becomes simply $y_1 = a_0$. If, moreover, n is any one of the even integers $2, 4, \ldots$, so that $n = 2m$ $(m = 1, 2, \ldots)$, then $a_{2m} \neq 0$ and $a_{2(m+1)} = a_{2(m+2)} = \cdots = 0$. Series (1) thus reduces to a polynomial whose degree is $2m$, or n. On the other hand, if $n = 1$, we see that $y_2 = a_1 x$; and if n is any one of the odd integers $n = 2m + 1$, then $a_{2m+1} \neq 0$ and $a_{2(m+1)+1} = a_{2(m+2)+1} = \cdots = 0$. Hence series (2) becomes a polynomial of degree n if n is odd.

Thus, *if λ has any one of the values* (3), *solution* (1) *reduces to the polynomial*

(5) $$y_1 = a_0 + a_2 x^2 + \cdots + a_n x^n \qquad (a_n \neq 0)$$

when n is even; and solution (2) *becomes*

(6) $$y_2 = a_1 x + a_3 x^3 + \cdots + a_n x^n \qquad (a_n \neq 0)$$

when n is odd. The coefficients a_0 and a_1 are arbitrary nonzero constants, and the others are determined by successive applications of relation (4). Observe that when n is even, solution (2) remains an infinite series and that when n is odd, the same is true of solution (1).

If n is even, it is customary to assign a value to a_0 such that when the coefficients $a_2, \ldots, a_n$ in expression (5) are determined by means of relation (4), the final coefficient a_n has the value

(7) $$a_n = \frac{(2n)!}{2^n (n!)^2}.$$

The reason for this requirement is that the polynomial (5) will then have the value unity when $x = 1$, as will be shown in Sec. 75. The precise value of a_0 that is needed is not important to us here. Using the convention that $0! = 1$, we note that $a_0 = 1$ if $n = 0$. In that case, $y_1 = 1$. If n is odd, we choose a_1 so that the final coefficient in expression (6) is also given by equation (7). The reason for this choice is similar to the one above regarding the value assigned to a_0. Note that $y_2 = x$ if $n = 1$, since $a_1 = 1$ for that value of n.

When $n = 2, 3, \ldots$, relation (4) can be used to write all the coefficients that precede a_n in expressions (5) and (6) in terms of a_n. To accomplish this, we first observe that the numerator on the right-hand side of relation (4) can be written

$$j(j + 1) - n(n + 1) = -\left[(n^2 - j^2) + (n - j)\right] = -(n - j)(n + j + 1).$$

We then solve for a_j; the result is

(8)
$$a_j = -\frac{(j + 2)(j + 1)}{(n - j)(n + j + 1)} a_{j+2}.$$

To express a_{n-2k} in terms of a_n, we now use relation (8) to write the following k equations:

$$a_{n-2} = -\frac{(n)(n - 1)}{(2)(2n - 1)} a_n,$$

$$a_{n-4} = -\frac{(n - 2)(n - 3)}{(4)(2n - 3)} a_{n-2},$$

$$\vdots$$

$$a_{n-2k} = -\frac{(n - 2k + 2)(n - 2k + 1)}{(2k)(2n - 2k + 1)} a_{n-2k+2}.$$

Equating the product of the left-hand sides of these equations to the product of their right-hand sides and then canceling the common factors a_{n-2}, $a_{n-4}, \ldots, a_{n-2k+2}$ on each side of the resulting equation, we find that

(9)
$$a_{n-2k} = \frac{(-1)^k}{2^k k!} \cdot \frac{n(n - 1) \cdots (n - 2k + 1)}{(2n - 1)(2n - 3) \cdots (2n - 2k + 1)} a_n.$$

Then, upon substituting expression (7) for a_n into equation (9) and combining various terms into the appropriate factorials (see Problem 5, Sec. 75), we arrive at the desired expression:

(10)
$$a_{n-2k} = \frac{1}{2^n} \cdot \frac{(-1)^k}{k!} \cdot \frac{(2n - 2k)!}{(n - 2k)!(n - k)!}.$$

As usual, $0! = 1$.

In view of equation (10), the polynomials (5) and (6), when the nonzero constants a_0 and a_1 are such that a_n has values (7), can be written

(11) $$P_n(x) = \frac{1}{2^n} \sum_{k=0}^{m} \frac{(-1)^k}{k!} \cdot \frac{(2n - 2k)!}{(n - 2k)!(n - k)!} x^{n-2k} \quad (n = 0, 1, 2, \ldots),$$

where $m = n/2$ if n is even and $m = (n - 1)/2$ if n is odd. Another expression for $P_n(x)$ will be given in Sec. 75. Note that since $P_n(x)$ is a polynomial containing only even powers of x if n is even and only odd powers if n is odd, it is an even or an odd function, depending on whether n is even or odd; that is,

(12) $$P_n(-x) = (-1)^n P_n(x) \qquad (n = 0,1,2,\dots).$$

The polynomial $P_n(x)$ is called the *Legendre polynomial of degree n.* For the first several values of n, expression (11) becomes

$$P_0(x) = 1, \qquad P_1(x) = x, \qquad P_2(x) = \frac{1}{2}(3x^2 - 1),$$

$$P_3(x) = \frac{1}{2}(5x^3 - 3x), \qquad P_4(x) = \frac{1}{8}(35x^4 - 30x^2 + 3),$$

$$P_5(x) = \frac{1}{8}(63x^5 - 70x^3 + 15x).$$

Observe that the value of each of these six polynomials is unity when $x = 1$, as anticipated. See Fig. 65, where the first four are displayed graphically on the interval $-1 \leqslant x \leqslant 1$.

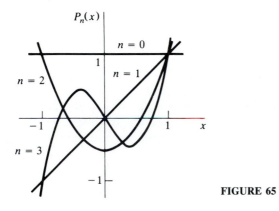

FIGURE 65

We have just seen that Legendre's equation

(13) $$(1 - x^2)y''(x) - 2xy'(x) + n(n + 1)y(x) = 0 \qquad (n = 0,1,2,\dots)$$

always has the polynomial solution $y = P_n(x)$, which is solution (5) (n even) or solution (6) (n odd) when appropriate values are assigned to the arbitrary constants a_0 and a_1 in those solutions. Details regarding the standard form of the accompanying series solution, which is denoted by $Q_n(x)$ and is called a *Legendre function of the second kind*, are left to the problems. We, of course, know from the statement in italics at the end of Sec. 72 that the series representing $Q_n(x)$ is convergent only when $-1 < x < 1$. It will, however, be sufficient for us to know that $Q_n(x)$ and $Q_n'(x)$ fail to be a pair of continuous functions on the *closed* interval $-1 \leqslant x \leqslant 1$ (Problem 9, Sec. 76). Since $P_n(x)$ and $Q_n(x)$ are linearly independent, the general solution of equation (13) is

(14) $$y = C_1 P_n(x) + C_2 Q_n(x),$$

where C_1 and C_2 are arbitrary constants.

74. ORTHOGONALITY OF LEGENDRE POLYNOMIALS

Let $X(x)$ denote the dependent variable in Legendre's equation, with arbitrary λ:

(1) $$(1 - x^2)X''(x) - 2xX'(x) + \lambda X(x) = 0.$$

Writing this equation in its self-adjoint form (Sec. 41)

(2) $$\left[(1 - x^2)X'(x)\right]' + \lambda X(x) = 0,$$

we see that we have a special case of the Sturm-Liouville differential equation

$$[r(x)X'(x)]' + [q(x) + \lambda p(x)]X(x) = 0,$$

where $p(x) = 1$, $q(x) = 0$, and $r(x) = 1 - x^2$. The function $r(x)$ vanishes at $x = \pm 1$; thus, as already pointed out in Example 2, Sec. 42, equation (2) here, without boundary conditions, is a singular Sturm-Liouville problem on the interval $-1 \leqslant x \leqslant 1$, where X and X' are required to be continuous on that closed interval.

The following theorem provides us with all the solutions of this problem.

Theorem. The eigenvalues and corresponding eigenfunctions of the singular Sturm-Liouville problem (2), *on the interval* $-1 \leqslant x \leqslant 1$, *are*

(3) $$\lambda_n = n(n + 1), \qquad X_n = P_n(x) \qquad (n = 0, 1, 2, \ldots),$$

where the $P_n(x)$ *are the Legendre polynomials.*

We start the proof by recalling from Sec. 73 that $P_n(x)$ and $Q_n(x)$ are linearly independent solutions of equation (2) when λ has any one of the values $\lambda = n(n + 1)$ $(n = 0, 1, 2, \ldots)$. Since the polynomial $P_n(x)$ and its derivative are continuous on the entire interval $-1 \leqslant x \leqslant 1$ and since this is not true of the Legendre function $Q_n(x)$, it is clear that the continuity requirements on X and X' are met only when X is a constant multiple of $P_n(x)$. Hence the λ_n and X_n in the statement of the theorem are, in fact, eigenvalues and eigenfunctions. It remains to show that there are no other eigenvalues.

To accomplish this, we digress for a moment and observe that, since the eigenfunctions just noted all correspond to different eigenvalues, *the set* $\{P_n(x)\}$ $(n = 0, 1, 2, \ldots)$ *is orthogonal on the interval* $-1 < x < 1$, *with weight function* $p(x) = 1$. (See the theorem in Sec. 43.) That is,

(4) $$\int_{-1}^{1} P_m(x) P_n(x) \, dx = 0 \qquad (m \neq n).$$

In the notation used for inner products, property (4) reads $(P_m, P_n) = 0 \ (m \neq n)$. Later on (Sec. 76), there will be a theorem telling us that if a function f is piecewise smooth on the interval $-1 < x < 1$, then the generalized Fourier series for f with respect to the orthonormal set of functions

$$(5) \qquad\qquad \phi_n(x) = \frac{P_n(x)}{\|P_n\|} \qquad\qquad (n = 0, 1, 2, \ldots)$$

converges to $f(x)$ at all but possibly a finite number of points in the interval $-1 < x < 1$. The set $\{\phi_n(x)\}$ is, therefore, closed (Sec. 12) in the function space $C'_p(-1, 1)$ (see Sec. 17).

Suppose now that λ is another eigenvalue, different from those listed in the statement of the theorem, and let X denote an eigenfunction corresponding to λ. Because of the orthogonality of eigenfunctions corresponding to distinct eigenvalues, $(X, \phi_n) = 0$ $(n = 0, 1, 2, \ldots)$, where the functions ϕ_n are those in equation (5). But the fact that $\{\phi_n(x)\}$ is closed requires that X, which is continuous on the entire interval $-1 \leqslant x \leqslant 1$, have value zero for each x in that interval. Consequently, since an eigenfunction cannot be identically zero, X is not an eigenfunction. In view of this contradiction, there are no other eigenvalues; and the proof of the theorem is finished.

If the interval $0 \leqslant x \leqslant 1$, rather than $-1 \leqslant x \leqslant 1$, is used, the differential equation (2) along with either one of the boundary conditions $X'(0) = 0$, $X(0) = 0$ is also a singular Sturm-Liouville problem (Sec. 42).

Corollary. *The eigenvalues and corresponding eigenfunctions of the singular Sturm-Liouville problem consisting of the differential equation* (2), *on the interval* $0 < x < 1$, *and the boundary condition* $X'(0) = 0$ *are*

$$(6) \qquad \lambda_n = 2n(2n+1), \qquad\qquad X_n = P_{2n}(x) \qquad\qquad (n = 0, 1, 2, \ldots).$$

If the condition $X(0) = 0$ *is used instead, the eigenvalues and eigenfunctions are*

$$(7) \qquad \lambda_n = (2n+1)(2n+2), \qquad X_n = P_{2n+1}(x) \qquad\qquad (n = 0, 1, 2, \ldots).$$

To see how these solutions follow, we consider first the solutions in the theorem when the condition $X'(0) = 0$ is imposed on them. Since $P_n'(0) = 0$ only when n is an even integer (Problem 7, Sec. 75), the polynomials $P_{2n+1}(x)$ $(n = 0, 1, 2, \ldots)$ must be eliminated. This leaves the eigenvalues and eigenfunctions (6). If, on the other hand, the condition $X(0) = 0$ is imposed, the fact that $P_n(0) = 0$ only when n is an odd integer leads us to the eigenvalues and eigenfunctions (7).

The theorem in Sec. 43, regarding the orthogonality of eigenfunctions, tells us that *each of the sets* $\{P_{2n}(x)\}$ $(n = 0, 1, 2, \ldots)$ *and* $\{P_{2n+1}(x)\}$ $(n = 0, 1, 2, \ldots)$ *is orthogonal on the interval* $0 < x < 1$ *with weight function unity.* That is,

$$(8) \qquad\qquad \int_0^1 P_{2m}(x) P_{2n}(x)\, dx = 0 \qquad\qquad (m \neq n)$$

and

$$(9) \qquad \int_0^1 P_{2m+1}(x)\,P_{2n+1}(x)\,dx = 0 \qquad (m \neq n),$$

where $m = 0, 1, 2, \ldots$ and $n = 0, 1, 2, \ldots$. Valid representations of piecewise smooth functions on the interval $0 < x < 1$ will follow (Sec. 76) from representations on the interval $-1 < x < 1$ in terms of the set $\{P_n(x)\}$ $(n = 0, 1, 2, \ldots)$, just as Fourier cosine and sine series follow from Fourier series involving both cosines and sines. Hence the same argument, involving closed sets, that was used in the proof of the theorem above can be used to show that there are no other eigenvalues of the Sturm-Liouville problems in the corollary.

75. RODRIGUES' FORMULA AND NORMS

According to expression (11), Sec. 73,

$$(1) \qquad P_n(x) = \frac{1}{2^n n!} \sum_{k=0}^m (-1)^k \frac{n!}{k!(n-k)!} \cdot \frac{(2n-2k)!}{(n-2k)!} x^{n-2k},$$

where $m = n/2$ if n is even and $m = (n-1)/2$ if n is odd. Since

$$\frac{d^n}{dx^n} x^{2n-2k} = \frac{(2n-2k)!}{(n-2k)!} x^{n-2k} \qquad (0 \leqslant k \leqslant m)$$

and because of the linearity of the differential operator d^n/dx^n, expression (1) can be written

$$(2) \qquad P_n(x) = \frac{1}{2^n n!} \frac{d^n}{dx^n} \sum_{k=0}^m (-1)^k \frac{n!}{k!(n-k)!} x^{2n-2k}.$$

The powers of x in the sum here decrease in steps of 2 as the index k increases; and the lowest power is $2n - 2m$, which is n if n is even and $n + 1$ if n is odd. Evidently, then, the sum can be extended so that k ranges from 0 to n. For the additional polynomial that is introduced is of degree less than n, and its nth derivative is, therefore, zero. Since the resulting sum is the binomial expansion of $(x^2 - 1)^n$, it follows from equation (2) that

$$(3) \qquad P_n(x) = \frac{1}{2^n n!} \frac{d^n}{dx^n} (x^2 - 1)^n \qquad (n = 0, 1, 2, \ldots).$$

This is *Rodrigues' formula* for the Legendre polynomials.

Various useful properties of Legendre polynomials are readily obtained from Rodrigues' formula with the aid of *Leibnitz' rule* for the nth derivative $D^n[f(x)g(x)]$ of the product of two functions:

$$(4) \qquad D^n(fg) = \sum_{k=0}^n \frac{n!}{k!(n-k)!} D^k(f) D^{n-k}(g),$$

where it is understood that all the required derivatives exist and that the zero-order derivative of a function is the function itself.

We note, for example, that if we write $u = x^2 - 1$, so that

$$u^n = (x^2 - 1)^n = (x + 1)^n (x - 1)^n,$$

it follows from Leibnitz' rule that

$$D^n u^n = \sum_{k=0}^{n} \frac{n!}{k!(n-k)!} D^k \big[(x+1)^n\big] D^{n-k}\big[(x-1)^n\big].$$

Now the first term in this sum is

$$D^0 \big[(x+1)^n\big] D^n \big[(x-1)^n\big] = (x+1)^n n!,$$

and the remaining terms all contain the factor $(x - 1)$ to some positive power. Hence the value of the sum when $x = 1$ is $2^n n!$, and it follows from Rodrigues' formula (3) that

$$(5) \qquad\qquad\qquad\qquad P_n(1) = 1 \qquad\qquad\qquad (n = 0, 1, 2, \ldots).$$

Observe how it follows from this and the relation $P_n(-x) = (-1)^n P_n(x)$ $(n = 0, 1, 2, \ldots)$, obtained in Sec. 73, that

$$(6) \qquad\qquad\qquad\qquad P_n(-1) = (-1)^n \qquad\qquad\qquad (n = 0, 1, 2, \ldots).$$

For another application of Rodrigues' formula (3), we use it to write

$$2^{n+1}(n + 1)! P_{n+1}(x) = D^{n+1} u^{n+1} = D^{n-1}(D^2 u^{n+1}),$$

where $u = x^2 - 1$. But

$$Du^{n+1} = 2(n + 1) x u^n,$$

and so

$$\begin{aligned}
D^2 u^{n+1} &= 2(n + 1)(u^n + 2nx^2 u^{n-1}) \\
&= 2(n + 1)\big[u^n + 2n(x^2 - 1)u^{n-1} + 2nu^{n-1}\big] \\
&= 2(n + 1)\big[(2n + 1)u^n + 2nu^{n-1}\big].
\end{aligned}$$

Consequently,

$$2^n n! P_{n+1}(x) = (2n + 1) D^{n-1} u^n + 2n D^{n-1} u^{n-1}.$$

Substituting $2^{n-1}(n - 1)! P_{n-1}(x)$ for $D^{n-1} u^{n-1}$ here, we find that

$$(7) \qquad\qquad P_{n+1}(x) - P_{n-1}(x) = \frac{2n + 1}{2^n n!} D^{n-1} u^n.$$

On the other hand, Leibnitz' rule (4) enables us to write

$$P_{n+1}(x) = \frac{D^n(Du^{n+1})}{2^{n+1}(n+1)!} = \frac{D^n(xu^n)}{2^n n!} = \frac{xD^n u^n + nD^{n-1}u^n}{2^n n!};$$

and, since $D^n u^n = 2^n n! P_n(x)$,

$$(8) \qquad P_{n+1}(x) - xP_n(x) = \frac{n}{2^n n!} D^{n-1} u^n.$$

Elimination of $D^{n-1}u^n$ between this and equation (7) gives the *recurrence relation*

$$(9) \qquad (n+1)P_{n+1}(x) + nP_{n-1}(x) = (2n+1)xP_n(x) \qquad (n = 1,2,\dots).$$

Note, too, that the relation

$$(10) \qquad P'_{n+1}(x) - P'_{n-1}(x) = (2n+1)P_n(x) \qquad (n = 1,2,\dots)$$

is an immediate consequence of equation (7).

We now show how relation (9) and its form

$$(11) \qquad nP_n(x) + (n-1)P_{n-2}(x) = (2n-1)xP_{n-1}(x) \qquad (n = 2,3,\dots),$$

obtained by replacing n by $n-1$, can be used to find the norms $\|P_n\| = (P_n, P_n)^{1/2}$ of the orthogonal polynomials P_n. Keeping in mind that $(P_{n+1}, P_{n-1}) = 0$ and $(P_{n-2}, P_n) = 0$, we find from equations (9) and (11), respectively, that

$$(12) \qquad n(P_{n-1}, P_{n-1}) = (2n+1)(xP_n, P_{n-1})$$

and

$$(13) \qquad n(P_n, P_n) = (2n-1)(xP_{n-1}, P_n).$$

The integrals representing (xP_n, P_{n-1}) and (xP_{n-1}, P_n) are identical, and we need only eliminate those quantities from equations (12) and (13) to see that

$$(2n+1)(P_n, P_n) = (2n-1)(P_{n-1}, P_{n-1}),$$

or

$$(14) \qquad (2n+1)\|P_n\|^2 = (2n-1)\|P_{n-1}\|^2 \qquad (n = 2,3,\dots).$$

It is easy to verify directly that equation (14) is also valid when $n = 1$.

Next, we let n be any fixed positive integer and use equation (14) to write the following n equations:

$$(2n+1)\|P_n\|^2 = (2n-1)\|P_{n-1}\|^2,$$
$$(2n-1)\|P_{n-1}\|^2 = (2n-3)\|P_{n-2}\|^2,$$
$$\vdots$$
$$(5)\|P_2\|^2 = (3)\|P_1\|^2,$$
$$(3)\|P_1\|^2 = (1)\|P_0\|^2.$$

Setting the product of the left-hand sides of these equations equal to the product of their right-hand sides and then canceling appropriately, we arrive at the result

$$(2n+1)\|P_n\|^2 = \|P_0\|^2 \qquad (n = 1,2,\dots).$$

Since $\|P_0\|^2 = 2$, this means that

$$(15) \qquad\qquad \|P_n\| = \sqrt{\frac{2}{2n+1}} \qquad\qquad (n = 0, 1, 2, \dots).$$

The set of polynomials

$$(16) \qquad\qquad \phi_n(x) = \sqrt{\frac{2n+1}{2}}\, P_n(x) \qquad\qquad (n = 0, 1, 2, \dots)$$

is, therefore, *orthonormal* on the interval $-1 < x < 1$.

PROBLEMS

1. From the orthogonality of the set $\{P_n(x)\}$, state why

   (a) $\displaystyle\int_{-1}^{1} P_n(x)\, dx = 0 \ (n = 1, 2, \dots)$;

   (b) $\displaystyle\int_{-1}^{1} (Ax + B)P_n(x)\, dx = 0 \ (n = 2, 3, \dots)$, where A and B are constants.

2. Verify directly that the Legendre polynomials

   $$P_0(x) = 1, \qquad P_1(x) = x, \qquad P_2(x) = \frac{1}{2}(3x^2 - 1), \qquad P_3(x) = \frac{1}{2}(5x^3 - 3x)$$

   form an orthogonal set on the interval $-1 < x < 1$. Show that their graphs are as indicated in Fig. 65 (Sec. 73).

3. Use the fact that the set $\{\phi_n(x)\}$ defined by equation (16), Sec. 75, is orthonormal on the interval $-1 < x < 1$ to show that the following sets are orthonormal on the interval $0 < x < 1$:

   (a) $\left\{\sqrt{4n+1}\, P_{2n}(x)\right\} \ (n = 0, 1, 2, \dots)$;

   (b) $\left\{\sqrt{4n+3}\, P_{2n+1}(x)\right\} \ (n = 0, 1, 2, \dots)$.

   Suggestion: Use the suggestion with Problem 2, Sec. 11, modified so as to apply when the even function f there is defined on the interval $-1 < x < 1$.

4. From recurrence relation (10), Sec. 75, obtain the integration formula

   $$\int_{a}^{1} P_n(x)\, dx = \frac{1}{2n+1}[P_{n-1}(a) - P_{n+1}(a)] \qquad (n = 1, 2, \dots).$$

5. Give details showing how expression (10) in Sec. 73 for the coefficients a_{n-2k} in the Legendre polynomials are obtained from equations (7) and (9) there.

   Suggestion: Observe that the factorials in equation (7), Sec. 73, can be written

   $$(2n)! = (2n)(2n-1)(2n-2) \cdots (2n-2k+1)(2n-2k)!,$$
   $$n! = n(n-1) \cdots (n-2k+1)(n-2k)!,$$
   $$n! = n(n-1) \cdots (n-k+1)(n-k)!.$$

6. With the aid of expression (11), Sec. 73, for $P_n(x)$, show that when $n = 2, 3, \dots$, the constants a_0 and a_1 in equations (5) and (6) in that section must have the following

values in order for the final constant a_n to have the value specified in equation (7) there:

$$a_0 = (-1)^{n/2} \frac{(1)(3)(5) \cdots (n-1)}{(2)(4)(6) \cdots (n)} \qquad (n = 2, 4, \ldots),$$

$$a_1 = (-1)^{(n-1)/2} \frac{(1)(3)(5) \cdots (n)}{(2)(4)(6) \cdots (n-1)} \qquad (n = 3, 5, \ldots).$$

7. Establish these properties of Legendre polynomials, where $n = 0, 1, 2, \ldots$:

(a) $P_{2n}(0) = (-1)^n \dfrac{(2n)!}{2^{2n}(n!)^2}$;   (b) $P'_{2n}(0) = 0$;   (c) $P_{2n+1}(0) = 0$;

(d) $P'_{2n+1}(0) = (2n+1)P_{2n}(0)$.

   Suggestion: For parts (a) and (d), refer to Problem 6.

8. Legendre's equation (1), Sec. 72, is often written

$$(1-x^2)y''(x) - 2xy'(x) + \nu(\nu+1)y(x) = 0,$$

where ν is an unrestricted complex number. Show that when $\lambda = \nu(\nu+1)$, recurrence relation (4), Sec. 72, can be put in the form

$$a_j = -\frac{(\nu - j + 2)(\nu + j - 1)}{j(j-1)}a_{j-2} \qquad (j = 2, 3, \ldots).$$

Then, by proceeding as we did in solving Bessel's equation (Sec. 59), use this relation to obtain the following linearly independent solutions of Legendre's equation:

$$y_1 = a_0\Bigg\{1 + \sum_{k=1}^{\infty} (-1)^k$$

$$\times \frac{[\nu(\nu-2) \cdots (\nu - 2k + 2)][(\nu+1)(\nu+3) \cdots (\nu + 2k - 1)]}{(2k)!}x^{2k}\Bigg\},$$

$$y_2 = a_1\Bigg\{x + \sum_{k=1}^{\infty} (-1)^k$$

$$\times \frac{[(\nu-1)(\nu-3) \cdots (\nu - 2k + 1)][(\nu+2)(\nu+4) \cdots (\nu + 2k)]}{(2k+1)!}x^{2k+1}\Bigg\},$$

where a_0 and a_1 are arbitrary nonzero constants. (These two series converge when $-1 < x < 1$, according to Sec. 72.)

9. Show that if ν is the complex number

$$\nu = -\frac{1}{2} + i\alpha \qquad (\alpha > 0),$$

Legendre's equation in Problem 8 becomes

$$(1-x^2)y''(x) - 2xy'(x) - \left(\frac{1}{4} + \alpha^2\right)y(x) = 0.$$

Then show how it follows from the solutions obtained in Problem 8 that the functions

$$P_\alpha(x) = 1 + \sum_{k=1}^{\infty} \left[\alpha^2 + \left(\frac{1}{2}\right)^2\right]\left[\alpha^2 + \left(\frac{5}{2}\right)^2\right]\cdots\left[\alpha^2 + \left(\frac{4k-3}{2}\right)^2\right]\frac{x^{2k}}{(2k)!},$$

$$q_\alpha(x) = x + \sum_{k=1}^{\infty} \left[\alpha^2 + \left(\frac{3}{2}\right)^2\right]\left[\alpha^2 + \left(\frac{7}{2}\right)^2\right]\cdots\left[\alpha^2 + \left(\frac{4k-1}{2}\right)^2\right]\frac{x^{2k+1}}{(2k+1)!}$$

are linearly independent solutions of this differential equation, valid on the interval $-1 < x < 1$. These particular Legendre functions arise in certain boundary value problems in regions bounded by cones.

10. Note that the solutions y_1 and y_2 obtained in Problem 8 are solutions (5) and (6) in Sec. 72 when $\lambda = \nu(\nu + 1)$. They remain infinite series when $\nu = n = 1, 3, 5, \ldots$ and $\nu = n = 0, 2, 4, \ldots$, respectively. When $\nu = n = 2m$ $(m = 0, 1, 2, \ldots)$, the Legendre function Q_n of the second kind is defined as y_2, where

$$a_1 = \frac{(-1)^m 2^{2m}(m!)^2}{(2m)!};$$

and when $\nu = n = 2m + 1$ $(m = 0, 1, 2, \ldots)$, Q_n is defined as y_1, where

$$a_0 = -\frac{(-1)^m 2^{2m}(m!)^2}{(2m+1)!}.$$

Using the fact that

$$\ln \frac{1+x}{1-x} = 2\sum_{k=0}^{\infty} \frac{x^{2k+1}}{2k+1} \qquad (-1 < x < 1),$$

show that

$$Q_0(x) = \frac{1}{2}\ln\frac{1+x}{1-x} \quad \text{and} \quad Q_1(x) = \frac{x}{2}\ln\frac{1+x}{1-x} - 1 = xQ_0(x) - 1.$$

11. Use mathematical induction on the integer n to verify Leibnitz' rule (4), Sec. 75.

12. Write

$$F(x,t) = (1 - 2xt + t^2)^{-1/2},$$

where $|x| \leqslant 1$ and t is as yet unrestricted.

(a) Note that $x = \cos\theta$ for some uniquely determined value of θ $(0 \leqslant \theta \leqslant \pi)$, and show that

$$F(x,t) = (1 - e^{i\theta}t)^{-1/2}(1 - e^{-i\theta}t)^{-1/2}.$$

Then, using the fact that $(1 - z)^{-1/2}$ has a valid Maclaurin series expansion when $|z| < 1$, point out why the functions $(1 - e^{\pm i\theta}t)^{-1/2}$, considered as functions of t, can be represented by Maclaurin series which are valid when $|t| < 1$. It follows that the product of those two functions also has such a representation when $|t| < 1.^\dagger$ That is, there are functions $f_n(x)$ $(n = 0, 1, 2, \ldots)$ such that

$$F(x, t) = \sum_{n=0}^{\infty} f_n(x) t^n \qquad\qquad (|t| < 1).$$

(b) Show that the function $F(x, t)$ satisfies the identity

$$(1 - 2xt + t^2) \frac{\partial F}{\partial t} = (x - t)F,$$

and use this result to show that the functions $f_n(x)$ in part (a) satisfy the recurrence relation

$$(n + 1)f_{n+1}(x) + nf_{n-1}(x) = (2n + 1)xf_n(x) \qquad (n = 1, 2, \ldots).$$

(c) Show that the first two functions $f_0(x)$ and $f_1(x)$ in part (a) are 1 and x, respectively, and notice that the recurrence relation obtained in part (b) can then be used to determine $f_n(x)$ when $n = 2, 3, \ldots$. Compare that relation with relation (9), Sec. 75, and conclude that the functions $f_n(x)$ are, in fact, the Legendre polynomials $P_n(x)$; that is, show that

$$(1 - 2xt + t^2)^{-1/2} = \sum_{n=0}^{\infty} P_n(x) t^n \qquad (|x| \leqslant 1, |t| < 1).$$

The function F is thus a *generating function* for the Legendre polynomials.

13. Give an alternative proof of the property (Sec. 75) $P_n(1) = 1$ $(n = 0, 1, 2, \ldots)$, using (a) recurrence relation (9), Sec. 75, and mathematical induction; (b) the generating function obtained in Problem 12(c).

76. LEGENDRE SERIES

In Sec. 75, we saw that the set of polynomials

$$(1) \qquad\qquad \phi_n(x) = \sqrt{\frac{2n + 1}{2}}\, P_n(x) \qquad\qquad (n = 0, 1, 2, \ldots)$$

is orthonormal on the interval $-1 < x < 1$. The Fourier constants (Sec. 12) with respect to that set, for a function f defined on the interval $-1 < x < 1$, are

$$c_n = (f, \phi_n) = \sqrt{\frac{2n + 1}{2}} \int_{-1}^{1} f(x) P_n(x)\, dx;$$

and the generalized Fourier series corresponding to $f(x)$ is

$$\sum_{n=0}^{\infty} c_n \phi_n(x) = \sum_{n=0}^{\infty} \frac{2n + 1}{2} P_n(x) \int_{-1}^{1} f(s) P_n(s)\, ds.$$

---

† For a discussion of this point, see, for example, the authors' book (1990, pp. 161–162), listed in the Bibliography.

That is,

(2) $$f(x) \sim \sum_{n=0}^{\infty} A_n P_n(x) \qquad (-1 < x < 1),$$

where

(3) $$A_n = \frac{2n+1}{2} \int_{-1}^{1} f(x) P_n(x)\, dx \qquad (n = 0, 1, 2, \ldots).$$

Series (2), with coefficients (3), is a *Legendre series*. We state here, without proof, a representation theorem that is applicable to piecewise smooth functions.[†]

Theorem. *Let f denote a function that is piecewise smooth on the interval* $-1 < x < 1$, *and suppose that* $f(x)$ *at each point of discontinuity of f in that interval is defined as the mean value of the one-sided limits* $f(x+)$ *and* $f(x-)$. *Then*

(4) $$f(x) = \sum_{n=0}^{\infty} A_n P_n(x) \qquad (-1 < x < 1),$$

where the coefficients A_n *are given by equation* (3).

According to Sec. 73, $P_{2n}(x)$ $(n = 0, 1, 2, \ldots)$ is even and $P_{2n+1}(x)$ $(n = 0, 1, 2, \ldots)$ is odd. That is,

$$P_{2n}(-x) = P_{2n}(x) \qquad \text{and} \qquad P_{2n+1}(-x) = -P_{2n+1}(x).$$

Evidently, then, if the function f in the statement of theorem is *even*, the product $f(x)P_{2n+1}(x)$ is odd and the graph of $y = f(x)P_{2n+1}(x)$ is symmetric with respect to the origin. On the other hand, $f(x)P_{2n}(x)$ is even, and the graph of $y = f(x)P_{2n}(x)$ is symmetric with respect to the y axis. Consequently,

$$\int_{-1}^{1} f(x) P_{2n+1}(x)\, dx = 0 \qquad \text{and} \qquad \int_{-1}^{1} f(x) P_{2n}(x)\, dx = 2\int_{0}^{1} f(x) P_{2n}(x)\, dx.$$

Hence it follows from expression (3) that the coefficients in representation (4) become $A_{2n+1} = 0$ $(n = 0, 1, 2, \ldots)$ and

(5) $$A_{2n} = (4n+1) \int_{0}^{1} f(x) P_{2n}(x)\, dx \qquad (n = 0, 1, 2, \ldots).$$

Thus if we apply the theorem to the even extension of a function f that is piecewise smooth on the interval $0 < x < 1$ and whose value $f(x)$ at each point

---

[†] The proof, which is rather lengthy, can be found in, for example, the book by Kreider, Kuller, Ostberg, and Perkins (1966, pp. 425–432), listed in the Bibliography. A simplified proof of a special case of the theorem appears in the book by Rainville (1971, pp. 177–179), also listed there.

of discontinuity is the mean value of $f(x+)$ and $f(x-)$, we find that

$$(6) \qquad\qquad f(x) = \sum_{n=0}^{\infty} A_{2n} P_{2n}(x) \qquad\qquad (0 < x < 1),$$

where the coefficients A_{2n} have the values (5).

Similarly, if f is an *odd* function, the products $f(x)P_{2n}(x)$ and $f(x)P_{2n+1}(x)$ are odd and even, respectively. It is then easy to show that when the value $f(x)$ of a piecewise smooth function f on $0 < x < 1$ is defined as the mean value of $f(x+)$ and $f(x-)$ at each point of discontinuity, $f(x)$ has the representation

$$(7) \qquad\qquad f(x) = \sum_{n=0}^{\infty} A_{2n+1} P_{2n+1}(x) \qquad\qquad (0 < x < 1),$$

where

$$(8) \qquad\qquad A_{2n+1} = (4n + 3) \int_0^1 f(x) P_{2n+1}(x)\, dx \qquad (n = 0, 1, 2, \ldots).$$

EXAMPLE. Let us expand the function $f(x) = 1$ $(0 < x < 1)$ in a series of type (7), involving Legendre polynomials of odd degree. According to expression (8),

$$A_{2n+1} = (4n + 3) \int_0^1 P_{2n+1}(x)\, dx \qquad (n = 0, 1, 2, \ldots).$$

The integral here is readily evaluated with the aid of the integration formula (Problem 4, Sec. 75)

$$\int_a^1 P_n(x)\, dx = \frac{1}{2n+1}\left[P_{n-1}(a) - P_{n+1}(a)\right] \qquad (n = 1, 2, \ldots),$$

which tells us that

$$(9) \qquad\qquad A_{2n+1} = P_{2n}(0) - P_{2(n+1)}(0).$$

Thus

$$(10) \qquad\qquad 1 = \sum_{n=0}^{\infty} \left[P_{2n}(0) - P_{2n+2}(0)\right] P_{2n+1}(x) \qquad (0 < x < 1).$$

Since [Problem 7(a), Sec. 75]

$$P_{2n}(0) = (-1)^n \frac{(2n)!}{2^{2n}(n!)^2} \qquad (n = 0, 1, 2, \ldots),$$

the coefficients (9) can also be written as

$$(11) \qquad\qquad A_{2n+1} = (-1)^n \left(\frac{4n+3}{2n+2}\right) \frac{(2n)!}{2^{2n}(n!)^2}.$$

This alternative form of representation (10) is then obtained:

$$(12) \qquad 1 = \sum_{n=0}^{\infty} (-1)^n \left(\frac{4n + 3}{2n + 2} \right) \frac{(2n)!}{2^{2n}(n!)^2} P_{2n+1}(x) \qquad (0 < x < 1).$$

PROBLEMS

1. Let F denote the odd extension of the function $f(x) = 1$ $(0 < x < 1)$ to the interval $-1 < x < 1$, where $F(0) = 0$. Also, let g be the function defined by means of the equations

$$g(x) = \begin{cases} 0 & \text{when } -1 < x < 0, \\ 1 & \text{when } 0 < x < 1, \end{cases}$$

and $g(0) = \frac{1}{2}$. Then, by observing that

$$g(x) = \frac{1}{2} + \frac{1}{2}F(x) \qquad\qquad (-1 < x < 1)$$

and referring to expansion (10), Sec. 76, show that

$$g(x) = \frac{1}{2}P_0(x) + \frac{1}{2}\sum_{n=0}^{\infty} [P_{2n}(0) - P_{2n+2}(0)]P_{2n+1}(x) \quad (-1 < x < 1).$$

2. Let f denote the function defined by the equations

$$f(x) = \begin{cases} 0 & \text{when } -1 < x \leqslant 0, \\ x & \text{when } 0 < x < 1. \end{cases}$$

(a) State why $f(x)$ is represented by its Legendre series (4), Sec. 76, at each point of the interval $-1 < x < 1$.

(b) Show that $A_{2n+1} = 0$ $(n = 1, 2, \dots)$ in the series in part (a).

(c) Find the first four nonvanishing terms of the series in part (a) to show that

$$f(x) = \frac{1}{4}P_0(x) + \frac{1}{2}P_1(x) + \frac{5}{16}P_2(x) - \frac{3}{32}P_4(x) + \cdots \quad (-1 < x < 1).$$

3. Verify that, for all x,

(a) $x^2 = \frac{1}{3}P_0(x) + \frac{2}{3}P_2(x)$; (b) $x^3 = \frac{3}{5}P_1(x) + \frac{2}{5}P_3(x)$.

4. Obtain the first three nonzero terms in the series of Legendre polynomials of even degree representing the function $f(x) = x$ $(0 < x < 1)$ to show that

$$x = \frac{1}{2}P_0(x) + \frac{5}{8}P_2(x) - \frac{3}{16}P_4(x) + \cdots \qquad (0 < x < 1).$$

Point out why this expansion remains valid when $x = 0$, and state what function the series represents on the interval $-1 < x < 1$.

5. By applying the corollary in Sec. 16 to the Fourier constants c_n in Sec. 76, state why

$$\lim_{n \to \infty} \sqrt{2n + 1} \int_{-1}^{1} f(x) P_n(x)\, dx = 0$$

when f is piecewise continuous on the interval $-1 < x < 1$.

6. Let f denote a function that is piecewise smooth on the interval $0 < x < 1$, and suppose that $f(x)$ at each point of discontinuity there is the mean value of the one-sided limits $f(x +)$ and $f(x -)$.

(a) By finding the Fourier constants for f with respect to the orthonormal set $\{\sqrt{4n + 1}\, P_{2n}(x)\}$ $(n = 0, 1, 2, \ldots)$ [Problem 3(a), Sec. 75], derive the coefficients (5), Sec. 76, appearing in expansion (6) in that section.

(b) Apply the corollary in Sec. 16 to the Fourier constants in part (a) to show that (compare Problem 5)

$$\lim_{n \to \infty} \sqrt{4n + 1} \int_{0}^{1} f(x) P_{2n}(x)\, dx = 0.$$

7. (a) By recalling that $P_m(x)$ is a polynomial of degree m containing only alternate powers of x (Sec. 73), state why

$$x^m = c P_m(x) + c_{m-2} x^{m-2} + c_{m-4} x^{m-4} + \cdots,$$

where the coefficients are constants. Apply the same argument to x^{m-2}, etc., to conclude that x^m is a finite linear combination of the polynomials $P_m(x), P_{m-2}(x), P_{m-4}(x), \ldots$.

(b) With the aid of the result in part (a), point out why

$$\int_{-1}^{1} P_n(x) p(x)\, dx = 0,$$

where $P_n(x)$ is a Legendre polynomial of degree n $(n = 1, 2, \ldots)$ and $p(x)$ is any polynomial whose degree is less than n.

8. Let n have any one of the values $n = 1, 2, \ldots$.

(a) By recalling the result in Problem 1(a), Sec. 75, state why $P_n(x)$ must change sign at least once in the open interval $-1 < x < 1$. Then let $x_1, x_2, \ldots, x_k$ denote the totality of distinct points in that interval at which $P_n(x)$ changes sign. Since any polynomial of degree n has at most n distinct zeros, we know that $1 \leqslant k \leqslant n$.

(b) *Assume* that the number of points $x_1, x_2, \ldots, x_k$ in part (a) is such that $k < n$, and consider the polynomial

$$p(x) = (x - x_1)(x - x_2) \cdots (x - x_k).$$

Use the result in Problem 7(b) to show that the integral

$$\int_{-1}^{1} P_n(x) p(x)\, dx$$

has value zero; and, after noting that $P_n(x)$ and $p(x)$ change sign at precisely the same points in the interval $-1 < x < 1$, state why the value of the integral cannot be zero. Having reached this contradiction, conclude that $k = n$ and hence that *the zeros of a Legendre polynomial $P_n(x)$ are all real and distinct and lie in the open interval* $-1 < x < 1$.

9. Show in the following way that, for each value of n $(n = 0, 1, 2, \ldots)$, the Legendre function of the second kind $Q_n(x)$ (Sec. 73) and its derivative $Q_n'(x)$ fail to be a pair of continuous functions on the closed interval $-1 \leqslant x \leqslant 1$. Suppose that there is an integer N such that $Q_N(x)$ and $Q_N'(x)$ are continuous on that interval. The functions $Q_N(x)$ and $P_n(x)$ $(n \neq N)$ are, then, eigenfunctions corresponding to different eigenvalues of the Sturm-Liouville problem (2), Sec. 74. Point out how it follows that

$$\int_{-1}^{1} Q_N(x) P_n(x)\, dx = 0 \qquad\qquad (n \neq N),$$

and then use the theorem in Sec. 76 to show that $Q_N(x) = A_N P_N(x)$, where A_N is some constant. This is, however, impossible since $P_N(x)$ and $Q_N(x)$ are linearly independent.

77.   DIRICHLET PROBLEMS IN SPHERICAL REGIONS

For our first application of Legendre series, we shall determine the harmonic function u in the region $r < c$ such that u assumes prescribed values $F(\theta)$ on the spherical surface $r = c$ (Fig. 66). Here r, ϕ, and θ are spherical coordinates, and u is independent of ϕ. Thus u satisfies Laplace's equation (Sec. 4)

$$(1) \qquad\qquad r \frac{\partial^2}{\partial r^2}(ru) + \frac{1}{\sin\theta} \frac{\partial}{\partial\theta}\left(\sin\theta \frac{\partial u}{\partial\theta}\right) = 0 \qquad (r < c, 0 < \theta < \pi)$$

and the condition

$$(2) \qquad\qquad u(c, \theta) = F(\theta) \qquad\qquad (0 < \theta < \pi).$$

The function u and its partial derivatives of the first and second order are to be continuous throughout the interior $(0 \leqslant r < c, 0 \leqslant \theta \leqslant \pi)$ of the sphere.

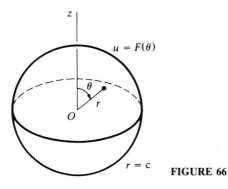

FIGURE 66

Physically, the function u may denote steady temperatures in a solid sphere $r \leqslant c$ whose surface temperatures depend only on θ; that is, the surface temperatures are uniform over each circle $r = c$, $\theta = \theta_0$. Also, u represents electrostatic potential in the space $r < c$, which is free of charges, when $u = F(\theta)$ on the boundary $r = c$.

Consider now a solution of equation (1) of the form $u = R(r)\Theta(\theta)$ that satisfies the stated continuity requirements. Separation of variables show that, for some constant λ,

$$\frac{1}{\sin\theta\,\Theta}\frac{d}{d\theta}\left(\sin\theta\frac{d\Theta}{d\theta}\right) = -\frac{r}{R}\frac{d^2}{dr^2}(rR) = -\lambda.$$

Consequently, R must satisfy the ordinary differential equation

$$(3) \qquad\qquad r\frac{d^2}{dr^2}(rR) - \lambda R = 0 \qquad\qquad (r < c)$$

and be continuous when $0 \leqslant r < c$. Also, for the same constant λ,

$$(4) \qquad\qquad \frac{1}{\sin\theta}\frac{d}{d\theta}\left(\sin\theta\frac{d\Theta}{d\theta}\right) + \lambda\Theta = 0 \qquad (0 < \theta < \pi),$$

where Θ and Θ' are to be continuous on the closed interval $0 \leqslant \theta \leqslant \pi$.

If, in equation (4), we make the substitution $x = \cos\theta$, so that

$$\sin\theta\frac{d\Theta}{d\theta} = (1 - \cos^2\theta)\frac{1}{\sin\theta}\frac{d\Theta}{d\theta} = -(1 - x^2)\frac{d\Theta}{dx},$$

it follows readily that

$$(5) \qquad\qquad \frac{d}{dx}\left[(1 - x^2)\frac{d\Theta}{dx}\right] + \lambda\Theta = 0 \qquad (-1 < x < 1),$$

where Θ and its derivative with respect to x are continuous on the entire closed interval $-1 \leqslant x \leqslant 1$. Equation (5) is Legendre's equation in self-adjoint form; and we know from the theorem in Sec. 74 that λ must be one of the eigenvalues $\lambda_n = n(n + 1)$ $(n = 0, 1, 2, \ldots)$ and that the corresponding eigenfunction is $\Theta_n = P_n(x)$. The functions $\Theta_n(\theta) = P_n(\cos\theta)$ $(n = 0, 1, 2, \ldots)$ thus satisfy equation (4) when $\lambda = n(n + 1)$:

$$(6) \qquad\qquad \frac{1}{\sin\theta}\frac{d}{d\theta}\left[\sin\theta\frac{d}{d\theta}P_n(\cos\theta)\right] + n(n + 1)P_n(\cos\theta) = 0$$

$$(n = 0, 1, 2, \ldots).$$

Writing equation (3) in the form

$$r^2 R'' + 2rR' - \lambda R = 0,$$

we see that it is a Cauchy-Euler equation, which reduces to a differential equation with constant coefficients after the substitution $r = \exp s$ is made (see Problem 3, Sec. 35). When $\lambda = n(n + 1)$, its general solution is

$$R = C_1 r^n + C_2 r^{-n-1},$$

as is easily verified. The continuity of R at $r = 0$ requires that $C_2 = 0$, and so the desired functions of r are $R_n(r) = r^n$ $(n = 0, 1, 2, \ldots)$.

The functions $u_n = r^n P_n(\cos\theta)$ $(n = 0, 1, 2, \ldots)$, therefore, satisfy Laplace's equation (1) and the continuity conditions accompanying it. Formally, their generalized linear combination

$$
(7) \qquad u(r, \theta) = \sum_{n=0}^{\infty} B_n r^n P_n(\cos \theta)
$$

is a solution of our boundary value problem if the constants B_n are such that $u(c, \theta) = F(\theta)$, or

$$
(8) \qquad F(\theta) = \sum_{n=0}^{\infty} B_n c^n P_n(\cos \theta) \qquad\qquad (0 < \theta < \pi).
$$

To find these constants, we introduce the new function

$$
(9) \qquad f(x) = F(\cos^{-1} x) \qquad\qquad (-1 < x < 1),
$$

where principal values of the inverse cosine are taken. Then if we make the substitutions $A_n = B_n c^n$ and $\theta = \cos^{-1} x$ in equation (8), that equation becomes

$$
(10) \qquad f(x) = \sum_{n=0}^{\infty} A_n P_n(x) \qquad\qquad (-1 < x < 1);
$$

and, according to the theorem in Sec. 76,

$$
(11) \qquad A_n = \frac{2n+1}{2} \int_{-1}^{1} f(x) P_n(x)\, dx \qquad (n = 0, 1, 2, \dots).
$$

We assume that f is piecewise smooth on the interval $-1 < x < 1$, so that expansion (10) is valid for each point x at which f is continuous.

In view of definition (9) of the function $f(x)$, the substitution $x = \cos \theta$ in integral (11) enables us to write A_n in terms of the original function $F(\theta)$:

$$
(12) \qquad A_n = \frac{2n+1}{2} \int_{0}^{\pi} F(\theta) P_n(\cos \theta) \sin \theta\, d\theta \qquad (n = 0, 1, 2, \dots).
$$

Since $B_n = A_n / c^n$, the required values of the constants in expression (7) are thus obtained; and the formal solution of our Dirichlet problem can be written in terms of the coefficients (12) as

$$
(13) \qquad u(r, \theta) = \sum_{n=0}^{\infty} A_n \left(\frac{r}{c}\right)^n P_n(\cos \theta) \qquad\qquad (r \leqslant c).
$$

We note that *the harmonic function v in the unbounded region $r > c$,* exterior to the spherical surface $r = c$, which assumes the values $F(\theta)$ on that surface and which is bounded as $r \to \infty$ can be found in like manner. Here $C_1 = 0$ in our solution $R = C_1 r^n + C_2 r^{-n-1}$ of equation (3) if R is to remain bounded as $r \to \infty$; and the solutions of equation (1) are $v_n = r^{-n-1} P_n(\cos \theta)$ $(n = 0, 1, 2, \dots)$. Thus

$$
(14) \qquad v(r, \theta) = \sum_{n=0}^{\infty} \frac{B_n}{r^{n+1}} P_n(\cos \theta) \qquad\qquad (r \geqslant c),
$$

where the B_n are this time related to the coefficients (12) by means of the equation $A_n = B_n/c^{n+1}$. That is,

$$(15) \qquad v(r, \theta) = \sum_{n=0}^{\infty} A_n \left(\frac{c}{r}\right)^{n+1} P_n(\cos \theta) \qquad (r \geqslant c).$$

78. STEADY TEMPERATURES IN A HEMISPHERE

The base $r < 1$, $\theta = \pi/2$ of a solid hemisphere $r \leqslant 1$, $0 \leqslant \theta \leqslant \pi/2$, part of which is shown in Fig. 67, is insulated. The flux of heat inward through the hemispherical surface is kept at prescribed values $F(\theta)$. In order that temperatures be steady, those values are such that the resultant rate of flow through the hemispherical surface is zero. That is, F satisfies the condition

$$\int_0^{\pi/2} F(\theta) 2\pi \sin \theta \, d\theta = 0,$$

which, in terms of the function

$$(1) \qquad f(x) = F(\cos^{-1} x) \qquad (0 < x < 1),$$

can also be written

$$(2) \qquad \int_0^1 f(x) \, dx = 0.$$

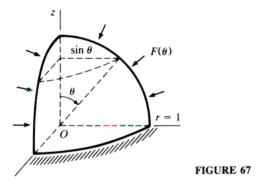

FIGURE 67

  If u denotes temperatures as a function of r and θ, then the condition that the base be insulated is, according to Problem 13, Sec. 4,

$$(3) \qquad u_\theta\left(r, \frac{\pi}{2}\right) = 0 \qquad (0 < r < 1).$$

The boundary value problem in $u(r, \theta)$ consists of Laplace's equation

(4) $$r\frac{\partial^2}{\partial r^2}(ru) + \frac{1}{\sin \theta}\frac{\partial}{\partial \theta}\left(\sin \theta \frac{\partial u}{\partial \theta}\right) = 0 \qquad \left(r < 1, 0 < \theta < \frac{\pi}{2}\right),$$

condition (3), and the flux condition (see Sec. 3)

(5) $$Ku_r(1, \theta) = F(\theta) \qquad \left(0 < \theta < \frac{\pi}{2}\right),$$

where K is thermal conductivity. We assume that the function f, defined by equation (1), is piecewise smooth on the interval $0 < x < 1$. Also, u must satisfy the usual continuity conditions when $0 \leqslant r < 1$ and $0 \leqslant \theta \leqslant \pi/2$.

Writing $u = R(r)\Theta(\theta)$ and separating variables in equations (3) and (4), we obtain the conditions

(6) $$r(rR)'' - \lambda R = 0 \qquad (r < 1),$$

where R must be continuous when $0 \leqslant r < 1$, and

(7) $$\frac{1}{\sin \theta}\frac{d}{d\theta}\left(\sin \theta \frac{d\Theta}{d\theta}\right) + \lambda \Theta = 0, \qquad \Theta'\left(\frac{\pi}{2}\right) = 0 \qquad \left(0 < \theta < \frac{\pi}{2}\right),$$

where Θ and Θ' are to be continuous when $0 \leqslant \theta \leqslant \pi/2$.

The substitution $x = \cos \theta$ transforms equations (7) into the singular Sturm-Liouville problem consisting of Legendre's equation (see Sec. 77)

$$\frac{d}{dx}\left[(1 - x^2)\frac{d\Theta}{dx}\right] + \lambda \Theta = 0 \qquad (0 < x < 1)$$

and the condition that

$$\frac{d\Theta}{dx} = 0 \qquad\qquad \text{when } x = 0,$$

where Θ and $d\Theta/dx$ are to be continuous when $0 \leqslant x \leqslant 1$. According to the corollary in Sec. 74, this problem has eigenvalues $\lambda_n = 2n(2n + 1)$ $(n = 0, 1, 2, \ldots)$ and eigenfunctions $\Theta_n = P_{2n}(x)$; hence $\Theta_n(\theta) = P_{2n}(\cos \theta)$. The corresponding bounded solution of the Cauchy-Euler equation (6) is $R_n = r^{2n}$.

Formally, then,

$$u(r, \theta) = \sum_{n=0}^{\infty} B_n r^{2n} P_{2n}(\cos \theta)$$

if the constants B_n are such that condition (5) is satisfied. That condition requires that

(8) $$\sum_{n=1}^{\infty} 2KnB_n P_{2n}(x) = f(x) \qquad (0 < x < 1),$$

where $x = \cos \theta$. This is the representation for $f(x)$ on the interval $0 < x < 1$ in a series of Legendre polynomials of even degree (Sec. 76) if $2KnB_n = A_{2n}$,

where

(9) $$A_{2n} = (4n + 1) \int_0^1 f(x) P_{2n}(x) \, dx \qquad (n = 1, 2, \ldots),$$

and if f is such that the condition $A_0 = 0$, which is precisely condition (2), is satisfied. Thus B_0 is left arbitrary; and

(10) $$u(r, \theta) = B_0 + \frac{1}{2K} \sum_{n=1}^{\infty} \frac{1}{n} A_{2n} r^{2n} P_{2n}(\cos \theta) \qquad \left(r \leqslant 1, 0 \leqslant \theta \leqslant \frac{\pi}{2} \right),$$

where the coefficients A_{2n} have the values (9).

The constant B_0 is the temperature at the origin $r = 0$. Solutions of such problems with just Neumann conditions (Sec. 7) are determined only up to such an arbitrary additive constant because all the boundary conditions prescribe only values of derivatives of the harmonic functions.

PROBLEMS

1. Suppose that u is harmonic throughout the regions $r < c$ and $r > c$, that $u \to 0$ as $r \to \infty$, and that $u = 1$ on the spherical surface $r = c$. Show from results found in Sec. 77 that $u = 1$ when $r \leqslant c$ and $u = c/r$ when $r \geqslant c$.

2. Suppose that, for all ϕ, the steady temperatures $u(r, \theta)$ in a solid sphere $r \leqslant 1$ are such that $u(1, \theta) = F(\theta)$, where

$$F(\theta) = \begin{cases} 1 & \text{when } 0 < \theta < \dfrac{\pi}{2}, \\ 0 & \text{when } \dfrac{\pi}{2} < \theta < \pi. \end{cases}$$

Derive the expression

$$u(r, \theta) = \frac{1}{2} + \frac{1}{2} \sum_{n=0}^{\infty} [P_{2n}(0) - P_{2n+2}(0)] r^{2n+1} P_{2n+1}(\cos \theta)$$

for those temperatures.

3. The base $r < 1$, $\theta = \pi/2$ of a solid hemisphere $r \leqslant 1$, $0 \leqslant \theta \leqslant \pi/2$ is kept at temperature $u = 0$, while $u = 1$ on the hemispherical surface $r = 1$, $0 < \theta < \pi/2$. Derive the expression

$$u(r, \theta) = \sum_{n=0}^{\infty} (-1)^n \left(\frac{4n + 3}{2n + 2} \right) \frac{(2n)!}{2^{2n}(n!)^2} r^{2n+1} P_{2n+1}(\cos \theta)$$

for the steady temperatures in that solid.

4. The base $r < c$, $\theta = \pi/2$ of a solid hemisphere $r \leqslant c$, $0 \leqslant \theta \leqslant \pi/2$ is insulated. The temperature distribution on the hemispherical surface is $u = F(\theta)$. Derive the expression

$$u(r, \theta) = \sum_{n=0}^{\infty} (4n + 1) \left(\frac{r}{c} \right)^{2n} P_{2n}(\cos \theta) \int_0^1 f(s) P_{2n}(s) \, ds,$$

where $f(x) = F(\cos^{-1} x)$ $(0 < x < 1)$, for the steady temperatures in the solid. Also, show that $u(r, \theta) = 1$ when $F(\theta) = 1$.

5. A function u is harmonic and bounded in the unbounded region $r > c$, $0 \leqslant \phi \leqslant 2\pi$, $0 \leqslant \theta < \pi/2$. Also, $u = 0$ everywhere on the flat boundary surface $r > c$, $\theta = \pi/2$ and $u = F(\theta)$ on the hemispherical boundary surface $r = c$, $0 < \theta < \pi/2$. Derive the expression

$$ u(r, \theta) = \sum_{n=0}^{\infty} (4n + 3) \left(\frac{c}{r}\right)^{2n+2} P_{2n+1}(\cos\theta) \int_0^1 f(s) P_{2n+1}(s)\, ds, $$

where $f(x) = F(\cos^{-1} x)$ $(0 < x < 1)$.

6. The flux of heat $Ku_r(1, \theta)$ into a solid sphere at its surface $r = 1$ is a prescribed function $F(\theta)$, where F is such that the net time rate of flow of heat into the solid is zero. Thus (see Sec. 78)

$$ \int_{-1}^{1} f(x)\, dx = 0, $$

where $f(x) = F(\cos^{-1} x)$ $(-1 < x < 1)$. Assuming that $u = 0$ at the center $r = 0$, derive the expression

$$ u(r, \theta) = \frac{1}{2K} \sum_{n=1}^{\infty} \frac{2n + 1}{n} r^n P_n(\cos\theta) \int_{-1}^{1} f(s) P_n(s)\, ds $$

for the steady temperatures throughout the entire sphere $0 \leqslant r \leqslant 1$.

7. Let $u(r, \theta)$ denote steady temperatures in a hollow sphere $a \leqslant r \leqslant b$ when

$$ u(a, \theta) = F(\theta) \qquad \text{and} \qquad u(b, \theta) = 0 \qquad (0 < \theta < \pi). $$

Derive the expression

$$ u(r, \theta) = \sum_{n=0}^{\infty} A_n \frac{b^{2n+1} - r^{2n+1}}{b^{2n+1} - a^{2n+1}} \left(\frac{a}{r}\right)^{n+1} P_n(\cos\theta), $$

where

$$ A_n = \frac{2n + 1}{2} \int_0^{\pi} F(\theta) P_n(\cos\theta) \sin\theta\, d\theta \qquad (n = 0, 1, 2, \dots). $$

8. Let $u(x, t)$ represent the temperatures in a nonhomogeneous insulated bar $-1 \leqslant x \leqslant 1$ along the x axis, and suppose that the thermal conductivity is proportional to $1 - x^2$. The heat equation takes the form

$$ \frac{\partial u}{\partial t} = b \frac{\partial}{\partial x}\left[(1 - x^2)\frac{\partial u}{\partial x}\right] \qquad (b > 0). $$

Here b is constant since we assume that the product of the physical constants σ and δ used in Sec. 2 and in Problem 8, Sec. 4, is constant. Note that the ends $x = \pm 1$ are insulated because the conductivity vanishes there. Assuming that $u(x, 0) = f(x)$ $(-1 < x < 1)$, derive the expression

$$ u(x, t) = \sum_{n=0}^{\infty} \frac{2n + 1}{2} \exp[-n(n + 1)bt] P_n(x) \int_{-1}^{1} f(s) P_n(s)\, ds. $$

9. Show that if $f(x) = x^2$ $(-1 < x < 1)$ in Problem 8, then

$$u(x,t) = \frac{1}{3} + \left(x^2 - \frac{1}{3}\right)\exp(-6bt).$$

10. Heat is generated at a steady and uniform rate throughout a solid hemisphere $0 \leqslant r \leqslant 1$, $0 < \theta \leqslant \pi/2$, and the entire surface is kept at temperature zero. Thus the steady temperatures $u = u(r, \theta)$ satisfy the nonhomogeneous differential equation

$$\frac{1}{r}\frac{\partial^2}{\partial r^2}(ru) + \frac{1}{r^2 \sin \theta}\frac{\partial}{\partial \theta}\left(\sin \theta \frac{\partial u}{\partial \theta}\right) + q_0 = 0 \qquad \left(0 < r < 1, 0 < \theta < \frac{\pi}{2}\right)$$

and the boundary conditions $u(1, \theta) = 0$, $u(r, \pi/2) = 0$. Also, $u(r, \theta)$ is continuous at $r = 0$. Point out how Problem 3 suggests seeking a solution of the form

$$u(r, \theta) = \sum_{n=0}^{\infty} B_n(r) P_{2n+1}(\cos \theta)$$

and applying the method of variation of parameters, which was first used in Sec. 33. Follow the steps below to find the solution by that method.

(*a*) Observe how it follows immediately from equation (6), Sec. 77, that

$$\frac{1}{\sin \theta}\frac{d}{d\theta}\left[\sin \theta \frac{d}{d\theta}P_{2n+1}(\cos \theta)\right] = -(2n+1)(2n+2)P_{2n+1}(\cos \theta)$$

$$(n = 0, 1, 2, \dots).$$

Then, with the aid of this identity and expansion (10), Sec. 76, obtain the initial value problem

$$r^2 B_n''(r) + 2rB_n'(r) - (2n+1)(2n+2)B_n(r) = -q_0 A_{2n+1}r^2, \quad B_n(1) = 0$$

$$(n = 0, 1, 2, \dots),$$

where $A_{2n+1} = P_{2n}(0) - P_{2n+2}(0)$ and where $B_n(r)$ is to be continuous on the interval $0 \leqslant r \leqslant 1$.

(*b*) Solve the differential equation in part (*a*) by adding a particular solution of it to the general solution of the complementary equation (compare Problem 13, Sec. 38). Then apply the required conditions on $B_n(r)$, stated in part (*a*), to complete the solution of the initial value problem in ordinary differential equations there. Thus arrive at the desired temperature function:

$$u(r, \theta) = \frac{q_0}{2}\sum_{n=0}^{\infty} \frac{P_{2n}(0) - P_{2n+2}(0)}{(2n-1)(n+2)}(r^2 - r^{2n+1})P_{2n+1}(\cos \theta).$$

Suggestion: Observe that the differential equation in part (*a*) has a particular solution of the form $B_n(r) = ar^2$, where a is a constant. Also, note that the complementary equation in part (*b*) is of Cauchy-Euler type, and solve it by the method described in Problem 3, Sec. 35.

11. Give a physical interpretation of the following boundary value problem in spherical coordinates for a harmonic function $u(r, \theta)$:

$$r \frac{\partial^2}{\partial r^2}(ru) + \frac{1}{\sin \theta} \frac{\partial}{\partial \theta}\left(\sin \theta \frac{\partial u}{\partial \theta}\right) = 0 \quad (1 < r < b, \theta_1 < \theta < \theta_2),$$

$$u(1, \theta) = 0, \qquad u(b, \theta) = 0,$$

$$u(r, \theta_1) = f(r), \qquad u(r, \theta_2) = 0,$$

where $0 < \theta_1 < \theta_2 < \pi$. Then, using the normalized eigenfunctions found in Problem 11, Sec. 45, and the functions p_α and q_α in Problem 9, Sec. 75, derive the expression

$$u(r, \theta) = \frac{1}{\sqrt{r}} \sum_{n=1}^{\infty} B_n \frac{F_n(\theta)}{F_n(\theta_1)} \sin(\alpha_n \ln r),$$

where

$$\alpha_n = \frac{n\pi}{\ln b}, \qquad B_n = \frac{2}{\ln b} \int_1^b \frac{f(r)}{\sqrt{r}} \sin(\alpha_n \ln r) \, dr,$$

and

$$F_n(\theta) = p_{\alpha_n}(\cos \theta) q_{\alpha_n}(\cos \theta_2) - p_{\alpha_n}(\cos \theta_2) q_{\alpha_n}(\cos \theta).$$

UNIQUENESS
OF SOLUTIONS

In this chapter, we examine in greater detail the question of verifying solutions of boundary value problems of certain types and, more particularly, the question of establishing that a solution of a given problem is the only possible solution. A multiplicity of solutions may actually arise when the statement of the problem does not demand adequate continuity or boundedness of a solution and its derivatives. This was illustrated in Problem 16, Sec. 58.

The theorem in Sec. 79 below enables us to establish uniform convergence of solutions obtained in the form of series and is useful in both verifying a solution and proving that it is unique. The remaining theorems in the chapter give conditions under which a solution is unique. They apply only to specific types of problems, and their applications are further limited because they require a rather high degree of regularity of the functions involved.

79. ABEL'S TEST FOR UNIFORM
CONVERGENCE

We begin with some needed background on uniform convergence. Let $s_n(x)$ denote the sum of the first n terms of a series of functions $X_i(x)$ which converges to the sum $s(x)$:

$$(1) \qquad s_n(x) = \sum_{i=1}^{n} X_i(x), \qquad s(x) = \lim_{n \to \infty} s_n(x).$$

Suppose that the series converges uniformly with respect to x for all x in some interval. Then (see Sec. 22), for each positive number ε, there exists a positive integer n_ε, independent of x, such that

$$|s(x) - s_n(x)| < \frac{\varepsilon}{2} \qquad\qquad \text{whenever } n > n_\varepsilon$$

for every x in the interval. Let j denote any positive integer. Then

$$|s_{n+j} - s_n| = |s_{n+j} - s + s - s_n| \leqslant |s - s_{n+j}| + |s - s_n| < \varepsilon,$$

provided $n > n_\varepsilon$. Thus *a necessary condition for uniform convergence of the series is that, for all positive integers* j,

(2) $$|s_{n+j}(x) - s_n(x)| < \varepsilon \qquad\qquad \text{whenever } n > n_\varepsilon.$$

Condition (2) is also a *sufficient* condition, known as the *Cauchy criterion*, for convergence of the series for each fixed x even if n_ε is not independent of x. Hence it implies that the sum $s(x)$ exists. Then, for any fixed n and x and for the given number ε, a positive integer $j_\varepsilon(x)$ exists such that

(3) $$|s(x) - s_{n+j}(x)| < \varepsilon \qquad\qquad \text{whenever } j > j_\varepsilon(x).$$

To show that condition (2) is sufficient for *uniform* convergence, let n denote any fixed integer greater than n_ε, where n_ε corresponds to the given number ε in the sense that condition (2) is satisfied for all x. Then, for each fixed x, condition (3) is satisfied when $j > j_\varepsilon(x)$; and, since

$$|s - s_n| = |s - s_{n+j} + s_{n+j} - s_n| \leqslant |s - s_{n+j}| + |s_{n+j} - s_n|,$$

it follows from conditions (3) and (2) that

(4) $$|s(x) - s_n(x)| < 2\varepsilon$$

when $j > j_\varepsilon(x)$ and $n > n_\varepsilon$. Thus $|s(x) - s_n(x)|$, which is independent of j, is arbitrarily small for each x when $n > n_\varepsilon$. Since n_ε is independent of x, uniform convergence is now established.

Note that x here may equally well denote elements $(x_1, x_2, \ldots, x_N)$ of some set in N-dimensional space. The uniform convergence is then with respect to all N variables $x_1, x_2, \ldots, x_N$ together.

We now derive a test for the uniform convergence of infinite series whose terms are products of certain types of functions. Its application in verifying formal solutions of boundary value problems was illustrated in Sec. 28. The test, known as *Abel's test*, involves functions in a sequence $T_i(t)$ $(i = 1, 2, \ldots)$ which is *uniformly bounded* for all points t in an interval. That is, there exists a constant M, independent of i, such that

(5) $$|T_i(t)| < M \qquad\qquad (i = 1, 2, \ldots)$$

for all t in the interval. The sequence is, moreover, *monotonic with respect to* i. Thus, for every t in the interval, either

(6) $$T_{i+1}(t) \leqslant T_i(t) \qquad\qquad (i = 1, 2, \ldots)$$

or

(7) $$T_{i+1}(t) \geqslant T_i(t) \qquad\qquad (i = 1, 2, \ldots).$$

We state the test as a theorem which shows that when the terms of a uniformly convergent series are multiplied by functions $T_i(t)$ of the type just described, the new series is also uniformly convergent.

Theorem. *The series*

(8) $$\sum_{i=1}^{\infty} X_i(x)T_i(t)$$

converges uniformly with respect to the two variables x and t together in a region R of the xt plane if the series

$$\sum_{i=1}^{\infty} X_i(x)$$

converges uniformly with respect to x for all x such that (x, t) is in R and if the functions $T_i(t)$ are uniformly bounded and monotonic with respect to i $(i = 1, 2, \ldots)$ for all t such that (x, t) is in R.

To start the proof, we let S_n denote partial sums of series (8):

$$S_n(x, t) = \sum_{i=1}^{n} X_i(x)T_i(t).$$

As indicated above, the uniform convergence of that series will be established if we prove that to each positive number ε there corresponds a positive integer n_ε, independent of x and t, such that

$$|S_m(x, t) - S_n(x, t)| < \varepsilon \qquad\qquad \text{whenever } n > n_\varepsilon,$$

for all integers $m = n + 1, n + 2, \ldots$ and for all points (x, t) in R.

We write the partial sum

$$s_n(x) = X_1(x) + X_2(x) + \cdots + X_n(x).$$

Then, for each pair of integers m and n $(m > n)$, $S_m - S_n$ can be written

$$X_{n+1}T_{n+1} + X_{n+2}T_{n+2} + \cdots + X_m T_m$$

$$= (s_{n+1} - s_n)T_{n+1} + (s_{n+2} - s_{n+1})T_{n+2} + \cdots + (s_m - s_{m-1})T_m$$

$$= (s_{n+1} - s_n)T_{n+1} + (s_{n+2} - s_n)T_{n+2} - (s_{n+1} - s_n)T_{n+2}$$

$$+ \cdots + (s_m - s_n)T_m - (s_{m-1} - s_n)T_m.$$

By pairing alternate terms here, we find that

(9) $$S_m - S_n = (s_{n+1} - s_n)(T_{n+1} - T_{n+2}) + (s_{n+2} - s_n)(T_{n+2} - T_{n+3})$$

$$+ \cdots + (s_{m-1} - s_n)(T_{m-1} - T_m) + (s_m - s_n)T_m.$$

Suppose now that the functions T_i are nonincreasing with respect to i, so that they satisfy condition (6), and that they also satisfy the uniform boundedness condition (5). Then the factors $T_{n+1} - T_{n+2}, T_{n+2} - T_{n+3}$, etc., in equation (9) are nonnegative, and $|T_i(t)| < M$. Since the series with terms $X_i(x)$ converges uniformly, an integer n_ε exists such that

$$|s_{n+j}(x) - s_n(x)| < \frac{\varepsilon}{3M} \qquad \text{whenever } n > n_\varepsilon,$$

for all positive integers j, where ε is any given positive number and n_ε is independent of x. Then if $n > n_\varepsilon$ and $m > n$, it follows from equation (9) that

$$|S_m - S_n| < \frac{\varepsilon}{3M}[(T_{n+1} - T_{n+2}) + (T_{n+2} - T_{n+3}) + \cdots + |T_m|]$$

$$= \frac{\varepsilon}{3M}(T_{n+1} - T_m + |T_m|) \leqslant \frac{\varepsilon}{3M}(|T_{n+1}| + 2|T_m|).$$

Therefore,

$$|S_m(x, t) - S_n(x, t)| < \varepsilon \qquad \text{whenever } m > n > n_\varepsilon;$$

and the uniform convergence of series (8) is established.

The proof is similar when the functions T_i are nondecreasing with respect to i.

When x is kept fixed, the series with terms X_i is a series of constants; and the only requirement placed on that series is that it be convergent. Then the theorem shows that when T_i are bounded and monotonic, the series of terms $X_i T_i(t)$ is uniformly convergent with respect to t.

Extensions of the theorem to cases in which X_i are functions of x and t, or both X_i and T_i are functions of several variables, become evident when it is observed that our proof rests on the uniform convergence of the series of terms X_i and the bounded monotonic nature of the functions T_i.

80. UNIQUENESS OF SOLUTIONS OF THE HEAT EQUATION

Let D denote the domain consisting of all points interior to a closed surface S; and let $\overline{D}$ be the closure of that domain, consisting of all points in D and all points on S. We assume always that the closed surface S is *piecewise smooth*. That is, it is a continuous surface consisting of a finite number of parts over each of which the outward unit normal vector exists and varies continuously from point to point. Then if U is a function of x, y, and z which is continuous in $\overline{D}$, together with its partial derivatives of the first and second order, a special case of Green's identity that we shall need here states that

(1)
$$\iint_S U \frac{dU}{dn} \, dA = \iiint_D \left(U \, \nabla^2 U + U_x^2 + U_y^2 + U_z^2 \right) dV.$$

Here dA is the area element on S, dV represents $dx\,dy\,dz$, and dU/dn is the derivative in the direction of the outward unit normal to S.[†]

Consider a homogeneous solid whose interior is the domain D and whose temperatures at time t are denoted by $u(x, y, z, t)$. A fairly general problem in heat conduction is the following:

$$(2) \qquad u_t = k\nabla^2 u + q(x, y, z, t) \qquad\qquad [(x, y, z) \text{ in } D, t > 0],$$

$$(3) \qquad u(x, y, z, 0) = f(x, y, z) \qquad\qquad\qquad [(x, y, z) \text{ in } \overline{D}],$$

$$(4) \qquad u = g(x, y, z, t) \qquad\qquad\qquad [(x, y, z) \text{ on } S, t \geqslant 0].$$

This is the problem of determining temperatures in a body, with prescribed initial temperatures $f(x, y, z)$ and surface temperatures $g(x, y, z, t)$, interior to which heat may be generated continuously at a rate per unit volume proportional to $q(x, y, z, t)$.

Suppose that the problem has two solutions

$$u = u_1(x, y, z, t), \qquad u = u_2(x, y, z, t),$$

where both u_1 and u_2 are continuous functions in the closed region $\overline{D}$ when $t \geqslant 0$, while their derivatives of the first order with respect to t and of the first and second orders with respect to x, y, and z are continuous in $\overline{D}$ when $t > 0$. Since u_1 and u_2 satisfy the linear equations (2), (3), and (4), their difference

$$U(x, y, z, t) = u_1(x, y, z, t) - u_2(x, y, z, t)$$

satisfies the homogeneous problem

$$(5) \qquad\qquad U_t = k\nabla^2 U \qquad\qquad [(x, y, z) \text{ in } D, t > 0],$$

$$(6) \qquad\qquad U(x, y, z, 0) = 0 \qquad\qquad [(x, y, z) \text{ in } \overline{D}],$$

$$(7) \qquad\qquad U = 0 \qquad\qquad [(x, y, z) \text{ on } S, t \geqslant 0].$$

Moreover, U and its derivatives have the continuity properties of u_1 and u_2 assumed above.

We shall now show that $U = 0$ in D when $t > 0$, so that the two solutions u_1 and u_2 are identical. That is, not more than one solution of the boundary value problem in u can exist if the solution is required to satisfy the stated continuity conditions.

The continuity of U with respect to x, y, z, and t together in the closed region $\overline{D}$ when $t \geqslant 0$ implies that the integral

$$(8) \qquad\qquad I(t) = \frac{1}{2}\iiint_D [U(x, y, z, t)]^2 \, dV$$

is a continuous function of t when $t \geqslant 0$; and, according to equation (6), $I(0) = 0$. Also, in view of the continuity of U_t when $t > 0$, we may use equation

---

[†] Identity (1) is found by applying Gauss's divergence theorem to the vector field U grad U. See the book by Taylor and Mann (1983, pp. 492–493), listed in the Bibliography.

(5) to write

$$I'(t) = \iiint_D UU_t \, dV = k \iiint_D U \, \nabla^2 U \, dV \qquad (t > 0).$$

Identity (1) applies to the last integral here because of the continuity of the derivatives of U when $t > 0$. Thus

(9) $$\iiint_D U \, \nabla^2 U \, dV = \iint_S U \frac{du}{dn} \, dA - \iiint_D \left(U_x^2 + U_y^2 + U_z^2 \right) dV$$

when $t > 0$. But $U = 0$ on S, and $k > 0$; consequently,

$$I'(t) = -k \iiint_D \left(U_x^2 + U_y^2 + U_z^2 \right) dV \leqslant 0.$$

The mean value theorem for derivatives applies to $I(t)$. That is, for each positive t, a number t_1 $(0 < t_1 < t)$ exists such that

$$I(t) - I(0) = tI'(t_1);$$

and, since $I(0) = 0$ and $I'(t_1) \leqslant 0$, it follows that $I(t) \leqslant 0$. However, definition (8) of the integral shows that $I(t) \geqslant 0$. Therefore,

$$I(t) = 0 \qquad (t \geqslant 0);$$

and so the nonnegative integrand U^2 cannot have a positive value at any point in D. For if it did, the continuity of U^2 would require that U^2 be positive throughout some neighborhood of the point, and that would mean $I(t) > 0$. Consequently,

$$U(x, y, z, t) = 0 \qquad \left[(x, y, z) \text{ in } \overline{D}, t \geqslant 0 \right];$$

and we arrive at the following theorem on uniqueness.

Theorem 1. Let u satisfy these conditions of regularity: (a) It is a continuous function of the variables x, y, z, and t together when the point (x, y, z) is in the closed region $\overline{D}$ and $t \geqslant 0$; (b) the derivatives of u appearing in the heat equation (2) are continuous in the same sense when $t > 0$. Then if u is a solution of the boundary value problem (2)–(4), it is the only possible solution satisfying conditions (a) and (b).

When conditions (a) and (b) in Theorem 1 are added to the requirement that u is to satisfy the heat equation and the boundary conditions, our boundary value problem is completely stated, provided it has a solution; for that will be the only possible solution.

The condition that u be continuous in $\overline{D}$ when $t = 0$ restricts the usefulness of our theorem. It is clearly not satisfied if the initial temperature function f in condition (3) fails to be continuous throughout $\overline{D}$, or if at some point on S the initial value $g(x, y, z, 0)$ of the prescribed surface temperature differs from the value $f(x, y, z)$. The continuity requirement at $t = 0$ can be relaxed in some cases.[†]

The proof of Theorem 1 required that the integral

$$\iint_S U \frac{dU}{dn} \, dA$$

in equation (9) either vanish or have a negative value. It vanished because $U = 0$ on S. But it is never positive if condition (4) is replaced by the boundary condition

(10) $$\frac{du}{dn} + hu = g(x, y, z, t) \quad [(x, y, z) \text{ on } S, t > 0],$$

where $h \geqslant 0$. For, in that case, $dU/dn = -hU$ on S; and $U \, dU/dn \leqslant 0$. Thus our theorem can be modified as follows.

Theorem 2. The conclusion in Theorem 1 is true if boundary condition (4) is replaced by condition (10), or if condition (4) is satisfied on part of the surface S and condition (10) is satisfied on the rest.

EXAMPLE. In the problem of temperature distribution in a slab with insulated faces $x = 0$ and $x = c$ and initial temperatures $f(x)$ (Secs. 27 and 28), write $c = \pi$ and assume that f is continuous and f' is piecewise continuous on the interval $0 \leqslant x \leqslant \pi$. Then the Fourier cosine series for f converges uniformly to $f(x)$ on that interval (Sec. 22). Let $u(x, t)$ denote the sum of the series

(11) $$\frac{a_0}{2} + \sum_{n=1}^{\infty} a_n e^{-n^2 kt} \cos nx \quad (0 \leqslant x \leqslant \pi, t \geqslant 0),$$

obtained as a formal solution of the problem, a_0 and a_n $(n = 1, 2, \dots)$ being the coefficients in the Fourier cosine series for f.

As noted in Sec. 27, this temperature problem for a slab is the same as the problem for a bar of uniform cross section whose bases, in the planes $x = 0$ and $x = \pi$, and whose lateral surface, parallel to the x axis, are all insulated $(du/dn = 0)$. Let the domain D consist of the interior points of the bar.

We can see from Abel's test (Sec. 79) that series (11) converges uniformly with respect to x and t together in the region $0 \leqslant x \leqslant \pi, t \geqslant 0$ of the xt plane; thus u is continuous there. When $t \geqslant t_0$, where t_0 is any positive number, the series obtained by differentiating series (11) term by term any number of times with respect to x or t is uniformly convergent, according to the Weierstrass M-test (Sec. 22). Consequently, we now know that u satisfies all the equations in

---

[†] Integral transforms can sometimes be used to prove uniqueness of solutions of certain types of boundary value problems. This is illustrated in the book by Churchill (1972, sec. 79), listed in the Bibliography.

the boundary value problem (compare Sec. 28) and also that u_t, u_x, and u_{xx} are continuous functions in the region $0 \leqslant x \leqslant \pi$, $t > 0$. Thus u satisfies the regularity conditions (a) and (b) stated in Theorem 1, and Theorem 2 applies to show that the sum $u(x, t)$ of series (11) is the only solution that satisfies those conditions.

81. SOLUTIONS OF LAPLACE'S OR POISSON'S EQUATION

Let U be a harmonic function in a domain D of three-dimensional space bounded by a continuous closed surface S that is piecewise smooth. Assume also that U and its partial derivatives of the first order are continuous in the closure $\bar{D}$ of the domain. Then, since

$$(1) \qquad \nabla^2 U(x, y, z) = 0 \qquad [(x, y, z) \text{ in } D],$$

Green's identity (1), Sec. 80, becomes

$$(2) \qquad \iint_S U \frac{dU}{dn} \, dA = \iiint_D \left(U_x^2 + U_y^2 + U_z^2 \right) dV.$$

This equation is valid for our function U even though we have required the derivatives of the second order to be continuous only in D, and not in the closed region $\bar{D}$. It is not difficult to prove that, because $\nabla^2 U = 0$, modification of the usual conditions in the divergence theorem from which Green's identity follows is possible.[†]

Suppose that $U = 0$ at all points on the surface S. Then the first integral in equation (2), and therefore the second, vanishes. But the integrand of the second is nonnegative and continuous in $\bar{D}$. Hence it must vanish throughout $\bar{D}$; that is,

$$(3) \qquad U_x = U_y = U_z = 0 \qquad [(x, y, z) \text{ in } \bar{D}].$$

Consequently, $U(x, y, z)$ is constant; but it is zero on S and continuous in $\bar{D}$, and so $U = 0$ throughout $\bar{D}$.

Suppose that dU/dn, instead of U, vanishes on S; or, to make the condition more general, suppose that

$$(4) \qquad \frac{dU}{dn} + hU = 0 \qquad [(x, y, z) \text{ on } S],$$

where $h \geqslant 0$ and h is either a constant or a function of x, y, and z. Then

$$U \frac{dU}{dn} = -hU^2 \leqslant 0$$

on S, so that the first integral in equation (2) is less than or equal to zero. But

---

[†] See the book by Kellogg (1953, p. 119), listed in the Bibliography.

the second integral is greater than or equal to zero. Hence it must vanish; and, again, conditions (3) follow. So U is constant in $\overline{D}$.

If U vanishes over part of S and satisfies condition (4) on the rest of that surface, our argument still shows that U is constant in $\overline{D}$. In such a case, the constant must be zero.

Now let u denote a function which is continuous in $\overline{D}$, together with its partial derivatives of the first order. Suppose that it also has continuous derivatives of the second order in D and satisfies these conditions:

$$(5) \qquad \nabla^2 u(x, y, z) = f(x, y, z) \qquad [(x, y, z) \text{ in } D],$$

$$(6) \qquad p\frac{du}{dn} + hu = g \qquad [(x, y, z) \text{ on } S].$$

Here f, p, h, and g denote prescribed constants or functions of x, y, and z. We assume that $p \geqslant 0$ and $h \geqslant 0$.

Equation (5) is known as Poisson's equation and is a generalization of Laplace's equation, which occurs when $f(x, y, z)$ is identically zero throughout D. It was encountered in Chaps. 1 and 4. Boundary condition (6) includes important special cases. When $p = 0$ on S, or on part of S, the value of u is assigned there. When $h = 0$, the value of du/dn is assigned. Of course, p and h must not vanish simultaneously.

If $u = u_1(x, y, z)$ and $u = u_2(x, y, z)$ are two solutions of this problem, their difference

$$U(x, y, z) = u_1(x, y, z) - u_2(x, y, z)$$

satisfies Laplace's equation in D and the condition

$$p\frac{dU}{dn} + hU = 0$$

on S. Moreover, U satisfies the conditions of regularity required of u_1 and u_2. Thus it is harmonic in D, and U and its derivatives of the first order are continuous in $\overline{D}$. It follows from the results established above for harmonic functions that U must be constant throughout $\overline{D}$. Thus $dU/dn = 0$ on S. If $h \neq 0$ at some point on S, then U vanishes there; and $U = 0$ throughout $\overline{D}$. For the harmonic function U vanishes over the closed surface S and hence cannot have values other than zero interior to S.

We have now established the following uniqueness theorem for problems in electrostatic or gravitational potential, steady temperatures, and other boundary value problems involving Laplace's or Poisson's equation.

Theorem. *Let $u(x, y, z)$ satisfy these conditions of regularity in a domain D bounded by a closed surface S: (a) It is continuous, together with its derivatives of the first order, in $\overline{D}$; (b) its derivatives of the second order are continuous in D. Then if u is a solution of the boundary value problem (5)–(6), it is the only solution satisfying conditions (a) and (b), except possibly for $u + C$, where C is an* *arbitrary constant. Unless h = 0 at every point on S, C = 0 and the solution is unique.*

It is possible to show that this theorem also applies when D is the unbounded domain exterior to the closed surface S, provided u satisfies the additional requirement that the absolute values of ru, $r^2 u_{xx}$, $r^2 u_{yy}$, and $r^2 u_{zz}$ be bounded for all r greater than some fixed number, where r is the distance from (x, y, z) to the origin.[†] Then, since u vanishes as $r \to \infty$, the constant C is zero; and the solution is unique. But note that S is a closed surface, so that this extension of the theorem does not apply, for instance, to unbounded domains between two planes or inside a cylinder.

Condition (a) in the theorem is severe because it requires u and its derivatives of the first order to be continuous on the surface S. For problems in which $p = 0$ on S, so that u is prescribed on the entire boundary, the condition can be relaxed so as to require only the continuity of u itself in $\overline{D}$ if derivatives are continuous in D. This follows directly from a fundamental result in potential theory: *If a function other than a constant is harmonic in D and continuous in $\overline{D}$, then its maximum and minimum values are assumed at points on S, and never in D.*[‡]

EXAMPLE. To illustrate the use of the theorem, consider the problem in Example 1, Sec. 34, of determining steady temperatures $u(x, y)$ in a rectangular plate with three edges at temperature zero and an assigned temperature distribution on the fourth edge. The faces of the plate are insulated. For convenience, we shall consider the plate to be square, with edge length π. As long as $du/dn = 0$ on the faces, the thickness of the plate does not affect the problem.

The domain D is the interior of the finite region bounded by the planes $x = 0$, $x = \pi$, $y = 0$, $y = \pi$, $z = z_1$, and $z = z_2$, where z_1 and z_2 are constants. Then S is the boundary of that domain. The required function u is harmonic in D. It vanishes on the three parts $x = 0$, $x = \pi$, and $y = \pi$ of S; and $u = f(x)$ on the part $y = 0$. Also, $u_z = 0$ on the parts $z = z_1$ and $z = z_2$. Thus the theorem applies if u and its derivatives of the first order are continuous in $\overline{D}$.

First, suppose that u is independent of z. Then

$$(7) \qquad u_{xx}(x, y) + u_{yy}(x, y) = 0 \qquad\qquad (0 < x < \pi, 0 < y < \pi),$$

$$(8) \qquad u(0, y) = u(\pi, y) = 0 \qquad\qquad (0 \leqslant y \leqslant \pi),$$

$$(9) \qquad u(x, 0) = f(x), \qquad u(x, \pi) = 0 \qquad\qquad (0 \leqslant x \leqslant \pi).$$

[†] See the book by Kellogg (1953), referred to earlier in this section.

[‡] Physically, the result seems evident since it states that steady temperatures cannot have maximum or minimum values interior to a solid in which no heat is generated. For a proof in three dimensions, see the book by Kellogg (1953), cited earlier in this section; and, for two dimensions, see the authors' book (1990, sec. 42), which is also listed in the Bibliography.

The formal solution found in Sec. 34 becomes

$$(10) \qquad u(x, y) = \sum_{n=1}^{\infty} b_n \frac{\sinh n(\pi - y)}{\sinh n\pi} \sin nx,$$

where b_n are the coefficients in the Fourier sine series for f on the interval $0 < x < \pi$.

To show that the function (10) satisfies the regularity conditions, let us require that f and f' be continuous and f'' piecewise continuous and that $f(0) = f(\pi) = 0$. Results found in Secs. 22 and 23 then show that

$$(11) \qquad f(x) = \sum_{n=1}^{\infty} b_n \sin nx, \qquad f'(x) = \sum_{n=1}^{\infty} nb_n \cos nx$$

and that both of these series converge uniformly on the interval $0 \leqslant x \leqslant \pi$. The second series, obtained by differentiating the first, is the Fourier cosine series for f' on $0 < x < \pi$; and, since f' is continuous and f'' is piecewise continuous, not only is that series uniformly convergent, but also the series of absolute values $|nb_n|$ of its coefficients converges. Hence it follows from the Weierstrass M-test (Sec. 22) that the series

$$(12) \qquad \sum_{n=1}^{\infty} nb_n \sin nx \qquad\qquad (0 \leqslant x \leqslant \pi)$$

also converges uniformly with respect to x.

Let us show that, for each fixed y, the sequence of functions

$$(13) \qquad \frac{\sinh n(\pi - y)}{\sinh n\pi} \qquad\qquad (0 \leqslant y \leqslant \pi)$$

appearing in series (10) is monotonic and nonincreasing as n increases. This is evident when $y = 0$ and when $y = \pi$. It is also true when $0 < y < \pi$, provided that the function

$$T(t) = \frac{\sinh \beta t}{\sinh \alpha t} \qquad\qquad (t > 0, \, \alpha > \beta > 0)$$

decreases as t increases. To see that this is so, we write

$$T'(t) \sinh^2 \alpha t = \beta \sinh \alpha t \cosh \beta t - \alpha \sinh \beta t \cosh \alpha t$$

$$= -\frac{1}{2}(\alpha - \beta) \sinh(\alpha + \beta)t + \frac{1}{2}(\alpha + \beta) \sinh(\alpha - \beta)t$$

$$= -\frac{\alpha^2 - \beta^2}{2}\left[\frac{\sinh(\alpha + \beta)t}{\alpha + \beta} - \frac{\sinh(\alpha - \beta)t}{\alpha - \beta} \right]$$

$$= -\frac{\alpha^2 - \beta^2}{2} \sum_{n=0}^{\infty} \frac{(\alpha + \beta)^{2n} - (\alpha - \beta)^{2n}}{(2n + 1)!} t^{2n+1}.$$

Since the terms of this series are positive, $T'(t) < 0$; thus $T(t)$ decreases as t increases.

Likewise, the positive-valued functions

(14) $$\frac{\cosh n(\pi - y)}{\sinh n\pi} \qquad (0 \leqslant y \leqslant \pi),$$

arising in the series for u_y, never increase in value as n grows because their squares can be written

(15) $$\frac{1}{\sinh^2 n\pi} + \left[\frac{\sinh n(\pi - y)}{\sinh n\pi} \right]^2$$

and each term here is nonincreasing.

The values of the functions (13) clearly vary only from zero to unity for all n and y involved. The functions (14) are also uniformly bounded. Hence those functions can be used in Abel's test for uniform convergence (Sec. 79). From the uniform convergence of the series in equations (11) and of series (12), on the interval $0 \leqslant x \leqslant \pi$, we conclude not only that series (10) converges uniformly with respect to x and y together in the region $0 \leqslant x \leqslant \pi, 0 \leqslant y \leqslant \pi$ of the xy plane, but also that the uniform convergence holds true for the series obtained by differentiating series (10) once termwise with respect to either x or y.

Consequently, series (10) is differentiable with respect to x and y; also, its sum $u(x, y)$ and u_x and u_y are continuous in the closed region $0 \leqslant x \leqslant \pi$, $0 \leqslant y \leqslant \pi$. Clearly, $u(x, y)$ satisfies boundary conditions (8) and (9).

The derivatives of the second order, with respect to either x or y, of the terms in series (10) have absolute values not greater than

(16) $$n^2 |b_n| \frac{\sinh n(\pi - y_0)}{\sinh n\pi}$$

when $0 \leqslant x \leqslant \pi$ and $y_0 \leqslant y \leqslant \pi$, where $y_0 > 0$. Let M be chosen such that $|b_n| < M$ for all n. Then, from the inequalities

$$2 \sinh n(\pi - y_0) < \exp n(\pi - y_0) \qquad \text{and} \qquad 2 \sinh n\pi \geqslant e^{n\pi}(1 - e^{-2\pi}),$$

it follows that the numbers (16) are less than

$$\frac{M}{1 - \exp(-2\pi)} n^2 \exp(-ny_0).$$

The series with these terms converges, according to the ratio test, since $y_0 > 0$; and so the Weierstrass M-test ensures the uniform convergence of the series of second-order derivatives of terms in series (10) when $y_0 \leqslant y \leqslant \pi$. Thus series (10) is twice-differentiable; also, u_{xx} and u_{yy} are continuous in the region $0 \leqslant x \leqslant \pi, 0 < y \leqslant \pi$.

Since the terms in series (10) satisfy Laplace's equation (7), the same is true of the sum $u(x, y)$ of that series. Thus u is established as a solution of our boundary value problem. Moreover, u satisfies our regularity conditions, even with respect to z since it is independent of z and $u_z = 0$ everywhere and on the parts $z = z_1$ and $z = z_2$ of S, in particular. According to the above theorem, the function defined by series (10) is, then, the only possible solution that satisfies the regularity conditions.

82. SOLUTIONS OF A WAVE EQUATION

Consider the following generalization of the problem solved in Sec. 29 for the transverse displacements in a stretched string:

$$(1) \qquad\qquad y_{tt}(x,t) = a^2 y_{xx}(x,t) + \phi(x,t) \qquad (0 < x < c, t > 0),$$

$$(2) \qquad\qquad y(0,t) = p(t), \qquad y(c,t) = q(t) \qquad\qquad (t \geqslant 0),$$

$$(3) \qquad\qquad y(x,0) = f(x), \qquad y_t(x,0) = g(x) \qquad\qquad (0 \leqslant x \leqslant c).$$

But we now require y to be of class C^2 in the region $R: 0 \leqslant x \leqslant c$, $t \geqslant 0$, by which we shall mean that y and its derivatives of the first and second order, including y_{xt} and y_{tx}, are to be continuous functions in R. As indicated in Sec. 30, the prescribed functions ϕ, p, q, f, and g must be restricted if the problem is to have a solution of class C^2.

Suppose that there are two solutions $y_1(x,t)$ and $y_2(x,t)$ in that class. Then the difference

$$Y(x,t) = y_1(x,t) - y_2(x,t)$$

is of class C^2 in R and satisfies the homogeneous problem

$$(4) \qquad\qquad Y_{tt}(x,t) = a^2 Y_{xx}(x,t) \qquad\qquad (0 < x < c, t > 0),$$

$$(5) \qquad\qquad Y(0,t) = 0, \qquad Y(c,t) = 0 \qquad\qquad (t \geqslant 0),$$

$$(6) \qquad\qquad Y(x,0) = 0, \qquad Y_t(x,0) = 0 \qquad\qquad (0 \leqslant x \leqslant c).$$

We shall prove that $Y = 0$ throughout R; thus $y_1 = y_2$, as stated in the following theorem.

Theorem. *The boundary value problem* (1)–(3) *cannot have more than one solution of class C^2 in R.*

To start the proof, we note that the integrand of the integral

$$(7) \qquad\qquad I(t) = \frac{1}{2} \int_0^c \left(Y_x^2 + \frac{1}{a^2} Y_t^2 \right) dx \qquad\qquad (t \geqslant 0)$$

satisfies conditions such that

$$(8) \qquad\qquad I'(t) = \int_0^c \left(Y_x Y_{xt} + \frac{1}{a^2} Y_t Y_{tt} \right) dx.$$

Since $Y_{tt} = a^2 Y_{xx}$, the integrand here can be written as

$$Y_x Y_{tx} + Y_t Y_{xx} = \frac{\partial}{\partial x}(Y_x Y_t).$$

So in view of equations (5), from which it follows that

$$Y_t(0, t) = 0, \qquad Y_t(c, t) = 0,$$

we can write

(9) $I'(t) = Y_x(c, t)Y_t(c, t) - Y_x(0, t)Y_t(0, t) = 0.$

Hence $I(t)$ is a constant. But equation (7) shows that $I(0) = 0$ because $Y(x, 0) = 0$, and so $Y_x(x, 0) = 0$; also, $Y_t(x, 0) = 0$. Thus $I(t) = 0$. The nonnegative continuous integrand of that integral must, therefore, vanish; that is,

$$Y_x(x, t) = Y_t(x, t) = 0 \qquad\qquad (0 \leqslant x \leqslant c, t \geqslant 0).$$

So Y is constant. In fact, $Y(x, t) = 0$ since $Y(x, 0) = 0$; and the proof of the theorem is complete.

If y_x, instead of y, is prescribed at the end point in either or both of conditions (2), the proof of uniqueness is still valid because condition (9) is again satisfied.

The requirement of continuity on derivatives of y is severe. Solutions of many simple problems in a wave equation have discontinuities in their derivatives.

PROBLEMS

1. In Problem 8, Sec. 33, on temperatures $u(x, t)$ in a slab, initially at temperatures $f(x)$ and throughout which heat is generated at a constant rate per unit volume, assume that f is continuous and f' is piecewise continuous $(0 \leqslant x \leqslant \pi)$ and that $f(0) = f(\pi) = 0$. Prove that the function $u(x, t)$ obtained there is the only solution of the problem which satisfies the regularity conditions (a) and (b) stated in Theorem 1, Sec. 80.

2. Verify the solution of Problem 8, Sec. 32, and prove that it is the only possible solution satisfying the regularity conditions (a) and (b) stated in Theorem 1, Sec. 80. Note that, in this case, the Weierstrass M-test suffices for all proofs of uniform convergence.

3. In the Dirichlet problem for a rectangle in Example 1, Sec. 34, let f be piecewise smooth on the interval $0 < x < a$. Verify that the formal solution found there satisfies the condition $u(x, 0 +) = f(x)$ when $0 < x < a$ if $f(x)$ is defined as the mean of $f(x +)$ and $f(x -)$ at its points of discontinuity.

4. Formulate a complete statement of the boundary value problem for steady temperatures in a square plate with insulated faces when the edges $x = 0$, $x = \pi$, and $y = 0$ are insulated and the edge $y = \pi$ is kept at temperatures $u = f(x)$. Show that if f, f', and f'' are continuous on the interval $0 \leqslant x \leqslant \pi$ and $f'(0) = f'(\pi) = 0$, the problem

has the unique solution

$$u(x, y) = \frac{a_0}{2} + \sum_{n=1}^{\infty} a_n \frac{\cosh ny}{\cosh n\pi} \cos nx,$$

where

$$a_n = \frac{2}{\pi} \int_0^{\pi} f(x) \cos nx \, dx \qquad\qquad (n = 0, 1, 2, \ldots).$$

5. Replace the thin disk in the example in Sec. 40 by a cylinder bounded by the surfaces $\rho = 1$, $z = z_1$, and $z = z_2$, where $u_z = 0$ on the last two parts. Also, let $f(\phi)$ be a periodic function of period 2π with a continuous derivative of the second order everywhere. Then show that the function u given by equation (6), Sec. 40, is the unique solution, satisfying our conditions of regularity, of the problem in steady temperatures.

6. Use the uniqueness that was established in Sec. 82 to show that the solution of Problem 1, Sec. 30, is the only solution of class C^2 in the region $0 \leqslant x \leqslant 1$, $t \geqslant 0$ of the xt plane.

7. Show that the solution of Problem 3, Sec. 38, is unique in the class C^2.

8. In Sec. 30, let $f(x)$ be such that its odd periodic extension $F(x)$ has a continuous second derivative $F''(x)$ for all x $(-\infty < x < \infty)$. Then show that the solution (10) there is unique in the class C^2.

BIBLIOGRAPHY

The following list of books and articles for supplementary study of the various topics that have been introduced is far from exhaustive. Further references are given in the ones mentioned here.

Abramowitz, M., and I. A. Stegun (Eds.): *Handbook of Mathematical Functions with Formulas, Graphs, and Mathematical Tables*, Dover Publications, Inc., New York, 1972.

Apostol, T. M.: *Mathematical Analysis* (2d ed.), Addison-Wesley Publishing Company, Inc., Reading, MA, 1974.

Bell, W. W.: *Special Functions for Scientists and Engineers*, D. Van Nostrand Company, Ltd., London, 1968.

Berg, P. W., and J. L. McGregor: *Elementary Partial Differential Equations* (Holden-Day, Inc., San Francisco, 1966), McGraw-Hill, Inc., New York.

Birkhoff, G., and G.-C. Rota: *Introduction to Ordinary Differential Equations* (4th ed.), John Wiley & Sons, Inc., New York, 1989.

Bowman, F.: *Introduction to Bessel Functions*, Dover Publications, Inc., New York, 1958.

Boyce, W. E., and R. C. DiPrima: *Elementary Differential Equations* (5th ed.), John Wiley & Sons, Inc., New York, 1992.

Broman, A.: *Introduction to Partial Differential Equations*, Dover Publications, Inc., New York, 1989.

Brown, J. W., and F. Farris: On the Laplacian, *Math. Mag.*, vol. 59, no. 4, pp. 227–229, 1986.

Buck, R. C.: *Advanced Calculus* (3d ed.), McGraw-Hill, Inc., New York, 1978.

Budak, B. M., A. A. Samarskii, and A. N. Tikhonov: *A Collection of Problems on Mathematical Physics*, Dover Publications, Inc., New York, 1988.

Byerly, W. E.: *Fourier's Series and Spherical Harmonics*, Dover Publications, Inc., New York, 1959.

Cannon, J. T., and S. Dostrovsky: *The Evolution of Dynamics, Vibration Theory from 1687 to 1742*, Springer-Verlag, New York, 1981.

Carslaw, H. S.: *Introduction to the Theory of Fourier's Series and Integrals* (3d ed.), Dover Publications, Inc., New York, 1952.

—— and J. C. Jaeger: *Conduction of Heat in Solids* (2d ed.), Oxford University Press, London, 1959.

Churchill, R. V.: *Operational Mathematics* (3d ed.), McGraw-Hill, Inc., New York, 1972.

—— and J. W. Brown: *Complex Variables and Applications* (5th ed.), McGraw-Hill, Inc., New York, 1990.

Coddington, E. A.: *An Introduction to Ordinary Differential Equations*, Dover Publications, Inc., New York, 1989.

—— and N. Levinson: *Theory of Ordinary Differential Equations*, Krieger Publishing Co., Inc., Melbourne, FL, 1984.

Courant, R., and D. Hilbert: *Methods of Mathematical Physics*, vols. 1 and 2, John Wiley & Sons, Inc., New York, 1989.

Davis, H. F.: *Fourier Series and Orthogonal Functions*, Dover Publications, Inc., New York, 1989.

Erdélyi, A. (Ed.): *Higher Transcendental Functions*, vols. 1–3, Krieger Publishing Co., Inc., Melbourne, FL, 1981.

Farrell, O. J., and B. Ross: *Solved Problems in Analysis as Applied to Gamma, Beta, Legendre, and Bessel Functions*, Dover Publications, Inc., New York, 1971.

Fourier, J.: *The Analytical Theory of Heat*, translated by A. Freeman, Dover Publications, Inc., New York, 1955.

Franklin, P.: *A Treatise on Advanced Calculus*, Dover Publications, Inc., New York, 1964.

Grattan-Guinness, I.: *Joseph Fourier, 1768–1830*, The MIT Press, Cambridge, MA, and London, 1972.

Gray, Alfred, and M. A. Pinsky: Gibbs' Phenomenon for Fourier-Bessel Series, *Expo. Math.* (in press).

Gray, Andrew, and G. B. Mathews: *A Treatise on Bessel Functions and Their Applications to Physics* (2d ed.), with T. M. MacRobert, Dover Publications, Inc., New York, 1966.

Hanna, J. R., and J. H. Rowland: *Fourier Series, Transforms, and Boundary Value Problems* (2d ed.), John Wiley & Sons, Inc., New York, 1990.

Hayt, W. H., Jr.: *Engineering Electromagnetics* (5th ed.), McGraw-Hill, Inc., New York, 1989.

Herivel, J.: *Joseph Fourier: The Man and the Physicist*, Oxford University Press, London, 1975.

Hewitt, E., and R. E. Hewitt: The Gibbs-Wilbraham Phenomenon: An Episode in Fourier Analysis, *Arch. Hist. Exact Sci.*, vol. 21, pp. 129–160, 1979.

Hobson, E. W.: *The Theory of Spherical and Ellipsoidal Harmonics*, Chelsea Publishing Company, Inc., New York, 1955.

Hochstadt, H.: *The Functions of Mathematical Physics*, Dover Publications, Inc., New York, 1986.

Ince, E. L.: *Ordinary Differential Equations*, Dover Publications, Inc., New York, 1956.

Jackson, D.: *Fourier Series and Orthogonal Polynomials*, Carus Mathematical Monographs, no. 6, Mathematical Association of America, 1941.

Jahnke, E., F. Emde, and F. Lösch: *Tables of Higher Functions* (6th ed.), McGraw-Hill, Inc., New York, 1960.

Kaplan, W.: *Advanced Calculus* (4th ed.), Addison-Wesley Publishing Company, Inc., Reading, MA, 1991.

———: *Advanced Mathematics for Engineers*, Addison-Wesley Publishing Company, Inc., Reading, MA, 1981.

Kellogg, O. D.: *Foundations of Potential Theory*, Dover Publications, Inc., New York, 1953.

Kevorkian, J.: *Partial Differential Equations*, Wadsworth, Inc., Belmont, CA, 1990.

Körner, T. W.: *Fourier Analysis*, Cambridge University Press, Cambridge, England, 1988.

Kreider, D. L., R. G. Kuller, D. R. Ostberg, and F. W. Perkins: *An Introduction to Linear Analysis*, Addison-Wesley Publishing Company, Inc., Reading, MA, 1966.

Lanczos, C.: *Discourse on Fourier Series*, Hafner Publishing Company, New York, 1966.

Langer, R. E.: Fourier's Series: The Genesis and Evolution of a Theory, Slaught Memorial Papers, no. 1, *Am. Math. Monthly*, vol. 54, no. 7, part 2, pp. 1–86, 1947.

Lebedev, N. N.: *Special Functions and Their Applications*, Dover Publications, Inc., New York, 1972.

———, I. P. Skalskaya, and Y. S. Uflyand: *Worked Problems in Applied Mathematics*, Dover Publications, Inc., New York, 1979.

McLachlan, N. W.: *Bessel Functions for Engineers* (2d ed.), Oxford University Press, London, 1955.

MacRobert, T. M.: *Spherical Harmonics* (3d ed.), with I. N. Sneddon, Pergamon Press, Ltd., Oxford, England, 1967.

Magnus, W., F. Oberhettinger, and R. P. Soni: *Formulas and Theorems for the Special Functions of Mathematical Physics* (3d ed.), Springer-Verlag, New York, 1966.

Oberhettinger, F.: *Fourier Expansions*: *A Collection of Formulas*, Academic Press, New York and London, 1973.

Özişik, M. N.: *Boundary Value Problems of Heat Conduction*, Dover Publications, Inc., New York, 1989.

Papp, F. J.: Two Equivalent Properties for Orthonormal Sets of Functions in Complete Spaces, *Int. J. Math. Educ. Sci. Technol.*, vol. 22, pp. 147–149, 1991.

Pinsky, M. A.: *Partial Differential Equations and Boundary-Value Problems with Applications* (2d ed.), McGraw-Hill, Inc., New York, 1991.

Powers, D. L.: *Boundary Value Problems* (3d ed.), Academic Press, Orlando, FL, 1987.

Rainville, E. D.: *Special Functions*, Chelsea Publishing Company, Inc., New York, 1971.

——— and P. E. Bedient: *Elementary Differential Equations* (7th ed.), Macmillan Publishing Co., New York, 1989.

Rogosinski, W.: *Fourier Series* (2d ed.), Chelsea Publishing Company, Inc., New York, 1959.

Sagan, H.: *Boundary and Eigenvalue Problems in Mathematical Physics*, Dover Publications, Inc., New York, 1989.

Sansone, G., and A. H. Diamond: *Orthogonal Functions* (rev. ed.), Dover Publications, Inc., New York, 1991.

Seeley, R. T.: *An Introduction to Fourier Series and Integrals*, W. A. Benjamin, Inc., New York, 1966.

Smirnov, V. I.: *Complex Variables—Special Functions*, Addison-Wesley Publishing Company, Inc., Reading, MA, 1964.

Sneddon, I. N.: *Elements of Partial Differential Equations*, McGraw-Hill, Inc., New York, 1957.

——: *Fourier Transforms*, McGraw-Hill, Inc., New York, 1951.

——: *Special Functions of Mathematical Physics and Chemistry* (3d ed.), Longman, New York, 1980.

——: *The Use of Integral Transforms*, McGraw-Hill, Inc., New York, 1972.

Streeter, V. L., and E. B. Wylie: *Fluid Mechanics* (8th ed.), McGraw-Hill, Inc., New York, 1985.

Taylor, A. E., and W. R. Mann: *Advanced Calculus* (3d ed.), John Wiley & Sons, Inc., New York, 1983.

Timoshenko, S. P., and J. N. Goodier: *Theory of Elasticity* (3d ed.), McGraw-Hill, Inc., New York, 1970.

Titchmarsh, E. C.: *Eigenfunction Expansions Associated with Second-Order Differential Equations*, Oxford University Press, London, part I (2d ed.), 1962; part II, 1958.

——: *Introduction to the Theory of Fourier Integrals* (3d ed.), Chelsea Publishing Company, Inc., New York, 1986.

Tolstov, G. P.: *Fourier Series*, Dover Publications, Inc., New York, 1976.

Tranter, C. J.: *Bessel Functions with Some Physical Applications*, Hart Publishing Company, Inc., New York, 1969.

Trim, D. W.: *Applied Partial Differential Equations*, PWS-KENT Publishing Company, Boston, 1990.

Van Vleck, E. B.: The Influence of Fourier's Series upon the Development of Mathematics, *Science*, vol. 39, pp. 113–124, 1914.

Watson, G. N.: *A Treatise on the Theory of Bessel Functions* (2d ed.), Cambridge University Press, Cambridge, England, 1952.

Weinberger, H. F.: *A First Course in Partial Differential Equations*, John Wiley & Sons, Inc., New York, 1965.

Whittaker, E. T., and G. N. Watson: *A Course of Modern Analysis* (4th ed.), Cambridge University Press, Cambridge, England, 1963.

Zygmund, A.: *Trigonometric Series* (2d ed.), 2 vols. in one, Cambridge University Press, Cambridge, England, 1977.

INDEX

Abel, N. H., 117
Abel's test for uniform convergence, 117
  proof of, 322
Adjoint of operator, 170
Antiperiodic functions, 189, 192, 193, 211, 216
Approximation:
  best in mean, 62
  least squares, $62n$.
  of solutions, 123, 133
Average value of function, 61

Bernoulli, D., 128
Bessel functions, 242
  of first kind, 245, 249
    boundedness of, 256, 259
    derivatives of, 250, 256
    generating function for, 259
    graphs of, 246
    integral forms of, 255, 256
    integrals of, 250, 251
    modified, 253, 266, 286, 287

    norms of, 270–271
    orthogonal sets of, 263, 269
    orthonormal sets of, 270
    recurrence relations for, 249
    series of, 273
    zeros of, 259–263, 269
    of second kind, Weber's, 247
Bessel's equation, 242
  modified, 253
  self-adjoint form of, 264
Bessel's inequality, 64, 67
Bessel's integral form, 255
Best approximation in mean, 62
Bibliography, 336–339
Boundary conditions, 2
  Cauchy type, 26
  Dirichlet type, 26
  linear, 3
  linear homogeneous, 4
  Neumann type, 26, 317
  periodic, 163, 172
  Robin, 26
  separated, 169
  types of, 26